Functional Ceramic Coatings

Functional Ceramic Coatings

Editors

Bożena Pietrzyk
Sebastian Miszczak

MDPI • Basel • Beijing • Wuhan • Barcelona • Belgrade • Manchester • Tokyo • Cluj • Tianjin

Editors
Bożena Pietrzyk
Lodz University of Technology
Poland

Sebastian Miszczak
Lodz University of Technology
Poland

Editorial Office
MDPI
St. Alban-Anlage 66
4052 Basel, Switzerland

This is a reprint of articles from the Special Issue published online in the open access journal *Coatings* (ISSN 2079-6412) (available at: https://www.mdpi.com/journal/coatings/special_issues/ceram_coat).

For citation purposes, cite each article independently as indicated on the article page online and as indicated below:

LastName, A.A.; LastName, B.B.; LastName, C.C. Article Title. *Journal Name* **Year**, *Volume Number*, Page Range.

ISBN 978-3-0365-0732-3 (Hbk)
ISBN 978-3-0365-0733-0 (PDF)

© 2021 by the authors. Articles in this book are Open Access and distributed under the Creative Commons Attribution (CC BY) license, which allows users to download, copy and build upon published articles, as long as the author and publisher are properly credited, which ensures maximum dissemination and a wider impact of our publications.

The book as a whole is distributed by MDPI under the terms and conditions of the Creative Commons license CC BY-NC-ND.

Contents

About the Editors . . . . **vii**

Bożena Pietrzyk and Sebastian Miszczak
Functional Ceramic Coatings
Reprinted from: *Coatings* 2021, 11, 130, doi:10.3390/coatings11020130 . . . . **1**

Karol Kyzioł, Julia Oczkowska, Daniel Kottfer, Marek Klich, Łukasz Kaczmarek, Agnieszka Kyzioł and Zbigniew Grzesik
Physicochemical and Biological Activity Analysis of Low-Density Polyethylene Substrate Modified by Multi-Layer Coatings Based on DLC Structures, Obtained Using RF CVD Method
Reprinted from: *Coatings* 2018, 8, 135, doi:10.3390/coatings8040135 . . . . **5**

Vilma Jonauske, Sandra Stanionyte, Shih-Wen Chen, Aleksej Zarkov, Remigijus Juskenas, Algirdas Selskis, Tadas Matijosius, Thomas C. K. Yang, Kunio Ishikawa, Rimantas Ramanauskas and Aivaras Kareiva
Characterization of Sol-Gel Derived Calcium Hydroxyapatite Coatings Fabricated on Patterned Rough Stainless Steel Surface
Reprinted from: *Coatings* 2019, 9, 334, doi:10.3390/coatings9050334 . . . . **19**

Kui Wen, Min Liu, Xuezhang Liu, Chunming Deng and Kesong Zhou
Deposition of Photocatalytic TiO_2 Coating by Modifying the Solidification Pathway in Plasma Spraying
Reprinted from: *Coatings* 2017, 7, 169, doi:10.3390/coatings7100169 . . . . **33**

Bożena Pietrzyk, Katarzyna Porębska, Witold Jakubowski and Sebastian Miszczak
Antibacterial Properties of Zn Doped Hydrophobic SiO_2 Coatings Produced by Sol-Gel Method
Reprinted from: *Coatings* 2019, 9, 362, doi:10.3390/coatings9060362 . . . . **43**

Bo Yu, Guoyan Fu, Yanbin Cui, Xiaomeng Zhang, Yubo Tu, Yingchao Du, Gaohong Zuo, Shufeng Ye and Lianqi Wei
Influence of Silicon-Modified Al Powders (SiO_2@Al) on Anti-oxidation Performance of Al_2O_3-SiO_2 Ceramic Coating for Carbon Steel at High Temperature
Reprinted from: *Coatings* 2019, 9, 167, doi:10.3390/coatings9030167 . . . . **57**

Katarzyna Banaszek and Leszek Klimek
Ti(C, N) as Barrier Coatings
Reprinted from: *Coatings* 2019, 9, 432, doi:10.3390/coatings9070432 . . . . **71**

Haiyan Liu, Yueguang Wei, Lihong Liang, Yingbiao Wang, Jingru Song, Hao Long and Yanwei Liu
Microstructure Observation and Nanoindentation Size Effect Characterization for Micron-/Nano-Grain TBCs
Reprinted from: *Coatings* 2020, 10, 345, doi:10.3390/coatings10040345 . . . . **85**

Maximilian Grimm, Susan Conze, Lutz-Michael Berger, Gerd Paczkowski, Thomas Lindner and Thomas Lampke
Microstructure and Sliding Wear Resistance of Plasma Sprayed Al_2O_3-Cr_2O_3-TiO_2 Ternary Coatings from Blends of Single Oxides
Reprinted from: *Coatings* 2020, 10, 42, doi:10.3390/coatings10010042 . . . . **97**

Piotr Kula, Robert Pietrasik, Sylwester Pawęta and Adam Rzepkowski
Low Frictional MoS_2/WS_2/FineLPN Hybrid Layers on Nodular Iron
Reprinted from: *Coatings* **2020**, *10*, 293, doi:10.3390/coatings10030293 109

Bożena Pietrzyk, Sebastian Miszczak, Ye Sun and Marcin Szymański
Al_2O_3 + Graphene Low-Friction Composite Coatings Prepared By Sol–Gel Method
Reprinted from: *Coatings* **2020**, *10*, 858, doi:10.3390/coatings10090858 117

Barbara Burnat, Patrycja Olejarz, Damian Batory, Michal Cichomski, Marta Kaminska and Dorota Bociaga
Titanium Dioxide Coatings Doubly-Doped with Ca and Ag Ions as Corrosion Resistant, Biocompatible, and Bioactive Materials for Medical Applications
Reprinted from: *Coatings* **2020**, *10*, 169, doi:10.3390/coatings10020169 133

Hieronim Szymanowski, Katarzyna Olesko, Jacek Kowalski, Mateusz Fijalkowski, Maciej Gazicki-Lipman and Anna Sobczyk-Guzenda
Thin SiNC/SiOC Coatings with a Gradient of Refractive Index Deposited from Organosilicon Precursor
Reprinted from: *Coatings* **2020**, *10*, 794, doi:10.3390/coatings10080794 153

Yi Wang, Jian Sun, Bing Sheng and Haifeng Cheng
Deposition Mechanism and Thickness Control of CVD SiC Coatings on Nextel^{TM}440 Fibers
Reprinted from: *Coatings* **2020**, *10*, 408, doi:10.3390/coatings10040408 177

Noor irinah Omar, Santirraprahkash Selvami, Makoto Kaisho, Motohiro Yamada, Toshiaki Yasui and Masahiro Fukumoto
Deposition of Titanium Dioxide Coating by the Cold-Spray Process on Annealed Stainless Steel Substrate
Reprinted from: *Coatings* **2020**, *10*, 991, doi:10.3390/coatings10100991 187

About the Editors

Bożena Pietrzyk, PhD DSc is a professor at the Institute of Materials Science and Engineering of the Lodz University of Technology, Poland. She obtained a PhD in 2000 at the Centre of Molecular and Macromolecular Studies of the Polish Academy of Sciences where she conducted research on the electron properties of thin films of amorphous materials prepared by plasma-enhanced chemical vapor deposition. In 2013, she obtained her habilitation at the Lodz University of Technology. Her research interests include methods of producing thin coatings of ceramic materials, such as sol–gel or hybrid plasma-assisted sol–gel methods. She conducts research on functional properties of ceramic and composite (layered and with the participation of particles) coatings, such as anticorrosive, antibacterial, photocatalytic and antifriction properties. She is an author and co-author of over 60 scientific journal articles and numerous conference announcements and presentations, as well as three patents. Bożena Pietrzyk is an active academic teacher conducting classes and supervising BSc, MSc and PhD students. She is also a coordinator of two study programs. She acts as a member of the Polish Materials Science Society (PTM) and Polish Society for Biomaterials (PSB).

Sebastian Miszczak, PhD Eng is a materials scientist specializing in thin ceramic coatings. He is an assistant professor at the Faculty of Mechanical Engineering of the Lodz University of Technology (LUT), Poland. After he received an MSc Eng degree in materials science, he started working at the Institute of Materials Science and Engineering, LUT. His work focused on the study of sol–gel coatings deposited in order to improve the heat resistance of metal substrates. He obtained a PhD on the basis of a dissertation entitled "Multilayer heat-resistant oxide coatings produced by the sol–gel method". Following the results obtained in previous research work, he continued his study on extending the functional properties of oxide coatings. His latest work concerns the properties of composite oxide coatings produced by the sol–gel method and modified with molybdenum disulfide and graphene in order to reduce dry friction. To date, Sebastian Miszczak has authored more than 20 journal publications, multiple articles and posters in conference proceedings.

Editorial

Functional Ceramic Coatings

Bożena Pietrzyk * and Sebastian Miszczak *

Institute of Materials Science and Engineering, Lodz University of Technology, Stefanowskiego Str. 1/15, 90-924 Lodz, Poland
* Correspondence: bozena.pietrzyk@p.lodz.pl (B.P.); sebastian.miszczak@p.lodz.pl (S.M.)

Received: 20 January 2021; Accepted: 22 January 2021; Published: 26 January 2021

1. Introduction

Modern materials engineering, just like other areas of today's science and technology, requires a comprehensive and balanced approach that takes into account all factors that affect not only the design and functional properties of materials, but also their economic profitability and rational management of the available resources. A reflection of this trend is the continuous improvement of the performance, functionality and versatility of engineering designs, which entails the need for new and/or precisely tailored materials.

In many cases, the surface properties of materials or devices are crucial for their application and operation. Therefore, it is advisable to apply the methods of surface engineering, leading to modifications in the structure and properties of the surface layers of materials, or deposition of coatings on their surface.

The advantage of the latter solution is, above all, the possibility of wider use of advanced materials with unique functional properties, which, for various reasons, cannot or should not be used as the bulk. Particularly interesting in this case is the group of ceramic materials, which in form of the coatings can significantly improve the functionality and applications of classic engineering materials, such as metals (and their alloys) or polymers. Ceramics is an important group of coating materials because of its wide range of controllable features, such as mechanical strength and hardness, corrosion resistance, electrical conductivity, surface properties (surface energy, catalytic behavior, biocompatibility), optical properties, etc. Using these features, it is possible to create tailored substrate-coating systems. From this point of view, it is crucial to understand the relationships between the structures, morphology and the properties of ceramics in the form of coatings.

2. Functional Ceramic Coatings

The aim of the *"Functional Ceramic Coatings"* Special Issue was to present the progress in the development of ceramic coatings for various purposes, with an emphasis on the influence of their internal structure and morphology on functional properties and potential applications. The Special Issue consisted of fourteen fully refereed scientific publications, the topics of which included:

- Bioactive/biomedical applications [1–4]
- Barrier and protective properties [5–7]
- Mechanical and anti-wear properties [8–10]
- Optical and photocatalytic properties [11,12]
- Deposition and adhesion mechanisms [13,14]

In the first topic, there were four publications related to the biomedical applications of ceramic coatings. Kyzioł et al. [1] presented their work on the deposition of various multi-layer coatings based on diamond-like carbon structures deposited on low-density polyethylene (LDPE). The tailored plasma pre-treatment and deposition of thin DLC-like coatings resulted in an improvement of the mechanical properties (higher hardness with favourable gradient) and wettability of the LDPE surface (as an important property in osseointegration processes), while maintaining good biocompatibility of the surface without significant changes in cell viability in vitro. Jonauske et al. [2] presented the results of their research on the manufacturing of thin calcium hydroxyapatite ($Ca_{10}(PO_4)_6(OH)_2$; CHA) films on stainless steel using the sol-gel method. Using this method, the authors produced good-quality hydrophobic CHA coatings, which induced the formation of amorphous calcium phosphate (ACP) in the environment of simulated body fluid (SBF). This could improve osseointegration and promote bone cell proliferation for a better bone-implant connection and biocompatibility. In the next article, Burnat et al. [3] developed Ca and Ag ions doped TiO_2 coatings, deposited by sol-gel dip-coating method on M30NW biomedical steel. Comprehensive characterization of the obtained coatings showed an anatase-based structure and anticorrosive, hydrophilic, and bioactive properties. The results of the biological evaluation indicated that independently of the Ca and Ag ions molar ratio, the obtained coatings are biocompatible and do not reduce the proliferation ability of the osteoblasts cells. The last work on the biomedical/bioactive topic was an article by Pietrzyk et al. [4] describing SiO_2 coatings modified with hydrophobizers and zinc compounds. The coatings, produced by the sol-gel method, were characterized in terms of their morphology, chemical structure, and antibacterial properties. It was found that zinc compounds, unlike hydrophobizers, had a crucial impact on the antibacterial properties of the coatings, preventing formation of *Escherichia coli* biofilm.

The second leading topic of the Special Issue was the barrier and protective properties of ceramic coatings. Very interesting research results were presented by Yu et al. [5], who proposed an innovative method of producing a protective coating based on Al_2O_3-SiO_2 compound. The Al powders were successfully encapsulated by SiO_2 and the SiO_2@Al slurry was used to coat the surface of carbon J55 steel. As a result, a new type of Al_2O_3-SiO_2 ceramic coating (ASMA) was formed that exhibited excellent high temperature protective properties against oxidation. In the next article by Banaszek and Klimek [6], Ti(C,N) coatings with various chemical composition were deposited by magnetron sputtering method on Ni-Cr Heraenium NA alloy as a barrier coatings. In-depth studies of the release of Ni and Cr ions into aqueous solutions have shown very good barrier properties of these coatings, therefore they can significantly contribute to reducing potential cytotoxicity and improving the biocompatibility of prosthetic alloys. The last work on the barrier ceramic coatings was that of Liu et al. [7], about the microstructure and mechanical properties of 8% yttria-stabilized zirconia (8YSZ) thermal barrier coatings deposited on Ni-based superalloy (GH3128) by plasma spraying. The authors analyzed the effect of the parameter-dependent plasma spraying deposition process on the microstructure of the coatings and their micromechanical properties. Then the trans-scale mechanics theory was adopted to characterize the nanoindentation size effect for the micron-/nano-grain microstructure of the coatings. Highly convergent simulation and experimental results showed much better micromechanical properties of the nano-grain YSZ coatings.

The mechanical and anti-wear properties of ceramic coatings was the third topic of the Special Issue. This topic was taken up by Grimm et al. [8], who investigated the structure, phase composition, and mechanical properties (including wear) of the Al_2O_3-Cr_2O_3-TiO_2 coatings. The authors showed that even in multi-component blends of oxides the dominant single oxide has a significant influence on the coating properties. The use of powder blends deposited by atmospheric plasma spraying (APS) presents a promising approach to adapt or extend the property profile of plain oxide coatings. The next two papers were focused on the friction and wear properties of the coatings. In their work, Kula et al. [9] described the new concept of low frictional hybrid MoS_2/WS_2/FineLPN composite coatings on nodular cast iron. Researchers used a hybrid coating method combining MoS_2+WS_2 slurry deposition with two types of thermo-chemical treatment: FineLPN low pressure nitriding, and sulphonitriding. This unconventional approach resulted in coatings with a layered microstructure, very low dry friction coefficient (0.13) and low linear wear. Interesting results were also obtained by Pietrzyk et al. [10], who used the sol-gel method to produce Al_2O_3 + graphene composite coatings on stainless steel substrates. The authors analyzed the influence of two types of graphene nanoplatelets (GNPs) and the parameters of coatings deposition on their microstructures and basic tribological properties, obtaining in some variants the dry friction coefficient of 0.11.

The fourth topic covers the optical and photocatalytic properties of the ceramic coatings. The article by Wen et al. [11] presents the properties of photocatalytic TiO_2 coatings deposited with modified atmospheric plasma spraying (APS) method. Water injection into the plasma jet allowed the modification of the solidification of the molten TiO_2 particles and induced nucleation of the desired anatase phase. As a result, TiO_2 coatings with 5-times higher anatase content and better photocatalytic activity were produced. In the second article, Szymanowski et al. [12] presented a novel method for deposition SiNC/SiOC optical coatings with a gradient of refractive index. The deposition was performed using radio frequency plasma-enhanced chemical vapor deposition (RF PECVD) technology from organosilicon compounds with a variable composition of N_2/O_2 gas mixture, resulting in a single-layer optical coating with the gradient of refractive index. The method developed was used to manufacture a high quality "cold mirror" type of interference filter.

The last Special Issue topic was related to deposition and adhesion mechanisms of ceramic coatings. Deposition mechanisms and thickness control were the main topics of the work of Wang et al. [13] on the SiC coatings on Nextel ™440 fibers. The authors synthesized carbon-rich SiC coatings on Nextel fibers using the CVD method, proposed an empirical formula to calculate the coating thickness, and presented a detailed mechanism of the reactions taking place during the deposition of the coating on the surface of the fibers. The topic of coating-to-substrate adhesion was explored by Omar et al. [14] in their work on the deposition of TiO_2 coating on stainless steel substrate by the cold-spray process. The obtained results proved the important role of annealing of austenitic 304 steel on the bonding of sprayed TiO_2 particles to its surface. Adhesion between the brittle TiO_2 particles and the stainless steel surface was attributed to a high-velocity collision that caused limited amorphization and mixing of atoms in the interface zone.

As the editors, we encourage readers to read this Special Issue. We hope that this reading will show the possibilities and variety of applications of ceramic coatings and will inspire readers to deepen their interest in this topic. Finally, we would like to express our appreciation to all the authors for their contribution to the development of research in the field of ceramic functional coatings.

Funding: This research received no external funding.

Acknowledgments: As guest editors, we would like to thank Bunny Zou, Section Managing Editor, for her support. We also wish to thank all the contributing authors and reviewers for their excellent work.

Conflicts of Interest: The authors declare no conflict of interest.

References

1. Kyzioł, K.; Oczkowska, J.; Kottfer, D.; Klich, M.; Kaczmarek, Ł.; Kyzioł, A.; Grzesik, Z. Physicochemical and Biological Activity Analysis of Low-Density Polyethylene Substrate Modified by Multi-Layer Coatings Based on DLC Structures, Obtained Using RF CVD Method. *Coatings* **2018**, *8*, 135. [CrossRef]
2. Jonauske, V.; Stanionyte, S.; Chen, S.-W.; Zarkov, A.; Juskenas, R.; Selskis, A.; Matijosius, T.; Yang, T.C.K.; Ishikawa, K.; Ramanauskas, R.; et al. Characterization of Sol-Gel Derived Calcium Hydroxyapatite Coatings Fabricated on Patterned Rough Stainless Steel Surface. *Coatings* **2019**, *9*, 334. [CrossRef]

3. Burnat, B.; Olejarz, P.; Batory, D.; Cichomski, M.; Kaminska, M.; Bociaga, D. Titanium Dioxide Coatings Doubly-Doped with Ca and Ag Ions as Corrosion Resistant, Biocompatible, and Bioactive Materials for Medical Applications. *Coatings* **2020**, *10*, 169. [CrossRef]
4. Pietrzyk, B.; Porębska, K.; Jakubowski, W.; Miszczak, S. Antibacterial Properties of Zn Doped Hydrophobic SiO_2 Coatings Produced by Sol-Gel Method. *Coatings* **2019**, *9*, 362. [CrossRef]
5. Yu, B.; Fu, G.; Cui, Y.; Zhang, X.; Tu, Y.; Du, Y.; Zuo, G.; Ye, S.; Wei, L. Influence of Silicon-Modified Al Powders (SiO_2@Al) on Anti-Oxidation Performance of Al_2O_3-SiO_2 Ceramic Coating for Carbon Steel at High Temperature. *Coatings* **2019**, *9*, 167. [CrossRef]
6. Banaszek, K.; Klimek, L. Ti(C, N) as Barrier Coatings. *Coatings* **2019**, *9*, 432. [CrossRef]
7. Liu, H.; Wei, Y.; Liang, L.; Wang, Y.; Song, J.; Long, H.; Liu, Y. Microstructure Observation and Nanoindentation Size Effect Characterization for Micron-/Nano-Grain TBCs. *Coatings* **2020**, *10*, 345. [CrossRef]
8. Grimm, M.; Conze, S.; Berger, L.-M.; Paczkowski, G.; Lindner, T.; Lampke, T. Microstructure and Sliding Wear Resistance of Plasma Sprayed Al_2O_3-Cr_2O_3-TiO_2 Ternary Coatings from Blends of Single Oxides. *Coatings* **2020**, *10*, 42. [CrossRef]
9. Kula, P.; Pietrasik, R.; Pawęta, S.; Rzepkowski, A. Low Frictional MoS_2/WS_2/FineLPN Hybrid Layers on Nodular Iron. *Coatings* **2020**, *10*, 293. [CrossRef]
10. Pietrzyk, B.; Miszczak, S.; Sun, Y.; Szymański, M. Al_2O_3 + Graphene Low-Friction Composite Coatings Prepared by Sol–Gel Method. *Coatings* **2020**, *10*, 858. [CrossRef]
11. Wen, K.; Liu, M.; Liu, X.; Deng, C.; Zhou, K. Deposition of Photocatalytic TiO_2 Coating by Modifying the Solidification Pathway in Plasma Spraying. *Coatings* **2017**, *7*, 169. [CrossRef]
12. Szymanowski, H.; Olesko, K.; Kowalski, J.; Fijalkowski, M.; Gazicki-Lipman, M.; Sobczyk-Guzenda, A. Thin SiNC/SiOC Coatings with a Gradient of Refractive Index Deposited from Organosilicon Precursor. *Coatings* **2020**, *10*, 794. [CrossRef]
13. Wang, Y.; Sun, J.; Sheng, B.; Cheng, H. Deposition Mechanism and Thickness Control of CVD SiC Coatings on NextelTM440 Fibers. *Coatings* **2020**, *10*, 408. [CrossRef]
14. Omar, N.I.; Selvami, S.; Kaisho, M.; Yamada, M.; Yasui, T.; Fukumoto, M. Deposition of Titanium Dioxide Coating by the Cold-Spray Process on Annealed Stainless Steel Substrate. *Coatings* **2020**, *10*, 991. [CrossRef]

© 2021 by the authors. Licensee MDPI, Basel, Switzerland. This article is an open access article distributed under the terms and conditions of the Creative Commons Attribution (CC BY) license (http://creativecommons.org/licenses/by/4.0/).

Article

Physicochemical and Biological Activity Analysis of Low-Density Polyethylene Substrate Modified by Multi-Layer Coatings Based on DLC Structures, Obtained Using RF CVD Method

Karol Kyzioł [1,*], Julia Oczkowska [1], Daniel Kottfer [2], Marek Klich [3], Łukasz Kaczmarek [3], Agnieszka Kyzioł [4] and Zbigniew Grzesik [1]

1. Faculty of Materials Science and Ceramics, AGH University of Science and Technology, A. Mickiewicza Av. 30, 30 059 Kraków, Poland; juliao@student.agh.edu.pl (J.O.); grzesik@agh.edu.pl (Z.G.)
2. Department of Technologies and Materials, Faculty of Mechanical Engineering, Technical University in Kosice, Masiarska 74, 040 01 Kosice, Slovakia; daniel.kottfer@tuke.sk
3. Institute of Materials Science and Engineering, Łódz University of Technology, Stefanowskiego Str. 1/15, 90 924 Łódz, Poland; marek.klich@p.lodz.pl (M.K.); lukasz.kaczmarek@p.lodz.pl (Ł.K.)
4. Faculty of Chemistry, Jagiellonian University, Gronostajowa 2, 30 387 Kraków, Poland; kyziol@chemia.uj.edu.pl
* Correspondence: kyziol@agh.edu.pl; Tel.: +48-12-637-2465

Received: 6 March 2018; Accepted: 5 April 2018; Published: 10 April 2018

Abstract: In this paper, the surface properties and selected mechanical and biological properties of various multi-layer systems based on diamond-like carbon structure deposited on low-density polyethylene (LDPE) substrate were studied. Plasma etching and layers deposition (incl. DLC, N-DLC, Si-DLC) were carried out using the RF CVD (radio frequency chemical vapor deposition) method. In particular, polyethylene with deposited N-DLC and DLC layers in one process was characterized by a surface hardness ca. seven times (up to ca. 2.3 GPa) higher than the unmodified substrate. Additionally, its surface roughness was determined to be almost two times higher than the respective plasma-untreated polymer. It is noteworthy that plasma-modified LDPE showed no significant cytotoxicity in vitro. Thus, based on the current research results, it is concluded that a multilayer system (based on DLC coatings) obtained using plasma treatment of the LDPE surface can be proposed as a prospective solution for improving mechanical properties while maintaining biocompatibility.

Keywords: LDPE; RF CVD; doped DLC structure; wettability; biocompatibility

1. Introduction

Polymers such as polyethylene (PE) are the most widely used materials for medical applications due to their properties (i.e., high flexibility, low density, high chemical resistance, biocompatibility) [1,2]. However, its medical application often requires a surface modification and enhancement of the surface properties (i.e., low surface hardness). Therefore, special surface treatments must be applied to improve the physicochemical properties. This can be achieved using thin layer technology, including oxygen and nitrogen plasma discharge [3], laser irradiation [4,5], deposition of anti-wear and/or functional coatings (i.e., diamond-like carbon (DLC)) [6–8] and immobilization of biopolymers (e.g., chitosan and its derivatives) [9,10]. Since plasma treatment results in the generation of high-energy species such as radicals, ions, or molecules in an excited electronic state, this enables surface reactions to take place and leads to surface activation and modification [11]. Such plasma techniques can transform PE into a valuable material for medical applications due to the modification of the surface without any interference in the substrate interior [12].

In the case of biomedical applications, the pre-treated processes (i.e., plasma etching) have a huge impact on the coating properties, biocompatibility, and enhancement of cell attachment [13]. Plasma treatment is the most versatile surface modification technique and involves electron-induced excitations, ionization, and dissociation to facilitate the production of tailored surface properties such as wettability, roughness, and many more chemical, physical, and biological properties of the polymer surface. For enhanced adhesion, the surface free energy of the polymer should be larger than that of the substrate it will be bonded with. This is why surface modification of polyethylene is required to increase the surface free energy, which corresponds to a decrease of contact angle. In addition, it is very beneficial for many applications to obtain protective and gas barrier coatings (based on DLC structure) as well as obtaining the most biocompatible surfaces (based on DLC and chitosan structures) on the polymeric substrate [7,9,14–16]. DLC layers are characterized by high hardness (up to 30 GPa) and a high Young's modulus, but usually also high internal stresses (up to 7 GPa). These properties are related to the presence of a sp^3 C fraction in the structure [17]. However, the stresses can be reduced by incorporating other atoms into the structure (i.e., Si, O, N, F [18–20] or metals [21,22]). It is worth mentioning that the reduction in stresses is often associated with a reduction in hardness and elastic modulus of the layers [6,23].

In this paper, the RF CVD (radio frequency chemical vapor deposition) method was used to modify the physicochemical parameters of LDPE (low-density polyethylene) substrate. The experimental part consisted of different approaches to studying various multi-layer systems (i.e., DLC, N-DLC layers, or Si-DLC layers). Every time, the polymer substrate was first treated and functionalized by plasma etching using Ar$^+$ ions. Precise characterization before and after surface modification was performed, presented, and discussed. The surface properties of the modified PE were determined by scanning electron microscopy (SEM) with chemical composition analysis (EDX). Atomic structure and topography were examined by infrared spectroscopy (Fourier transform infrared-attenuated total reflectance (FTIR-ATR)) and atomic force microscopy (AFM), respectively. The nanoindentation method was applied to assess hardness and Young's modulus profiles. Cytotoxicity in vitro against the MG-63 cell line was evaluated by Alamar Blue assay. Additionally, the influence of surface modification on wettability and surface free energy of modified polyethylene substrate was also examined.

2. Materials and Methods

2.1. Sample Preparation and Surface Treatment

The material used in this study was low-density polyethylene prepared in the form of regular samples (width/length/height—10 mm/7 mm/4 mm). The average chemical composition of this material was 99.5 at.% carbon and 0.5 at.% oxygen (according to EDS analysis, EDAX Genesis, EDAX Inc., Mahwah, NJ, USA) and provided by Sigma-Aldrich (Karlsruhe, Germany). The oxygen content determined on the LDPE surface is probably the result of the adsorption of this element on the surface in air atmosphere. Before coating deposition on PE substrates, the samples were chemically purified in isopropanol and subjected to Ar$^+$ ion pre-treatment (etching process) in plasma conditions. The process of PE surface modification was performed in a RF CVD reactor (Elettrorava S.p.A., Turin, Italy), which generated plasma using radio-frequency discharge (13.56 MHz) under low-pressure conditions. The distance between the electrodes (cathode–anode) was 20 mm. All plasma processes (treatment and layers deposition) on the PE substrate were performed at room temperature (297 K) due to the low thermal resistance (melting point 383 K) of polyethylene. The treatment in the RF reactor of the polyethylene surface (ion etching) was conducted for all samples in order to prepare the substrates for further modification processes and surface purification from adsorbed gases (e.g., O$_2$). This stage of processing was conducted in Ar flow conditions (75 cm^3/min) under a pressure of 53 Pa and plasma density of 0.08 W/cm^2 for 10 min. The experiments consisted of four independent series of PE surface modification, including the deposition of DLC, N-DLC, and Si-DLC coatings, as well as multilayer systems. These processes were carried out in accordance with the technical parameters presented in Table 1.

Table 1. Technical parameters applied in surface modification of LDPE (low-density polyethylene) substrates using the RF CVD (radio frequency chemical vapor deposition) method.

Series	Process	Gas Mixture		Parameters		
		Gas	Flow (sccm)	$\rho_{Prf.}$ (W/cm^2)	p (Pa)	t (min)
PE_1	DLC deposition	Ar/H$_2$/CH$_4$	50/80/8	0.80	40	60
PE_2	N-DLC deposition	Ar/N$_2$/CH$_4$	75/84/10	0.60	40	60
PE_3	N-DLC deposition	Ar/N$_2$/CH$_4$	75/84/10	0.60	40	30
	DLC deposition	Ar/H$_2$/CH$_4$	50/80/8	0.80	40	30
PE_4	N-DLC deposition	Ar/N$_2$/CH$_4$	75/84/10	0.60	40	30
	Si-DLC deposition	Ar/CH$_4$/SiH$_4$	75/8/8	0.50	53	30

Note: $\rho_{Prf.}$—plasma density; p—pressure in the chamber; t—process time.

In addition, the unmodified substrate (PE_0 series) was also tested to compare the obtained test results. The obtained samples were characterized using techniques adequate for material engineering as well as the evaluation of biological activity.

2.2. Surface Characterization and Mechanical Study

The surface microstructure and chemical composition of unmodified and coated polyethylene was examined using scanning electron microscopy (NOVA NANO SEM 200, FEI, Hillsboro, OR, USA) with energy dispersive X-ray spectroscopy analysis (EDX). In the case of the detection of light elements (i.e., C, N, and O), an accelerating voltage of 5 eV was applied. In addition, the thickness of the obtained layers was examined based on the cross-section of the tested samples. Furthermore, the topography of tested samples was investigated using atomic force microscopy (AFM, Bruker, Santa Barbara, CA, USA) equipped with peak force tapping mode with a MultiMode 8 (Bruker, Santa Barbara, CA, USA) microscope with a Sb-doped silicon tip of 8 nm diameter. The chemical structure of the polymer surface was examined using FTIR-ATR (Fourier transform infrared) spectroscopy on a Bio-Rad FTS60 V device (Digilab Division, Cambridge, MA, USA). The spectra were measured within 400–4000 cm^{-1}, 275 scans, and 4 cm^{-1} resolution.

Contact angle measurements were conducted using the sessile drop technique performed on a DSA10Mk2 (Kruss, Hamburg, Germany) analyzer. Wettability and surface energy measurements were made using ultra-high-quality water (UHQ—water produced with the use of UHQ-PS, Elga, Buckinghamshire, UK) and diiodomethane (Aldrich, Taufkichen, Germany) droplets with a volume of 0.2 µL. For each tested sample, five independent measurements were carried out on the surface.

Indentation technique with G200 Nano Indenter® (MTS, Oak Ridge, TN, USA) equipped with Berkovich-type diamond tip was used to assess the mechanical properties of the surface. Profiles of hardness and modulus of elasticity were acquired by the continuous stiffness measurement (CSM) method, where the maximum penetration depth was set to 2 µm.

2.3. Cytotoxicity

Human osteoblast-like MG-63 cell lines (ATCC: CRL-1427) were cultured in DMEM (Dulbecco's Modified Eagle's Medium) medium (Immuniq, Żory, Poland) without phenol red with the addition of 10% fetal bovine serum (FBS) and 1% streptomycin/penicillin (Gibco-BRL, Life Technologies, Karlsruhe, Germany) in standard cell conditions (37 °C, 5% CO$_2$). Cytotoxicity assay was conducted for both pristine polyethylene and modified samples. The samples were sterilized overnight in ethanol vapor. Then, the samples were placed in a sterile 24-well culture plate and a cell suspension with a density of 50×10^4 cells/0.2 mL was added. Cells without the tested samples were treated as a reference sample. Treated and untreated cells were incubated for 72 h. Every 24 h, the medium was changed for a fresh one. After that time, cytotoxicity was determined by Alamar Blue assay (Sigma-Aldrich, Bornem, Belgium) according to the well-described protocol in [24]. In brief, cells were washed with phosphate-buffered saline (PBS) and incubated with resazurin sodium salt solution (25 µM in PBS) for 4 h at 37 °C in the dark. The fluorescence caused by the cellular metabolic activity was measured at 605 nm (excitation wavelength 560 nm) with a multimode microplate reader (Infinite 200M PRO NanoQuant, Tecan, Männedorf, Switzerland). Cytotoxicity was expressed as a percentage of viable cells after treatment with coated polyethylene samples in reference to untreated cells (control). The experiment was repeated three times. Cell morphology was examined using a fluorescence microscope (Olympus IX51, Olympus, Tokyo, Japan) with an excitation filter of 470/20 nm. After staining with acridine orange (viable cells) and propidium iodide (dead cells), cells were observed for morphological changes. At least five viewing fields containing ca. 100 cells each were analyzed. Photographs of CT26 cells after treatment with the tested alloys were taken using an inverted microscope equipped with a reflected fluorescence system (Olympus IX51, Olympus, Tokyo, Japan).

3. Results

3.1. Morphological Analysis

SEM and AFM techniques were used to obtain detailed information about morphological and topographical changes of polyethylene induced by plasma treatment and deposition of DLC-based coatings. It is worth noting that air and oxygen plasma treatments are more aggressive than argon plasma. Even though the ions of these gases are reactive and aggressive in contact with the surface layer of the polymeric substrate [25], these atmospheres are also used to clean the surface prior to the coating deposition. For instance, Rohrbeck et al. [26] applied an oxygen plasma cleaning process (10 min, 200 W), and after such treatment the initially smooth polymer surface turned out to be considerably rougher, and trenches and holes were more pronounced. However, in our work, the etching process with application of argon (less-reactive gas than oxygen and air) was carried out under the plasma power of 8 W, and in these conditions no significant negative influence of temperature on LDPE was observed. SEM analysis (Figure 1) revealed that each surface modification resulted in the formation of continuous and homogenous structures on the surface, without any cracks. Only in the case of modifications with DLC layer deposition (PE_1 series), could a more diverse microstructure be observed, with visible heterogeneities in the micro scale.

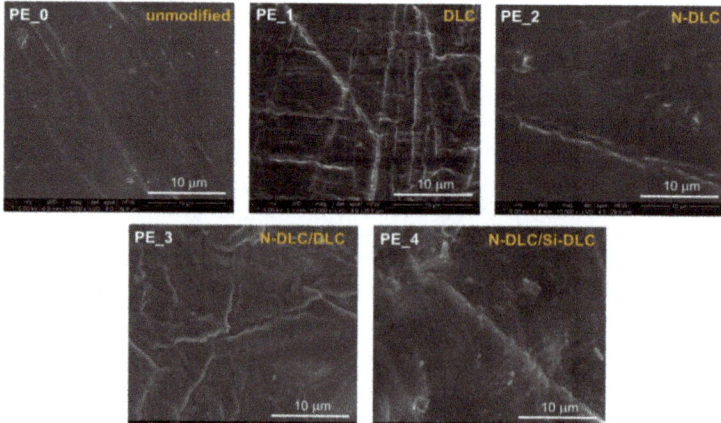

Figure 1. SEM images of unmodified polyethylene (PE) (PE_0 series) and selected modified PE surface after plasma processes: (PE_1) DLC deposition; (PE_2) N-DLC deposition; (PE_3) N-DLC and DLC deposition; (PE_4) N-DLC and Si-DLC deposition.

The new formed structures showed more details with atomic force microscopy, at the nanometric scale (Figure 2).

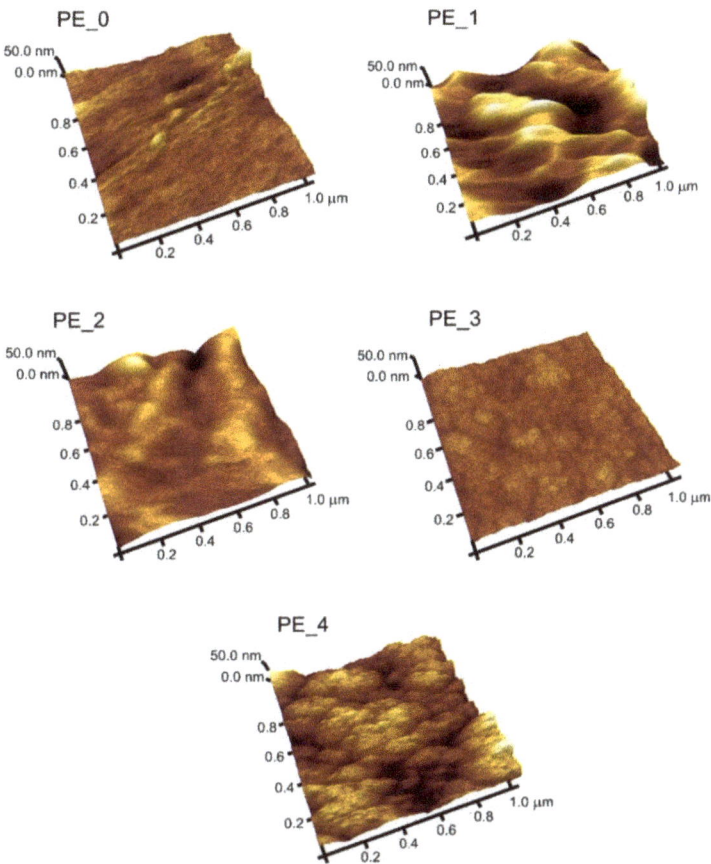

Figure 2. Atomic force microscopy (AFM) images of unmodified PE (PE_0 series) and modified polyethylene surface modification: (PE_1) DLC deposition; (PE_2) N-DLC deposition; (PE_3) N-DLC and DLC deposition; (PE_4) N-DLC and Si-DLC deposition.

Analysis of AFM images of modified substrates showed granular-like structures, which in the case of Si-DLC coatings were composed of agglomerated clusters (see Figure 2, sample PE_4).

A similar effect was observed Catena et al. when DLC layers were deposited on polyethylene [1]. In this respect, it is worth mentioning that plasma treatment of PE surfaces (with coating deposition) caused an increase in the surface area of tested samples, which is also beneficial for cell adhesion. The characteristic bulges (observed in sample PE_3 and PE_4) are similar to those presented by Catena et al. [27], caused by intrinsic stress release phenomena. More details concerning the surface roughness values of all samples, their chemical composition, and layer thicknesses are presented in Table 2.

It is important that determined roughness values (R_a) for samples after coatings deposition were similar in the case of one-layer modifications (ca. 24–30 nm) as well as for the two-layer ones (ca. 13–16 nm). These values were two to over three times higher than the value of this parameter for the unmodified polyethylene (ca. 9 nm). On one hand, the plasma treatment contributed to the increase in surface roughness of the PE substrate, and on the other hand it influenced the surface structure of the modified samples. The increases of surface roughness value after plasma processes are in good agreement with results obtained by Novotná et al. [28].

The chemical composition of tested samples (series PE_1–PE_4 in Table 2) confirms that the obtained coatings consisted of C, N, and Si elements, depending on the chemical composition of the gas mixture during plasma processes in the RF reactor. In the case of PE_2 series (with the N-DLC coating), nitrogen was incorporated into the structure to ca. 8 at.%, while Si atoms (for the PE_4 series) in the Si-DLC structure to ca. 27 at.%. It is worth noting that the addition of N and Si atoms to the diamond-like carbon structure caused a decrease in the value of internal stresses inside the obtained coatings as well as their hardness, which was also observed in other works [29–31]. However, the presence of silicon above (ca. 16 at.%) positively affected the anti-bacterial properties of the DLC coatings as well, which was also confirmed by Bociaga et al. [19]. The chemical composition studies of the tested samples revealed the presence of oxygen atoms in the structure, up to ca. 3 at.% in the case of the PE_1, PE_2, and PE_3 series.

Table 2. Layer thickness (d), surface roughness (R_a), and chemical composition of unmodified (PE_0 series) and modified polyethylene with obtained coatings.

Sample	d (μm)	R_a (nm)	Chemical Composition (at.%)			
			C	O	N	Si
PE_0	–	9 ± 2	99.5 ± 0.1	0.5 ± 0.1	–	–
PE_1	0.96 ± 0.06	24 ± 10	96.9 ± 0.1	3.1 ± 0.1	–	–
PE_2	0.77 ± 0.02	30 ± 4	89.7 ± 0.1	2.2 ± 0.1	8.1 ± 0.1	–
PE_3	0.83 ± 0.04	16 ± 2	95.0 ± 0.1	2.7 ± 0.1	2.3 ± 0.1	–
PE_4	1.95 ± 0.09 *	13 ± 3	51.3 ± 0.1	21.5 ± 0.1	0.0 **	27.2 ± 0.1

R_a: arithmetic average roughness (nm), measured using AFM; *: the thickness of Si-DLC and N-DLC layers was ca. 1.55 μm and ca. 0.40 μm, respectively; **: N-DLC layer was out of range of EDS analysis (thickness of Si-DLC above 1 μm).

The growth of oxygen concentration after plasma treatments was strongly affected by the creation of polar oxygen groups, which was also concluded by Novotná et al. [28]. In the case of PE_4, the thickness of Si-DLC layer was above 1 μm and the N-DLC layer was out of range of EDS analysis. This can be explained by the absence of nitrogen in the average content (at.%). In the case of PE_1, PE_2 and PE_3 series oxygen appeared in EDX analysis (up to ca. 3 at.%), possibly as the result of the adsorption of this element after the coating deposition process at ambient air conditions. In the case of the series with a Si-DLC layer (PE_4), the content of oxygen atoms was much higher (ca. 20 at.%), which can be associated with a large silicon content in the DLC structure, and therefore increased compliance for the incorporation of oxygen into the top surface of the modified PE substrate. The presence of oxygen was attributed to the surface oxidation. This process was also observed by Batory et al. [32]. The confirmation of this fact was by the presence of the Si–O atomic groups in the IR spectra as well as the highest range of surface hardness for obtained Si-DLC layers (the tested PE_4 sample, vide infra Figure 3 and see Section 3.4). This is mainly due to the very high binding energy for Si–O (ca. 532 eV), compared to the value for C–H (ca. 338.5 kJ/mol) and Si–H (ca. 298.7 kJ/mol) [33]. Additionally, Si–H bonds were less stable than C–H bonds, which also confirms the incorporation of oxygen into the Si-DLC structure.

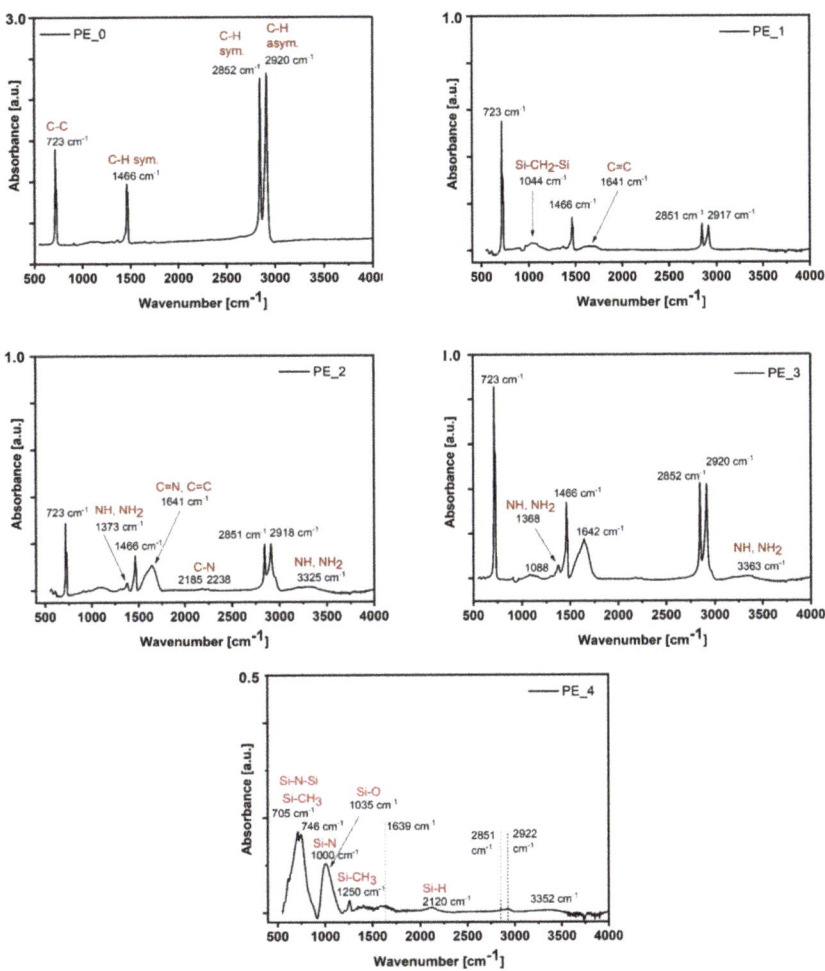

Figure 3. FTIR-ATR spectra of unmodified (PE_0) and surface-modified low-density polyethylene: (PE_1) DLC deposition; (PE_2) N-DLC deposition; (PE_3) N-DLC and DLC deposition; (PE_4) N-DLC and Si-DLC deposition.

3.2. FTIR-ATR Analysis

The modification of the PE surface led to significant changes in atomic structure, which was confirmed by the FTIR-ATR method. Obtained results are shown in Figure 3.

The IR spectra of unmodified PE (PE_0 series) showed two large peaks at 2920 and 2852 cm^{-1}, which correspond to C–H asymmetric and symmetric stretching vibrations in the CH$_2$ group, respectively. Two smaller absorption peaks at 1466 and 723 cm^{-1} can be identified as C–H symmetric and C–C bonds. It can be clearly concluded that this spectra is characteristic for unmodified polyethylene [2]. Deposition of DLC coating (see Figure 3—PE_1) noticeably changed the IR spectra of pure PE, the peaks assigned to C–H and C–C vibrations decreased, and the reordered spectra are typical for DLC structures. This modification also caused the appearance of a spectral line at 1641 cm^{-1} that was assigned to vibrations in C=C groups.

In the case of the next modification (PE_2 series), due to the obtained N-DLC layer, the new spectra lines were centered at 3325 cm^{-1} (assigned to NH and NH$_2$ groups, in the energy range 3300–3400 cm^{-1}) [34,35] and additionally confirmed by a peak at ca. 1373 cm^{-1} [36]. Furthermore, the relatively wide peak (in comparison to spectra for PE_1 series, centered at 1641 cm^{-1}) was also attributed to C=N bonds vibrations [37,38], while two weak spectral lines (for 2185 and 2238 cm^{-1}) were attributed to stretching vibrations in C–N groups [39].

In the case of PE substrate modification with the deposition of two layers (N-DLC/DLC, PE_3 series), the highest intensity (about two times higher than the IR spectra for PE_2 series) of spectral lines was assigned to C–H, C=C, and C–C vibration groups.

The last surface modification of polyethylene substrate (after deposition of N-DLC/Si-DLC layers, PE_4 series) resulted in significant changes to the obtained IR spectra. The spectra were dominated by various atomic groups containing Si atoms, including Si–H (2120 cm^{-1}) [40], Si–CH$_3$ (1250 cm^{-1}) [41], Si–N (750–1050 cm^{-1}) [36], and Si–CH$_2$–Si (1090–1020 cm^{-1}) vibrations [42]. In addition, in the range of 600 cm^{-1} to 850 cm^{-1}, many vibration modes, Si–C stretching, Si–N–Si asymmetrical stretching, CH$_3$–Si rocking-stretching, and Si–H bending [42,43] can be noticed. The high value of oxygen content in this case (ca. 21 at.%, based on EDS analysis) is probably associated with vibrations in the Si–O in Si–O–Si groups, which was assigned to the 1035 cm^{-1} spectral line [41]. It is noteworthy that in the case of modification with Si-DLC layers, the high dissociation energy of the Si–O bonding (798 kJ/mol) [44] resulted in a significant increase in the mechanical resistance of the modified surface.

3.3. Contact Angle and Surface Energy Analysis

Unmodified polyethylene is a low-energy hydrophobic material which must be modified in order to be useful in biomedical applications. Wettability is one of the most important surface parameters for biomedical applications, because hydrophilic material with higher surface energy favors cell adhesion and biocompatibility [4]. The contact angle value of untreated polyethylene (PE_0 series) was determined to be high (ca. 85°) for two measuring liquids (water and diiodomethane), see Figure 4.

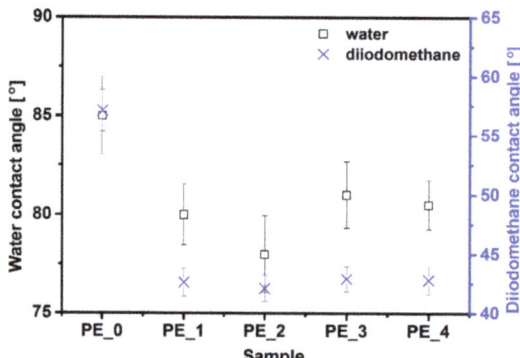

Figure 4. Contact angle values for water (black) and diiodomethane (blue) of PE surface before and after plasma modification.

A significant decrease in contact angle values after the deposition of DLC-based structures was observed for all tested series, while the contact angle for diiodomethane was lower than for water. A similar effect after the deposition of two different diamond-like carbon structures (flexible-DLC and robust-DLC) was described by Catena et al. [1]. This shows that in most cases plasmochemical treatment causes the contact angle to decrease, which was discussed broadly in many papers [2,10,45,46]. Polyethylene with N-DLC coating (PE_2 series) is probably the best for biomedical applications, because it exhibits low and comparable water and diiodomethane contact angles.

Figure 5 shows results of surface free energy obtained for the tested samples, including polar and dispersive components of SFE. The influence of plasma treatment for LDPE on surface free energy was also described by Pandiyaraj et al. [47,48].

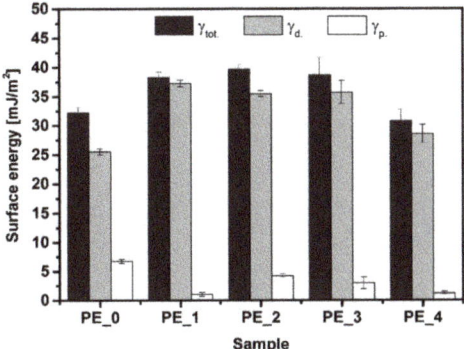

Figure 5. Surface free energy ($\gamma_{tot.}$—total surface energy, $\gamma_{d.}$—dispersive surface energy, $\gamma_{p.}$—polar surface energy) of unmodified and modified polyethylene, calculated using two different measuring liquids.

The authors concluded that usually in such processes oxygen flow results in an increase in the polar component (by incorporation of polar functional groups), without significantly changing the dispersive component. In our case, the performed experiments demonstrated that after deposition of DLC-based structures, the dispersive component increased, while the polar component decreased in relation to unmodified PE (PE_0 series). For example, in the case of modification with undoped DLC coatings (PE_1 series), the dispersive and polar components of surface free energy increased up to 37.5 mJ/m^2 ($\gamma_{d.}$) and decreased to 1.0 mJ/m^2 ($\gamma_{p.}$), respectively. Despite the fact that the total surface energy of all modified samples increased considerably, the best results (the highest $\gamma_{tot.}$ value and the lowest contact angle value) were obtained for the PE_2 series. This leads to the conclusion that the deposition of DLC layers (mainly N-DLC) can improve biocompatibility by increasing the surface energy of the substrate.

3.4. Mechanical Analysis

Surface modification of the LDPE surface under plasmochemical conditions improves its mechanical properties. The hardness and Young's modulus profiles in relation to displacement into the surface are shown in Figure 6.

Deposition of DLC-based coatings generally improves hardness (by up to nine times), especially at a distance of 600 nm from the surface (Figure 6a). The unmodified polyethylene (sample PE_0) was characterized by increased hardness only up to about 50 nm displacement into the surface (hardness of 0.3 GPa) and then stabilized at a value of ca. 0.1 GPa. In the case of the DLC layer obtained on the PE substrate (PE_1 series), we observed a hardness increase of up to ca. 2 GPa. The DLC structure doped with N atoms (N-DLC coating, PE_2 series) was characterized by lower hardness than the previous one (PE_1), the surface hardness achieved a value of ca. 1 GPa, and the strengthening remained at ca. 550 nm distance from the surface. The PE_3 series, corresponding to N-DLC/DLC multilayer, exhibited the highest surface hardness of up to 2.3 GPa, at a similar distance from the surface as in the case of the PE_2 series (ca. 600 nm). As a result, the addition of N to the structure of DLC caused a decrease in layer hardness compared to undoped DLC coatings. A comparable relationship was also observed for the addition of Si, but only at a distance of about 50 nm from the surface. Similar dependencies were observed by Ruijun et al. [49] and Wang et al. [50]. For the tested samples, the highest strengthening range (up to ca. 1750 nm from the surface) was shown in the PE_4 series, after deposition of the N-DLC/Si-DLC coating. In this case, the hardness increased to ca. 1.8 GPa.

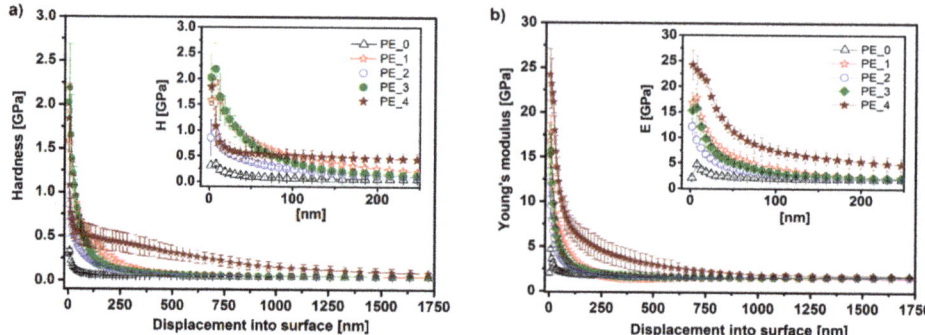

Figure 6. (a) Hardness and (b) Young's modulus profiles of tested samples.

Similar relationships were observed for the Young's modulus of tested samples (Figure 6b). Unmodified polyethylene reached a maximum value of 5 GPa near the surface, while in the interior it was about 2 GPa, which is a typical value for polyethylene. First modifications (DLC, N-DLC, and DLC/N-DLC corresponding to PE_1–PE_3) caused significant alterations in the Young's modulus of the samples (18 GPa for PE_1, 12 GPa for PE_2, and 16 GPa for PE_3), but these modifications increased these values only up to about 200 nm displacement into the surface. Again, the best mechanical properties were exhibited by the PE_4 series (N-DLC/Si-DLC modification), with a maximum Young's modulus value of 25 GPa on the surface. Increased E modulus in relation to unmodified PE remained escalated for about 1000 nm. This value is substantial, because it is the closest result to bone stiffness, which can be related to the enhanced biocompatibility required for implants.

3.5. Cytotoxicity Assay

In vitro cytotoxicity was evaluated on human osteoblast-like MG-63 cells after 72 h. No significant changes in cell viability were observed in all tested series. As shown in Figure 7, the decrease in cell survival after treatment for all tested samples was assessed to be not higher than 12%.

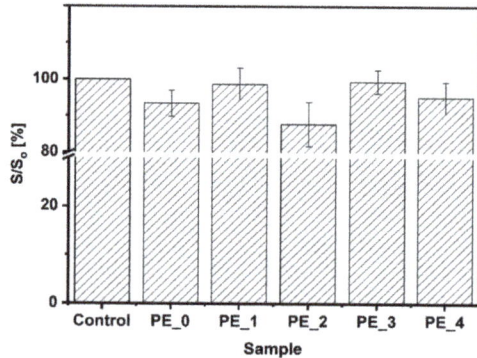

Figure 7. Cytotoxicity test results represented as survivability of cells (in %).

The most significant cytotoxicity (ca. 12%) showed only LDPE substrate after deposition of N-DLC layer on the top (series PE_2). This elevated cytotoxicity can be explained by a high content of N (ca. 8 at.%) in the modified surface.

Additionally, detailed analysis based on cellular morphology observation under fluorescent microscopy (Figure 8) did not reveal any significant changes in mitochondrial shape and size, and no apoptotic bodies were formed after this long incubation time.

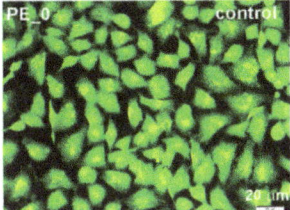 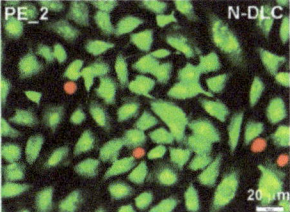 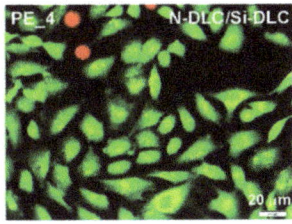

Figure 8. Live (green) and dead (red) human osteoblast-like MG-63 cells after 72 h of treatment. Cells were stained with acridine orange and propidium iodide, indicating viable and dead necrotic or late apoptotic cells, respectively.

Additionally, PE_3 modification (with the deposition of N-DLC/DLC layers) resulted in comparable biocompatibility to the control (untreated MG-63 cells) with significant improvement of mechanical properties (vide supra Figure 6).

The proposed DLC, N-DLC, or Si-DLC coatings on polymeric substrate were less cytotoxic than studied by us previously (DLC layers doped with N and Si atoms and deposited on titanium alloy). The latter coatings influenced the viability of the treated MG-63 cells, decreasing the cell surviving fraction by even up ca. 29% (modification C; after nitriding process and Si/N-DLC deposition) [51]. It can be concluded that the application of N-DLC or Si-DLC resulted in lower cytotoxic effect than the addition of N and Si atoms to the DLC structure during coating deposition.

4. Conclusions

On the basis of the performed experiments, it can be concluded that the modification of the LDPE in plasma conditions obtaining multi-layer coatings based on DLC structures resulted in profitable changes of physicochemical and biological properties for their medical applications. In particular, it allowed for a required hardness gradient on the modified surface to be obtained, and as a result made the polymer surface harder and less elastic. It was shown that all modifications in which DLC structures (doped and undoped with Si or N atoms) were obtained improved the wettability of the polymer surface, which is an important property in osseointegration processes. Importantly, in the case of N-DLC/DLC (PE_3 series) and DLC/Si-DLC (PE_4 series) multi-layers deposition on the LDPE surface, a significant improvement in hardness was observed—up to ca. 2.3 and 2.5 GPa, respectively. In addition, a high content of Si (ca. 27 at.%) in the top layer of PE_4 series enabled the incorporation of oxygen in the DLC structure after the modification process at ambient air conditions. All obtained surface modifications resulted in considerable increase in the mechanical properties of the tested surface by improving the PE substrate strengthening at a distance of ca. 1.7 μm from the surface. It is noteworthy that for all tested series no significant changes in cell viability in vitro were observed.

Acknowledgments: This work has been supported by the statutory research of Department of Physical Chemistry and Modeling of Faculty of Materials Science and Ceramics, AGH-UST (subject number 11.11.160.768).

Author Contributions: Karol Kyzioł conceived the article concept, designed and performed the experiments, and wrote the paper; Julia Oczkowska analyzed the data and prepared part of the manuscript; Daniel Kottfer, Marek Klich, Łukasz Kaczmarek and Zbigniew Grzesik analyzed the data, participated in discussion; Agnieszka Kyzioł designed and performed biological experiments and analyzed the obtained results.

Conflicts of Interest: The authors declare no conflict of interest.

References

1. Catena, A.; Agnello, S.; Rösken, L.M.; Bergen, H.; Recktenwald, E.; Bernsmann, F.; Busch, H.; Cannas, M.; Gelardi, F.M.; Hahn, B.; et al. Characteristics of industrially manufactured amorphous hydrogenated carbon (a-C:H) depositions on high-density polyethylene. *Carbon N. Y.* **2016**, *96*, 661–671. [CrossRef]
2. De Geyter, N.; Morent, R.; Leys, C. Surface characterization of plasma-modified polyethylene by contact angle experiments and ATR-FTIR spectroscopy. *Surf. Interface Anal.* **2008**, *40*, 608–611. [CrossRef]
3. Junkar, I.; Vesel, A.; Cvelbar, U.; Mozetič, M.; Strnad, S. Influence of oxygen and nitrogen plasma treatment on polyethylene terephthalate (PET) polymers. *Vacuum* **2009**, *84*, 83–85. [CrossRef]
4. Slepicka, P.; Kasalkova, N.S.; Siegel, J.; Kolska, Z.; Bacakova, L.; Svorcik, V. Nano-structured and functionalized surfaces for cytocompatibility improvement and bactericidal action. *Biotechnol. Adv.* **2015**, *33*, 1120–1129. [CrossRef] [PubMed]
5. Riveiro, A.; Soto, R.; Del Val, J.; Comesaña, R.; Boutinguiza, M.; Quintero, F.; Lusquiños, F.; Pou, J. Laser surface modification of ultra-high-molecular-weight polyethylene (UHMWPE) for biomedical applications. *Appl. Surf. Sci.* **2014**, *302*, 236–242. [CrossRef]
6. Love, C.A.; Cook, R.B.; Harvey, T.J.; Dearnley, P.A.; Wood, R.J.K. Diamond like carbon coatings for potential application in biological implants—A review. *Tribol. Int.* **2013**, *63*, 141–150. [CrossRef]
7. Fischer, C.B.; Rohrbeck, M.; Wehner, S.; Richter, M.; Schmeißer, D. Interlayer formation of diamond-like carbon coatings on industrial polyethylene: Thickness dependent surface characterization by SEM, AFM and NEXAFS. *Appl. Surf. Sci.* **2013**, *271*, 381–389. [CrossRef]
8. Choudhury, D.; Morita, T.; Sawae, Y.; Lackner, J.M.; Towler, M.; Krupka, I. A Novel functional layered diamond-like carbon coating for orthopedics applications. *Diam. Relat. Mater.* **2015**, *61*, 56–69. [CrossRef]
9. Xin, Z.; Hou, J.; Ding, J.; Yang, Z.; Yan, S.; Liu, C. Surface functionalization of polyethylene via covalent immobilization of O-stearoyl-chitosan. *Appl. Surf. Sci.* **2013**, *279*, 424–431. [CrossRef]
10. Pandiyaraj, K.N.; Ferraria, A.M.; Rego, A.M.B.D.; Deshmukh, R.R.; Su, P.; Halleluyah, J.M.; Halim, A.S. Low-pressure plasma enhanced immobilization of chitosan on low-density polyethylene for bio-medical applications. *Appl. Surf. Sci.* **2015**, *328*, 1–12. [CrossRef]
11. Holmes, C.; Tabrizian, M. *Surface Functionalization of Biomaterials*; Elsevier Inc.: New York, NY, USA, 2014; pp. 187–206.
12. Švorčík, V.; Kotál, V.; Slepička, P.; Macková, A.; Novotná, M.; Hnatowicz, V. Modification of surface properties of polyethylene by Ar plasma discharge. *Nucl. Instrum. Methods Phys. Res. Sect. B Beam Interact. Mater. Atoms.* **2006**, *244*, 365–372. [CrossRef]
13. Chung, T.W.; Liu, D.Z.; Wang, S.Y.; Wang, S.S. Enhancement of the growth of human endothelial cells by surface roughness at nanometer scale. *Biomaterials* **2003**, *24*, 4655–4661. [CrossRef]
14. Ray, S.C.; Mukherjee, D.; Sarma, S.; Bhattacharya, G.; Mathur, A.; Roy, S.S.; McLaughlin, J.A. Functional diamond like carbon (DLC) coatings on polymer for improved gas barrier performance. *Diam. Relat. Mater.* **2017**, *80*, 59–63. [CrossRef]
15. Dufils, J.; Faverjon, F.; Héau, C.; Donnet, C.; Benayoun, S.; Valette, S. Evaluation of a variety of a-C:H coatings on PEEK for biomedical implants. *Surf. Coat. Technol.* **2017**, *313*, 96–106. [CrossRef]
16. Ashtijoo, P.; Bhattacherjee, S.; Sutarto, R.; Hu, Y.; Yang, Q. Fabrication and characterization of adherent diamond-like carbon based thin films on polyethylene terephthalate by end hall ion beam deposition. *Surf. Coat. Technol.* **2016**, *308*, 90–97. [CrossRef]
17. Grill, A. Diamond-like carbon: State of the art. *Diam. Relat. Mater.* **1999**, *8*, 428–434. [CrossRef]
18. Zhang, T.F.; Wan, Z.X.; Ding, J.C.; Zhang, S.; Wang, Q.M.; Kim, K.H. Microstructure and high-temperature tribological properties of Si-doped hydrogenated diamond-like carbon films. *Appl. Surf. Sci.* **2018**, *435*, 963–973. [CrossRef]
19. Bociaga, D.; Kaminska, M.; Sobczyk-Guzenda, A.; Jastrzebski, K.; Swiatek, L.; Olejnik, A. Surface properties and biological behaviour of Si-DLC coatings fabricated by a multi-target DC-RF magnetron sputtering method for medical applications. *Diam. Relat. Mater.* **2016**, *67*, 41–50. [CrossRef]

20. Batory, D.; Jedrzejczak, A.; Kaczorowski, W.; Szymanski, W.; Kolodziejczyk, L.; Clapa, M.; Niedzielski, P. Influence of the process parameters on the characteristics of silicon-incorporated a-C:H:SiO$_x$ coatings. *Surf. Coat. Technol.* **2015**, *271*, 112–118. [CrossRef]
21. Bociaga, D.; Jakubowski, W.; Komorowski, P.; Sobczyk-Guzenda, A.; Jędrzejczak, A.; Batory, D.; Olejnik, A. Surface characterization and biological evaluation of silver-incorporated DLC coatings fabricated by hybrid RF PACVD/MS method. *Mater. Sci. Eng. C* **2016**, *63*, 462–474. [CrossRef] [PubMed]
22. Usman, A.; Rafique, M.S.; Khaleeq-ur-Rahman, M.; Siraj, K.; Anjuma, S.; Latifa, H.; Khan, T.M.; Mehmood, M. Growth and characterization of Ni:DLC composite films using pulsed laser deposition technique. *Mater. Chem. Phys.* **2011**, *126*, 649–654. [CrossRef]
23. Dean, J.; Aldrich-Smith, G.; Clyne, T.W. Use of nanoindentation to measure residual stresses in surface layers. *Acta Mater.* **2011**, *59*, 2749–2761. [CrossRef]
24. Vega-Avila, E.; Pugsley, M.K. An overview of colorimetric assay methods used to assess survival or proliferation of mammalian cells. *Proc. West. Pharmacol. Soc.* **2011**, *54*, 10–14. [CrossRef] [PubMed]
25. Couto, E.; Tan, I.H.; Demarquette, N.; Caraschi, J.C.; Leão, A. Oxygen plasma treatment of sisal fibers and polypropylene: Effects on mechanical properties of composites. *Polym. Eng. Sci.* **2002**, *42*, 790–797. [CrossRef]
26. Rohrbeck, M.; Körsten, S.; Fischer, C.B.; Wehner, S.; Kessler, B. Diamond-like carbon coating of a pure bioplastic foil. *Thin Solid Films* **2013**, *545*, 558–563. [CrossRef]
27. Catena, A.; Agnello, S.; Cannas, M.; Gelardi, F.M.; Wehner, S.; Fischer, C.B. Evolution of the sp^2 content and revealed multilayer growth of amorphous hydrogenated carbon (a-C:H) films on selected thermoplastic materials. *Carbon* **2017**, *117*, 351–359. [CrossRef]
28. Novotná, Z.; Rimpelová, S.; Ju, P.; Veselý, M.; Kolská, Z.; Václav, Š. The interplay of plasma treatment and gold coating and ultra-high molecular weight polyethylene: On the cytocompatibility. *Mater. Sci. Eng. C* **2017**, *71*, 125–131. [CrossRef] [PubMed]
29. Chouquet, C.; Gerbaud, G.; Bardet, M.; Barrat, S.; Billard, A.; Sanchette, F.; Ducros, C. Structural and mechanical properties of a-C:H and Si doped a-C:H thin films grown by LF-PECVD. *Surf. Coat. Technol.* **2010**, *204*, 1339–1346. [CrossRef]
30. Charitidis, C.A. Nanomechanical and nanotribological properties of carbon-based thin films: A review. *Int. J. Refract. Met. Hard Mater.* **2010**, *28*, 51–70. [CrossRef]
31. Ray, S.C.; Pong, W.F.; Papakonstantinou, P. Iron, nitrogen and silicon doped diamond like carbon (DLC) thin films: A comparative study. *Thin Solid Films* **2016**, *610*, 42–47. [CrossRef]
32. Batory, D.; Jedrzejczak, A.; Szymanski, W.; Niedzielski, P.; Fijalkowski, M.; Louda, P.; Kotela, I.; Hromadka, M.; Musil, J. Mechanical characterization of a-C:H:SiO$_x$ coatings synthesized using radio-frequency plasma-assisted chemical vapor deposition method. *Thin Solid Films* **2015**, *590*, 299–305. [CrossRef]
33. Ong, S.E.; Zhang, S.; Du, H.; Sun, D. Relationship between bonding structure and mechanical properties of amorphous carbon containing silicon. *Diam. Relat. Mater.* **2007**, *16*, 1628–1635. [CrossRef]
34. De Graaf, A.; Dinescu, G.; Longueville, J.L.; Van de Sanden, M.C.M.; Schram, D.C.; Dekempeneer, E.H.A.; Van Ijzendoorn, L.J. Amorphous hydrogenated carbon nitride films deposited via an expanding thermal plasma at high growth rates. *Thin Solid Films* **1998**, *333*, 29–34. [CrossRef]
35. Kyzioł, K.; Koper, K.; Kaczmarek, Ł.; Grzesik, Z. Plasmochemical modification of aluminum-zinc alloys using NH$_3$-Ar atmosphere with anti-wear coatings deposition. *Mater. Chem. Phys.* **2017**, *189*, 198–206. [CrossRef]
36. Giorgis, F.; Pirri, C.F.; Tresso, E. Structural properties of a-Si$_{1-x}$N$_x$: H films grown by plasma enhanced chemical vapour deposition by SiH$_4$ + NH$_3$ + H$_2$ gas mixtures. *Thin Solid Films* **1997**, *307*, 298–305. [CrossRef]
37. Szörényi, T.; Fuchs, C.; Fogarassy, E.; Hommet, J.; Le Normand, F. Chemical analysis of pulsed laser deposited a-CN$_x$ films by comparative infrared and X-ray photoelectron spectroscopies. *Surf. Coat. Technol.* **2000**, *125*, 308–312. [CrossRef]
38. Jonas, S.; Januś, M.; Jaglarz, J.; Kyzioł, K. Formation of Si$_x$N$_y$(H) and C:N:H layers by plasma-assisted chemical vapor deposition method. *Thin Solid Films* **2016**, *600*, 162–168. [CrossRef]
39. Wang, Y.H.; Moitreyee, M.R.; Kumar, R.; Shen, L.; Zeng, K.Y.; Chai, J.W.; Pan, J.S. A comparative study of low dielectric constant barrier layer, etch stop and hardmask films of hydrogenated amorphous Si-(C, O, N). *Thin Solid Films* **2004**, *460*, 211–216. [CrossRef]

40. Vassallo, E.; Cremona, A.; Ghezzi, F.; Dellera, F.; Laguardia, L.; Ambrosone, G.; Coscia, U. Structural and optical properties of amorphous hydrogenated silicon carbonitride films produced by PECVD. *Appl. Surf. Sci.* **2006**, *252*, 7993–8000. [CrossRef]
41. Kafrouni, W.; Rouessac, V.; Julbe, A.; Durand, J. Synthesis of PECVD a-SiC$_x$N$_y$: H membranes as molecular sieves for small gas separation. *J. Memb. Sci.* **2009**, *329*, 130–137. [CrossRef]
42. Kim, M.T.; Lee, J. Characterization of amorphous SiC:H films deposited from hexamethyldisilazane. *Thin Solid Films* **1997**, *303*, 173–179. [CrossRef]
43. Kyzioł, K.; Koper, K.; Środa, M.; Klich, M.; Kaczmarek, Ł. Influence of gas mixture during N$^+$ ion modification under plasma conditions on surface structure and mechanical properties of Al–Zn alloys. *Surf. Coat. Technol.* **2015**, *278*, 30–37. [CrossRef]
44. Lanigan, J.L.; Wang, C.; Morina, A.; Neville, A. Repressing oxidative wear within Si doped DLCs. *Tribol. Int.* **2016**, *93*, 651–659. [CrossRef]
45. Popelka, A.; Kronek, J.; Novák, I.; Kleinová, A.; Mičušík, M.; Špírková, M.; Omastová, M. Surface modification of low-density polyethylene with poly(2-ethyl-2-oxazoline) using a low-pressure plasma treatment. *Vacuum* **2014**, *100*, 53–56. [CrossRef]
46. Pandiyaraj, K.N.; RamKumar, M.C.; Arun Kumar, A.; Padmanabhan, P.V.A.; Deshmukh, R.R.; Bendavid, A.; Su, P.; Sachdev, A.; Gopinat, P. Cold atmospheric pressure (CAP) plasma assisted tailoring of LDPE film surfaces for enhancement of adhesive and cytocompatible properties: Influence of operating parameters. *Vacuum* **2016**, *130*, 34–47. [CrossRef]
47. Pandiyaraj, K.N.; Arun Kumar, A.; RamKumar, M.C.; Deshmukh, R.R.; Bendavid, A.; Su, P.; Kumar, S.U.; Gopinath, P. Effect of cold atmospheric pressure plasma gas composition on the surface and cyto-compatible properties of low density polyethylene (LDPE) films. *Curr. Appl. Phys.* **2016**, *16*, 784–792. [CrossRef]
48. Pandiyaraj, K.N.; Deshmukh, R.R.; Ruzybayev, I.; Shah, S.I.; Su, P.; Halleluyah, M., Jr.; Halim, A.S. Influence of non-thermal plasma forming gases on improvement of surface properties of low density polyethylene (LDPE). *Appl. Surf. Sci.* **2014**, *307*, 109–119. [CrossRef]
49. Ruijun, Z.; Hongtao, M. Nano-mechanical properties and nano-tribological behaviors of nitrogen-doped diamond-like carbon (DLC) coatings. *J. Mater. Sci.* **2006**, *41*, 1705–1709. [CrossRef]
50. Wang, J.; Pu, J.; Zhang, G.; Wang, L. Tailoring the structure and property of silicon-doped diamond-like carbon films by controlling the silicon content. *Surf. Coat. Technol.* **2013**, *235*, 326–332. [CrossRef]
51. Kyzioł, K.; Kaczmarek, Ł.; Brzezinka, G.; Kyzioł, A. Structure, characterization and cytotoxicity study on plasma surface modified Ti-6Al-4V and γ-TiAl alloys. *Chem. Eng. J.* **2014**, *240*, 516–526. [CrossRef]

© 2018 by the authors. Licensee MDPI, Basel, Switzerland. This article is an open access article distributed under the terms and conditions of the Creative Commons Attribution (CC BY) license (http://creativecommons.org/licenses/by/4.0/).

Article

Characterization of Sol-Gel Derived Calcium Hydroxyapatite Coatings Fabricated on Patterned Rough Stainless Steel Surface

Vilma Jonauske [1,*], Sandra Stanionyte [2], Shih-Wen Chen [3], Aleksej Zarkov [1], Remigijus Juskenas [2], Algirdas Selskis [2], Tadas Matijosius [2], Thomas C. K. Yang [3], Kunio Ishikawa [4], Rimantas Ramanauskas [2] and Aivaras Kareiva [1]

[1] Department of Inorganic Chemistry, Faculty of Chemistry and Geosciences, Vilnius University, Naugarduko str. 24, LT-03225 Vilnius, Lithuania; aleksej.zarkov@chf.vu.lt (A.Z.); aivaras.kareiva@chgf.vu.lt (A.K.)
[2] Center for Physical Sciences and Technology, Sauletekio av. 3, LT-10257 Vilnius, Lithuania; sandra.stanionyte@ftmc.lt (S.S.); remigijus.juskenas@ftmc.lt (R.J.); algirdas.selskis@ftmc.lt (A.S.); tadas.matijosius@ftmc.lt (T.M.); rimantas.ramanauskas@ftmc.lt (R.R.)
[3] Department of Chemical Engineering and Biotechnology, National Taipei University of Technology, 1, Sec. 3, Chung-Hsiao E. Road, Taipei 106, Taiwan; shihwenc@gmail.com (S.-W.C.); ckyang@mail.ntut.edu.tw (T.C.K.Y.)
[4] Department of Biomaterials, Faculty of Dental Science, Kyushu University, Maidashi, Higashi-Ku, Fukuoka 8120053, Japan; ishikawa@dent.kyushu-u.ac.jp
* Correspondence: vilma.ciuvasovaite@gmail.com; Tel.: +370-6-106-4808

Received: 16 April 2019; Accepted: 21 May 2019; Published: 24 May 2019

Abstract: Sol-gel derived calcium hydroxyapatite ($Ca_{10}(PO_4)_6(OH)_2$; CHA) thin films were deposited on stainless steel substrates with transverse and longitudinal patterned roughness employing a spin-coating technique. Each layer in the preparation of CHA multilayers was separately annealed at 850 °C in air. Fabricated CHA coatings were placed in simulated body fluid (SBF) for 2, 3, and 4 weeks and investigated after withdrawal. For the evaluation of obtained and treated with SBF coatings, diffuse reflectance infrared Fourier transform spectroscopy (DRIFTS), X-ray diffraction (XRD) analysis, Raman spectroscopy, XPS spectroscopy, scanning electron microscopy (SEM) analysis, and contact angle measurements were used. The tribological properties of the CHA coatings were also investigated in this study.

Keywords: calcium hydroxyapatite; sol-gel synthesis; thin films; spin coating; surface roughness; simulated body fluid

---

1. Introduction

For many years, 316L stainless steel has been widely used for orthopedic and orthodontic applications due to its excellent mechanical properties, good biocompatibility, and high corrosion resistance [1,2]. Metallic implants themselves are not bioactive; therefore, medical doctors and scientists are always concerned with achieving acceptable integration of implants into living bone. The main problem with metallic implants is the possible formation of fibrous tissue around the guest body implanted into bone [3]. Fibrous tissue formation is caused by the absorption of protein and other organic molecules on the hydrophobic surface of metals. Hydrophobic surfaces attract these organic molecules with enhanced formation of the above-mentioned biofilms [4,5]. It was suggested that formation of protein capsules may cause inflammation at the implant-bone interface, followed by the rejection of an implant [6]. Therefore, modified surfaces with antibacterial properties have been developed [7–9].

Implants are foreign bodies, and therefore, surface morphology and other properties have an impact on the behavior of bone cells that come into contact with implants [10,11]. Of course, the immune system of an individual patient is very important for the response to implanting biomaterials [12]. Bioglasses and other bioceramics such as calcium phosphates bond to natural bone and support the proliferation of new cells [2,13–15]. Surface wettability enhances osteoblast adhesion of human cells and also cell proliferation [16–18]. The problems associated with the application of metallic implants might be solved by the formation of calcium hydroxyapatite ($Ca_{10}(PO_4)_6(OH)_2$; CHA) on the surface [19,20]. CHA is a well-known biomaterial for its bioactivity, biocompatibility, and ability to induce growth of a new bone tissue. After implantation of a ceramic-coated, metal implant, the CHA forms strong chemical bonds with natural bone tissue due to its structural and chemical similarity to the bone [21,22].

Application of biphasic calcium phosphate in biocomposites or hydroxyapatite showed no fibrous encapsulation of the particles and composites after implantation allowing further infiltration of cells within the sample implants [23–25].

Previously we described sol-gel chemistry processing for the preparation of calcium hydroxyapatite films [26–28]. The CHA thin films were fabricated on substrates including silicon, quartz, and modified titanium. Both dip-coating and spin-coating techniques were compared. The same sol-gel method was used for the synthesis of CHA coatings on stainless steel substrates [29]. The quality of thin films obtained using the spin-coating technique was slightly better than when using dip-coating. However, the adhesion of CHA coatings obtained by both processes were less than desired. Recently, the effects of surface roughness on the adhesive properties of thin films were discussed and investigated [30]. In our study, the CHA thin films were applied to roughened stainless steel substrates using the spin-coating technique. Transverse and longitudinal roughness was patterned on the metal. The bioactivity of sol-gel derived calcium hydroxyapatite coatings in simulated body fluid are presented.

2. Experimental

For the sol-gel processing, calcium acetate monohydrate ($Ca(CH_3COO)_2 \cdot H_2O$; 99.9%; Fluka, Saint Louis, MO, USA, 5.2854 g (0.03 mol)) was dissolved in 40 mL of distilled water and mixed with 4 mL of 1,2-ethanediol (99.0%; Alfa Aesar, Haverhill, MA, USA). Then, a mixture of ethylenediaminetetraacetic acid (EDTA; 99.0%; Alfa Aesar, Haverhill, MA, USA, 9.6439 g (0.033 mol)) and triethanolamine (99.0%; Merck, Darmstadt, Germany, 24 mL) was added to the solution, which was continuously stirred for 10 h. Next, phosphoric acid (H_3PO_4; 85.0%; Reachem, Manhattan, NY, USA, 1.23 mL) was added with stirring for about 24 h. Finally, the obtained sol-gel solution was mixed with 3% solution of polyvinyl alcohol (PVA7200, 99.5%; Aldrich, Saint Louis, MA, USA) in the proportion 25 to 15 mL. This solution was used as sol-gel precursor solution for the synthesis of CHA coatings on prepared 316L stainless steel substrate. The surface of square plates of 10 mm × 10 mm × 0.5 mm was modified by formation of transverse and longitudinal patterned roughness (less than 0.2 μm) prior to the spin-coating of the Ca–P–O sol-gel solution (0.5 mL). A sandpaper (2500 grit) was used to obtain the pattern. Two pieces of sandpaper were lightly rubbed against each other to remove any possible large particles to protect the surface from gouging. Then the metal was roughened with the sandpaper applying medium pressure from an index finger.in two directions 90° apart. A schematic diagram of the preparation of the CHA coatings is presented in Figure 1.

The spinning procedure was performed for 1 min in air with a spinning rate of 2000 revolutions per minute. Each layer in the preparation of CHA multilayers was separately annealed at 850 °C for 5 h in air. CHA thin films with 15, 20, 25, and 30 layers were fabricated. To evaluate in vivo bioactivity of the sol-gel derived CHA coatings, the samples were immersed into simulated body fluid (SBF) for 2, 3, and 4 weeks. The SBF was prepared using the method proposed by Kokuba and Takadama [3]. The concentrations of ions in the prepared SBF solution are presented in Table 1.

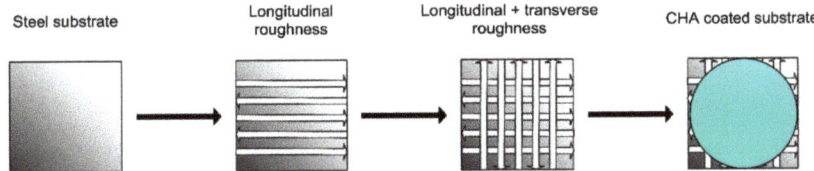

Figure 1. A schematic diagram of the preparation of CHA coatings.

Table 1. Nominal ion concentrations of simulated body fluid (SBF) in comparison with those in human blood plasma.

Ion	Ion Concentration (mM)	
	Blood Plasma	SBF
Na^+	142.0	142.0
K^+	5.0	5.0
Mg^{2+}	1.5	1.5
Ca^{2+}	2.5	2.5
Cl^-	103.0	147.8
HCO_3^-	27.0	4.2
HPO_4^{2-}	1.0	1.0
SO_4^{2-}	0.5	0.5
pH	7.2–7.4	7.40

X-ray diffraction (XRD) analysis was performed with a D8 Focus diffractometer (Bruker AXS Inc., Hamburg, Germany) with a LynxEye detector (Bruker, Billerica, MA, USA) using Cu Kα radiation. The changes in the layers were evaluated using diffuse reflectance infrared Fourier transform spectroscopy (DRIFTS) with a FTIR Spectrum BX II spectrometer (Perkin–Elmer, Waltham, MA, USA). The microstructure and morphology of the obtained samples were investigated using a SU-70 scanning electron microscope (SEM) (Hitachi, Tokyo, Japan). For the evaluation of the hydrophobicity of the CHA coatings, the contact angle measurements were performed using contact angle meter (CAM) (CAM 100, KSV, San Francisco, CA, USA). Distilled water was used for CAM measurements. Raman spectra were measured on Ramboss 500i micro spectrometer (Dongwo, Hsinchu, Taiwan). XPS measurements of the synthesized samples were carried out employing the JPS-9030 spectrometer (JEOL, Akishima, Japan). For tribological measurements, a TriTec SA CSM Tribometer (Anton Paar, Buchs, Switzerland) was used with a ball-on-plate linearly reciprocal configuration, as described by Matijosius et al. [31]. A corundum ball (RGP International Srl, Cinisello Balsamo MI, Italy) of 6 mm (outer diameter) was held stationary, and the load was 1 or 5 N. Stainless steel substrates coated with calcium hydroxyapatite were used as the moving part, which was mounted on a pre-installed tribometer module. Linear reciprocal motion within the amplitude of 2 mm was maintained resulting in a 4 mm range and 8 mm of the total distance for one reciprocal friction cycle. At a velocity of 2 cm/s, each friction cycle produced approx. 100 data points of the 'instantaneous' coefficient of friction (COF).

3. Results and Discussion

Figure 2 shows XRD patterns of CHA films synthesized using the spin-coating technique on a steel surface with patterned roughness.

The multilayer coatings were obtained by increasing the number of spinning and annealing cycles from 15 to 30. The dependence of CHA formation on the times of spinning and annealing is evident. With 15 spinning and annealing procedures, iron oxides (Fe_2O_3 and $Fe_{2.932}O_4$) were the dominant crystalline phases. The iron oxide phase was formed by heating an uncoated but roughened stainless steel substrate at the same temperature (see Figure 3).

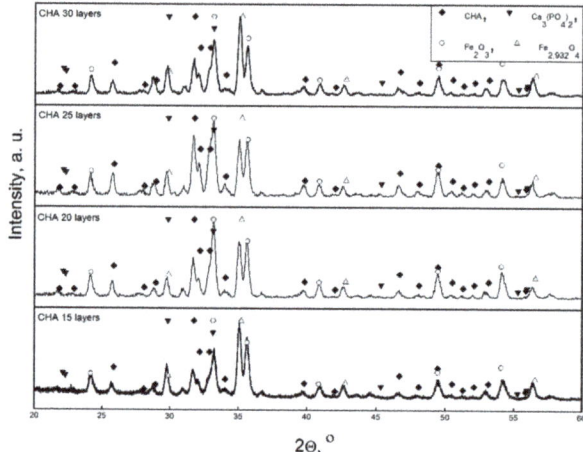

Figure 2. XRD patterns of calcium hydroxyapatite (CHA) coatings fabricated using different times of spinning and annealing procedures.

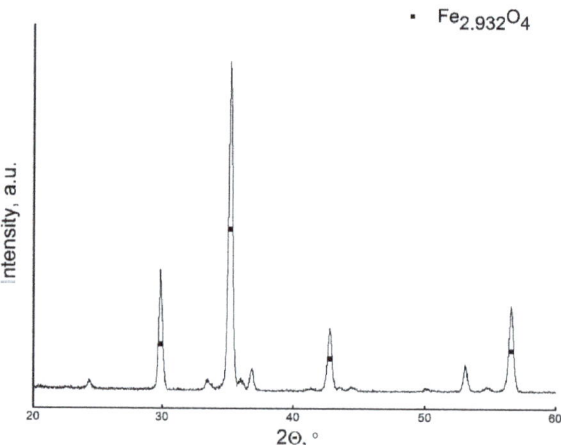

Figure 3. XRD pattern of uncoated rough 316 L stainless steel substrate annealed at 850 °C. $Fe_{2.932}O_4$ is the only dominant phase.

Even after 15 coating cycles, the calcium phosphate phases such as CHA and $Ca_3(PO_4)_2$ have also formed. The intensity of diffraction lines attributable to the CHA phase monotonically increased with increasing the number of layers to 20 and 25. Well resolved CHA diffraction peaks were seen and in the XRD pattern of the sample with 30 layers. However, after 30 spin-coating procedures the intensity of peaks attributable to calcium phosphate phases surprisingly decreased. These XRD results allow us to draw the initial conclusion that the most crystalline CHA sample was obtained with 25 CHA layers.

The FTIR spectra with the DRIFTS attachment are presented in Figure 4 for the CHA coatings.

The FTIR spectra were almost identical among all the samples with various number of layers. The characteristic calcium hydroxyapatite absorption bands in the range of 1100–950 cm^{-1} were clearly visible [32]. Strong bands located at 1035 and 1090 cm^{-1} resulted from the v_1 symmetric P–O stretching vibrations in PO_4^{3-} [32,33]. The broad band visible in the range of 3600–3300 cm^{-1} is associated with O–H stretch vibrations and attributed to adsorbed water [29,32,34]. In all spectra, weak and broad bands in the range of 1550–1370 cm^{-1} were also seen. The origin of these bands is related to the stretching and bending modes of CO_3^{2-} in CHA (C–O bond), despite the last band being slightly shifted to the region of lower wavenumbers. The existence of low-intensity bands at about 870 cm^{-1} may be ascribed to the ν_2 bending mode of CO_3^{2-} (C–O bond) in CHA and confirmed the presence of carbonated apatite in the samples [32]. No specific bands attributable to oxyhydroxyapatite $Ca_{10}(PO_4)_6(OH)_{2-2x}O_x$ were detected in our spectra; however, such bands were present for the CHA thin films deposited on Si substrate [35].

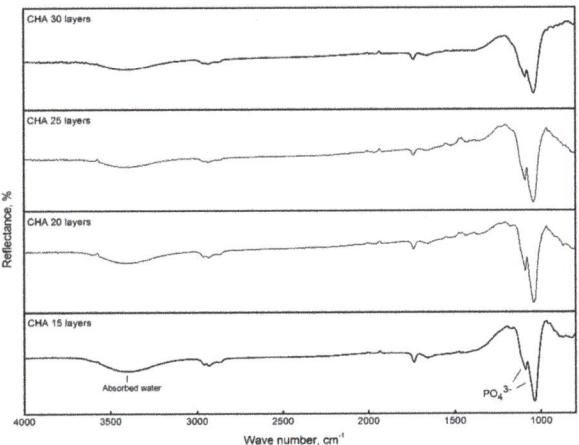

Figure 4. FTIR spectra of sol-gel derived films with 15, 20, 25, 30 layers of CHA.

Figure 5 shows Raman spectra in the 200–1400 cm^{-1} spectral region of the CHA samples deposited on rough stainless steel substrate.

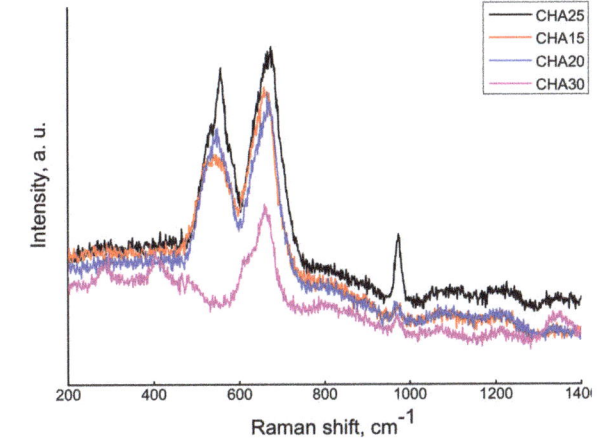

Figure 5. Raman spectra of the CHA samples containing 15, 20, 25, and 30 layers of Ca–P–O gel deposited on rough stainless-steel substrate in 200–1400 cm^{-1} spectral region.

The most intensive Raman bands were observed for the CHA sample having 25 layers. The intensity of the bands monotonically increased with the number of layers but decreased for the specimen having 30 layers. The most intense bands were observed at 580 and 640 cm^{-1}. These bands could be assigned to the triple degenerate (F_2 symmetry) asymmetric bending modes ν_4 of phosphate group in calcium hydroxyapatite [36]. Interestingly, the band at 580 cm^{-1} was not observed in Raman spectrum of the

CHA sample synthesized with 30 layers. The band at about 960 cm^{-1} is due to ν_1 (A$_1$) symmetric stretching vibration of the tetrahedral PO$_4^{3-}$ group. This band was the most intense in the Raman spectrum of the CHA sample synthesized on a silicon nitride substrate [35]. For the specimens in this study, the intensity of this band was substantially less. The peak position of this band confirms the exact stoichiometry (molar ratio Ca:P = 1.667) for calcium hydroxyapatite [37,38]. The Raman spectroscopy results are in good agreement with the XRD analysis data, because in both spectra the most intense peaks attributable to CHA were visible for the sample with 25 layers.

The XPS of samples was measured by calibrating the binding energy scale with C 1s peak as reference. All XPS survey spectra (Figure 6a) exhibited signals characteristic of calcium, oxygen, phosphorous, and carbon. The high-resolution O 1s XPS spectra (Figure 6b) showed a signal centered at 535.1 eV corresponding to O–P–O and OH bonding in hydroxyapatite [39]. Figure 6c shows P 2p XPS spectra, which consist of $2p^{3/2}$ and $2p^{1/2}$ components, corresponding to O–P–O bonding in the (PO$_4$) network of calcium hydroxyapatite and tricalcium phosphate (TCP) [40]. The high resolution Ca 2p spectra are shown in Figure 6d. That spectra consists of two peaks at 351.1 and 354.7 eV. These peaks are attributed to the spin-orbit splitting components of $2p^{3/2}$ and $2p^{1/2}$ with energy difference of 3.4 eV [41]. The elemental analysis, represented as Ca to P ratio, showed that the Ca to P ratio was lower than in stoichiometric calcium hydroxyapatite (1.67). This is not surprising, since the samples also contain TCP. The Ca:P varied between 1.40 and 1.51 depending on the number of sol-gel layers.

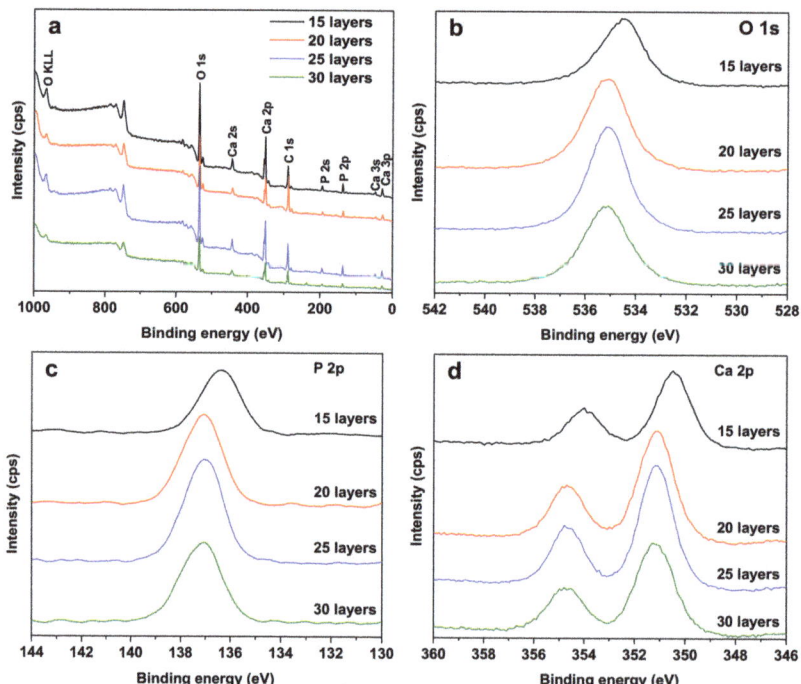

Figure 6. XPS survey spectra (**a**) and high-resolution O 1s (**b**), P 2p (**c**) and Ca 2p (**d**) XPS spectra taken from the surface of deposited and annealed CHA films with different numbers of layers.

The SEM micrographs of the surfaces of obtained CHA samples are shown in Figure 7.

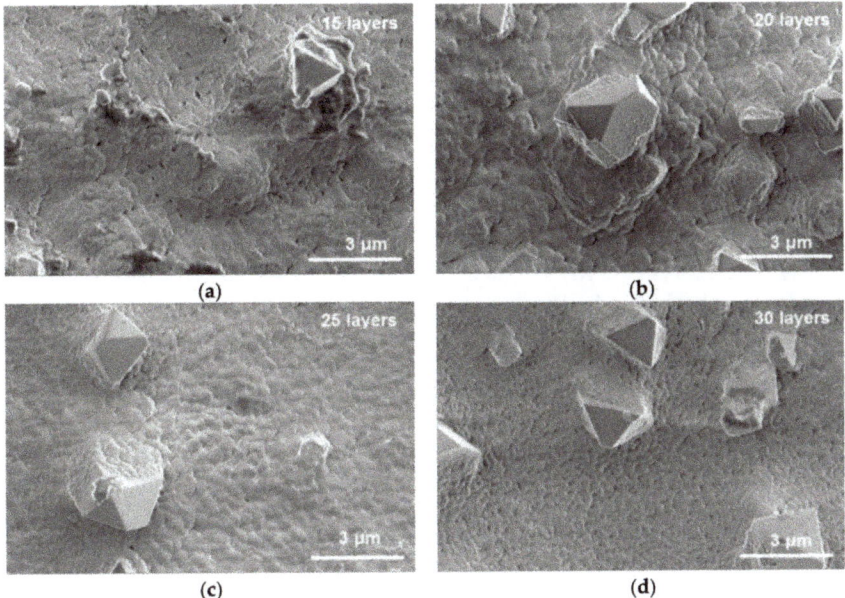

Figure 7. SEM micrographs of the CHA samples containing (**a**) 15, (**b**) 20, (**c**) 25, and (**d**) 30 layers. Magnification 15,000.

The surface of the specimen obtained after 15 dipping times is uneven with clearly pronounced asperities and pores. The surface smoothness was better when the number of layers was 20. The steel substrate after 25 spin-coating procedures was even more uniform. The surface with 25 layers of CHA was composed of homogeneously distributed well interconnected spherical grains about 250 nm in size. The layer was continuous and pore-free. Surprisingly, the morphological features of the sample with 30 layers changed, and the formation of nanospheres on the surface cannot be seen. Nano-sized pores were visible in the sample containing 30 layers of CHA. The irregularly shaped crystals are probably iron oxides. The results of the tribological measurements are presented in Figure 8. COF changes were observed after more loading cycles were performed for either a 1 or 5 N load.

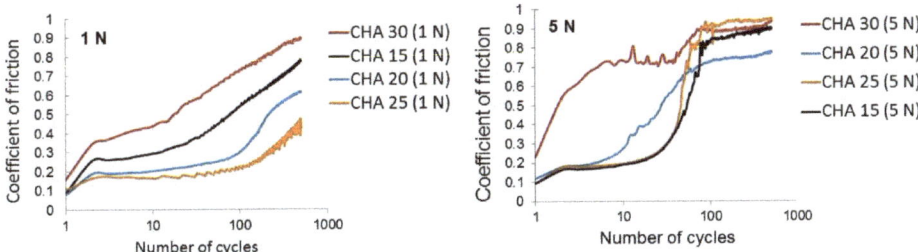

Figure 8. Friction dependence on the coating cycles of CHA layers deposited on the patterned rough surface of stainless steel substrate under (**a**) 1 and (**b**) 5 N loads. Number of layers: 15 layers—CHA 15, 20 layers—CHA 20, 25 layers—CHA 25, and 30 layers—CHA 30.

In most samples, the instantaneous COF values remained similar throughout the selected range of the reciprocal motion, i.e., the middle 80% of a segment. Tribological effectiveness (lowest COF) of sol-gel derived CHA layers coated on the rough stainless steel substrates was evaluated under 1 and 5 N loads. The layer thickness of CHA influenced friction from 15 to 30. Layers of 15 and 30 had the lowest friction by showing a gradual increase of COF after a few friction cycles under 1 N load. Samples with 30 coating cycles had the highest friction. This effect may be explained by rapid CHA layer deformation and degradation, producing wear debris particles that damaged the substrate. The CHA coating with 30 layers was thick enough to break off or delaminate into debris particles. Another possible mechanism to explain such an increase in friction could be related to changes in the surface morphology of these CHA specimens. The best tribological effectiveness (lowest COF) was CHACOF reduction below 0.2 for 100 friction cycles under 1 N load. Friction for the sample with 30 CHA layers evaluated under 5 N load was even higher; the COF value was 0.6 after just a few friction cycles. The samples with 15 and 25 layers had much lower COF, below 0.2 for 30 friction cycles. Despite lower durability (30 vs. 100 friction cycles) the CHA layers on the patterned surface resist abrasion under considerable loads. Thin layers, produced after 15 coating cycles, were tribologically effective under 5 N load, while under 1 N load this effect was not observed. Overall, CHA layers showed the best protective and anti-frictional properties when produced after 25 coating cycles under 1 and 5 N loads. SEM micrographs show that the steel substrate with 25 layers was evenly coated with homogeneously distributed spherical grains of about 250 nm in size.

The contact angle measured for the uncoated patterned stainless-steel substrate was 81.3°, as shown in Figure 9.

The contact angle values of all CHA coatings fell within a small range (113°–116°). The images of water drops on CHA surfaces coated with 15, 20, 25, or 30 sol-gel layers show the CHA coatings were slightly hydrophobic.

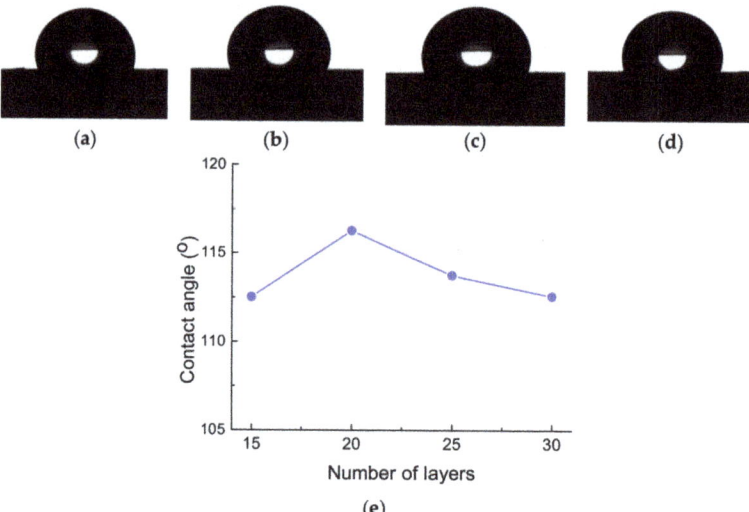

Figure 9. Variation of contact angle of CHA coatings with increasing number of layers (**e**) and the images of water drops on CHA surfaces coated with (**a**) 15, (**b**) 20, (**c**) 25, and (**d**) 30 sol-gel layers.

All CHA samples were immersed into simulated body fluid (SBF) for one month. The phases and morphology were examined after 2, 3, and 4 weeks. The XRD patterns of the CHA samples with 15, 20, 25 and 30 layers after immersion into SBF for 2, 3 and 4 weeks are presented in Figures 10–13. Once placed into SBF, TCP starts to dissolve and induce formation of amorphous calcium phosphate (ACP) and precipitation of CHA. However, the amount of precipitated ACP and CHA is probably too small to be detected using XRD [42]. ACP is the precursor phase of bone-like hydroxyapatite [43]. Loss of crystallinity due to lower peak intensity is visible in Figures 10 and 11 for samples with 15 and 20 layers respectively.

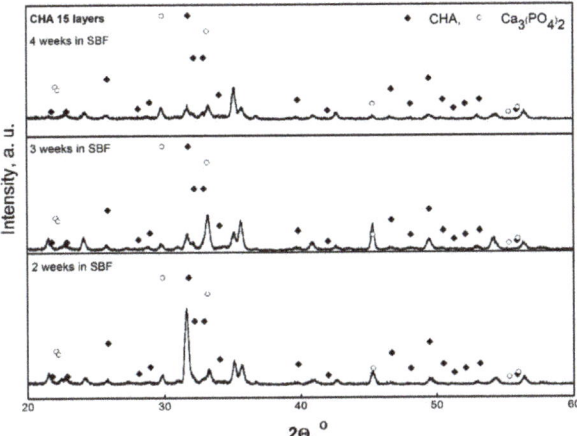

Figure 10. XRD patterns of CHA samples coated with 15 layers after immersion in SBF for 2, 3, and 4 weeks.

Phase changes during one month of soaking in SBF are not monotonic.

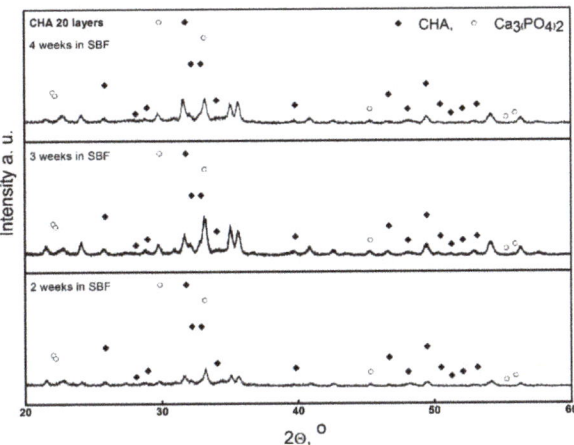

Figure 11. XRD patterns of CHA samples coated with 20 layers after immersion in SBF for 2, 3, and 4 weeks.

The situation changed with 25 layers (Figure 12).

After 1 month of soaking in SBF, a decrease in the intensity of the peaks attributable to both CHA and TCP phases can be seen in the XRD patterns. Amorphous precipitate of CHA and ACP dominated the surface of stainless steel. However, the presence of amorphous calcium phosphate has a very positive impact for bone repair [44]. With further increases in the number of CHA layers, the dissolution and precipitation process remains very similar (Figure 13).

Slightly different actions for similar samples in SBF solution have been observed in other studies. For example, Chen et al. determined that in the early soaking stage the dissolution action dominates [45]. Following this, new precipitates form, indicating that precipitation reaction is dominating. In addition, this process can be different when the soaking time is increased. Thus, the amorphous phosphate phase on the surface formed with increasing soaking time. The most obvious change occurs in the samples with 25 and 30 layers of CHA. Probably due to precipitation of ACP from SBF, the diffraction peaks of

CHA were no longer visible after soaking for more than 3 weeks. The XRD results clearly indicate that ACP formed, which is suitable as a bone substitute material, having sufficient cell proliferation and ALP activity [46].

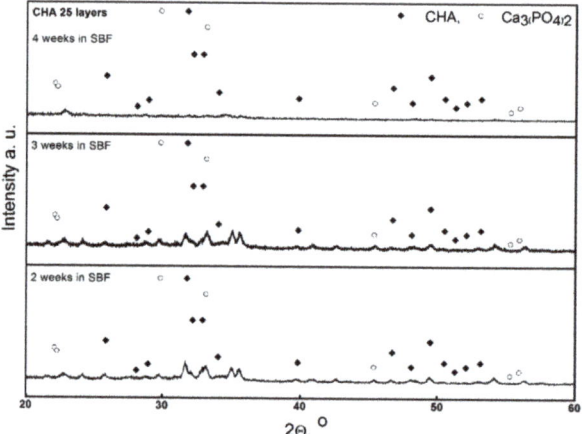

Figure 12. XRD patterns of CHA samples coated with 25 layers after immersion in SBF for 2, 3, and 4 weeks.

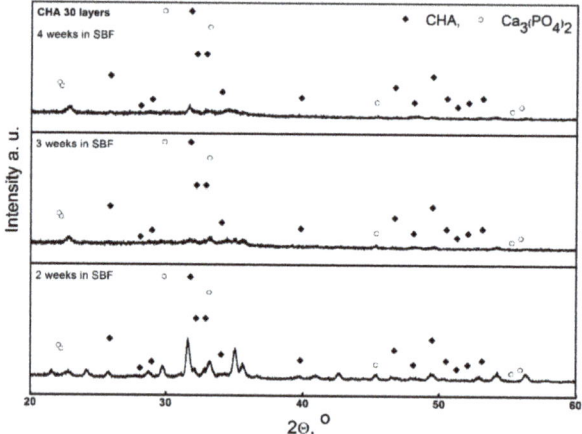

Figure 13. XRD patterns of CHA samples coated with 30 layers after immersion in SBF for 2, 3, and 4 weeks.

The SEM micrographs of the CHA samples with 15, 20, 25, and 30 layers obtained after immersion into SBF for 2, 3, and 4 weeks can be seen in Figure 14.

Interestingly, almost no changes in the surface morphology of CHA coatings were observed after soaking in SBF. The microstructure was not influenced by immersion time or by the number of layers on the substrate, despite the slight changes in phase composition and crystallinity during SBF immersion. Neither etching behavior, nor additional precipitates were observed. Probably, the dissolution and the precipitation processes proceeded simultaneously, without any domination [45]. However, SEM images obtained using secondary electron or backscattered electron modes give poorly distinguishable results among different calcium phosphate phases [46–49]. Loss of crystallinity and formation of various defects including twining, dislocations, stacking faults, and grain boundaries should be investigated using transmission electron microscopy [50].

The SEM results were partially confirmed by the contact angle measurements. All CHA coatings after soaking in SBF were more hydrophilic compared to the initial samples. The values of the contact angle were almost the same for all (100°–98°). Thus, wettability of the CHA samples containing 15–30 layers did not change significantly after 4 weeks of soaking in SBF.

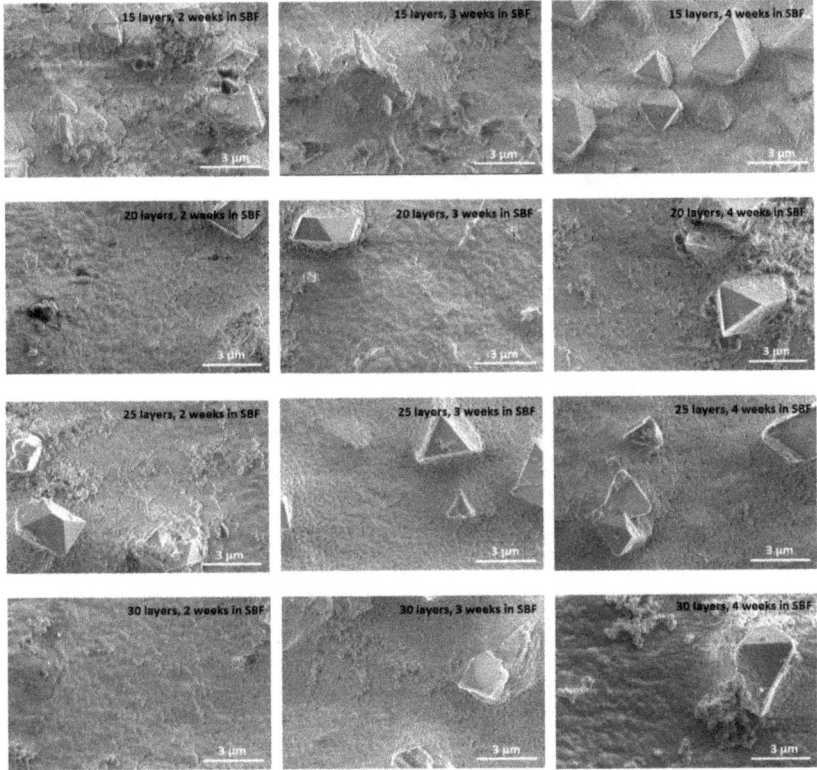

Figure 14. SEM micrographs of CHA samples coated with 15, 20, 25, and 30 layers after immersion in SBF for 2, 3, and 4 weeks.

4. Conclusions

An aqueous sol-gel method was used for the synthesis of calcium hydroxyapatite ($Ca_{10}(PO_4)_6(OH)_2$; CHA) thin films on medical grade stainless steel substrates with transverse and longitudinal patterned roughness. Each layer in the preparation of CHA multilayers (15, 20, 25, and 30) was separately annealed at 850 °C in air. According to the XRD analyses results, the most crystalline CHA sample was obtained with 25 spin-coating and annealing layers. The FTIR and Raman spectroscopy results were in good agreement with the XRD analysis data, confirming the formation of good quality CHA coatings on the rough surface of stainless steel. The formation of CHA coatings was also confirmed by XPS analysis. The chemical bonding such as P–O and Ca–O were noticed in the CHA thin films using XPS measurements. The samples with 25 layers had the lowest friction under both 1 and 5 N loads. The contact angle of coatings (113°–116°) showed the formation of hydrophobic layers of CHA. CHA coated samples were immersed in simulated body fluid (SBF) for 2, 3, and 4 weeks and formed amorphous calcium phosphate (ACP). However, the microstructure of SBF soaked CHA samples was not influenced by immersion time or by the number of layers on the substrate, despite the phase composition and crystallinity differences. All CHA coatings after soaking in SBF were more hydrophilic compared to the initial samples, which could improve osseointegration and promote bone cell proliferation for a better bone-implant connection. ACP is biocompatible and a precursor for CHA, indicating this coating material may be suitable to promote cell proliferation and ALP activity.

Author Contributions: Formal Analysis, V.J., A.Z. and A.K..; Investigation, V.J., S.S., S.W.C., R.J., A.S. and T.M.; Resources, T.C.K.Y., K.I. and R.R.; Data Curation, V.J.; Writing—Original Draft Preparation, V.J.; Writing—Review and Editing, A.K.; Visualization, V.J.; Supervision, S.W.C. and A.K.

Funding: This research was funded by The Japan Society for the Promotion of Science (JSPS). Fellow's ID No.: L12546.

Conflicts of Interest: The authors declare no conflict of interest.

References

1. Furkó, M.; Balázsi, K.; Balázsi, C. Comparative study on preparation and characterization of bioactive coatings for biomedical applications—A review on recent patents and literature. *Rev. Adv. Mater. Sci.* **2017**, *48*, 25–51.
2. Daud, N.M.; Al-Ashwal, R.H.; Kadir, M.R.A.; Saidin, S. Polydopamine-assisted chlorhexidine immobilization on medical grade stainless steel 316L: Apatite formation and in vitro osteoblastic evaluation. *Ann. Anat.* **2018**, *220*, 29–37. [CrossRef] [PubMed]
3. Kokubo, T.; Takadama, H. How useful is SBF in predicting in vivo bone bioactivity? *Biomaterials* **2006**, *27*, 2907–2915. [CrossRef]
4. Bekmurzayeva, A.; Duncansond, W.J.; Azevedo, H.S.; Kanayeva, D. Surface modification of stainless steel for biomedical applications: Revisiting a century-old material. *Mater. Sci. Eng. C* **2018**, *93*, 1073–1089. [CrossRef] [PubMed]
5. Dayan, A.; Lamed, R.; Benayahu, D.; Fleminger, G. RGD-modified dihydrolipoamide dehydrogenase as a molecular bridge for enhancing the adhesion of bone forming cells to titanium dioxide implant surfaces. *J. Biomed. Mater. Res. Part A* **2019**, *107*, 545–551. [CrossRef] [PubMed]
6. Liu, D.M.; Yang, Q.; Troczynski, T. Sol-gel hydroxyapatite coatings on stainless steel substrates. *Biomaterials* **2002**, *23*, 691–698. [CrossRef]
7. Shen, J.; Jin, B.; Qi, Y.C.; Jiang, Q.Y.; Gao, X.F. Carboxylated chitosan/silver-hydroxyapatite hybrid microspheres with improved antibacterial activity and cytocompatibility. *Mater. Sci. Eng. C* **2017**, *78*, 589–597. [CrossRef] [PubMed]
8. Iconaru, S.L.; Chifiriuc, M.C.; Groza, A. Structural and antimicrobial evaluation of silver doped hydroxyapatite-polydimethylsiloxane thin layers. *J. Nanomater.* **2017**, *2017*, 7492515. [CrossRef]
9. Predoi, D.; Popa, C.L.; Chapon, P.; Groza, A.; Iconaru, S. Evaluation of the antimicrobial activity of different antibiotics enhanced with silver-doped hydroxyapatite thin films. *Materials* **2016**, *9*, 778. [CrossRef] [PubMed]
10. Takechi, M.; Ishikawa, K.; Miyamoto, Y.; Nagayama, M.; Suzuki, K. Tissue responses to anti-washout apatite cement using chitosan when implanted in the rat tibia. *J. Mater. Sci. Mater. Med.* **2001**, *12*, 597–602. [CrossRef] [PubMed]
11. Shah, F.A.; Thomsen, P.; Palmquist, A. Osseointegration and current interpretations of the bone-implant interface. *Acta Biomater.* **2019**, *84*, 1–15. [CrossRef] [PubMed]
12. Sadtler, K.; Wolf, M.T.; Ganguly, S.; Moad, C.A.; Chung, L.; Majumdar, S.; Housseau, F.; Pardoll, D.M.; Elisseeff, J.H. Divergent immune responses to synthetic and biological scaffolds. *Biomaterials* **2019**, *192*, 405–415. [CrossRef] [PubMed]
13. Shariff, K.A.; Tsuru, K.; Ishikawa, K. Fabrication of dicalcium phosphate dihydrate-coated beta-TCP granules and evaluation of their osteoconductivity using experimental rats. *Mater. Sci. Eng. C* **2017**, *75*, 1411–1419. [CrossRef] [PubMed]
14. Fujioka-Kobayashi, M.; Tsuru, K.; Nagai, H.; Fujisawa, K.; Kudoh, T.; Ohe, G.; Ishikawa, K.; Miyamoto, Y. Fabrication and evaluation of carbonate apatite-coated calcium carbonate bone substitutes for bone tissue engineering. *J. Tissue Eng. Regener. Med.* **2018**, *12*, 2077–2087. [CrossRef] [PubMed]
15. Kargozar, S.; Hamzehlou, S.; Baino, F. Can bioactive glasses be useful to accelerate the healing of epithelial tissues? *Mater. Sci. Eng. C* **2019**, *97*, 1009–1020. [CrossRef] [PubMed]
16. Zhang, H.; Han, J.; Sun, Y.; Huang, Y.; Zhou, M. MC3T3-E1 cell response to stainless steel 316L with different surface treatments. *Mater. Sci. Eng. C* **2015**, *56*, 22–29. [CrossRef] [PubMed]

17. Akindoyo, J.O.; Beg, M.D.; Ghazali, S.; Alam, A.K.M.M.; Heim, H.P.; Feldmann, M. Synergized poly(lactic acid)-hydroxyapatite composites: Biocompatibility study. *J. Appl. Polym. Sci.* **2019**, *136*, 47400. [CrossRef]
18. Dai, G.; Wan, W.; Chen, J.; Wu, J.; Shuai, X.; Wang, Y. Enhanced osteogenic differentiation of MC3T3-E1 on rhBMP-2 immobilized titanium surface through polymer-mediated electrostatic interaction. *Appl. Surf. Sci.* **2019**, *471*, 986–998. [CrossRef]
19. Balla, V.K.; Das, M.; Bose, S.; Ram, G.J.; Manna, I. Laser surface modification of 316 L stainless steel with bioactive hydroxyapatite. *Mater. Sci. Eng. C* **2013**, *33*, 4594–4598. [CrossRef] [PubMed]
20. Prosolov, K.A.; Belyavskaya, O.A.; Muehle, U.; Sharkeev, Y.P. Thin bioactive Zn substituted hydroxyapatite coating deposited on ultrafine-grained titanium substrate: Structure analysis. *Front. Mater.* **2018**, *5*, 3. [CrossRef]
21. Xin, F.; Jian, C.; Zou, J.P.; Qian, W.; Zhou, Z.C.; Ruan, J.M. Bone-like apatite formation on HA/316L stainless steel composite surface in simulated body fluid. *Transac. Nonferr. Metals Soc. China* **2009**, *19*, 347–352. [CrossRef]
22. Sarkar, C.; Kumari, P.; Anuvrat, K.; Sahu, S.K.; Chakraborty, J.; Garai, S. Synthesis and characterization of mechanically strong carboxymethyl cellulose-gelatin-hydroxyapatite nanocomposite for load-bearing orthopedic application. *J. Mater. Sci.* **2018**, *53*, 230–246. [CrossRef]
23. Testori, T.; Iezzi, G.; Manzon, L.; Fratto, G.; Piattelli, A.; Weinstein, R.L. High temperature-treated bovine porous hydroxyapatite in sinus augmentation procedures: A case report. *Int. J. Periodont. Restor. Dent.* **2012**, *32*, 295–301.
24. Abueva, C.D.; Padalhin, A.R.; Min, Y.K.; Lee, B.T. Preformed chitosan cryogel-biphasic calcium phosphate: A potential injectable biocomposite for pathologic fracture. *J. Biomater. Appl.* **2015**, *30*, 182–192. [CrossRef] [PubMed]
25. Durham III, J.W.; Allen, M.J.; Rabiei, A. Preparation, characterization and in vitro response of bioactive coatings on polyether ether ketone. *J. Biomed. Mater. Res. Part B Appl. Biomater.* **2017**, *105*, 560–567. [CrossRef] [PubMed]
26. Malakauskaite-Petruleviciene, M.; Stankeviciute, Z.; Beganskiene, A.; Kareiva, A. Sol-gel synthesis of calcium hydroxyapatite thin films on quartz substrate using dip-coating and spin-coating techniques. *J. Sol Gel Sci. Technol.* **2014**, *71*, 437–446. [CrossRef]
27. Malakauskaite-Petruleviciene, M.; Stankeviciute, Z.; Niaura, G.; Prichodko, A.; Kareiva, A. Synthesis and characterization of sol-gel derived calcium hydroxyapatite thin films spin-coated on silicon substrate. *Ceram. Int.* **2015**, *41*, 7421–7428. [CrossRef]
28. Usinskas, P.; Stankeviciute, Z.; Beganskiene, A.; Kareiva, A. Sol-gel derived porous and hydrophilic calcium hydroxyapatite coating on modified titanium substrate. *Surf. Coat. Technol.* **2016**, *307*, 935–940. [CrossRef]
29. Jonauske, V.; Prichodko, A.; Skaudzius, R.; Kareiva, A. Sol-gel derived calcium hydroxyapatite thin films on 316L stainless steel substrate: Comparison of spin-coating and dip-coating techniques. *Chemija* **2016**, *27*, 192–201.
30. Rao, P.S.; Murmu, B.; Agarwal, S. Effects of surface roughness and non-newtonian micropolar fluid squeeze film between truncated conical bearings. *J. Nanofluids* **2019**, *8*, 1338–1344. [CrossRef]
31. Matijošius, T.; Ručinskienė, A.; Selskis, A.; Stalnionis, G.; Leinartas, K.; Asadauskas, S. Friction reduction by nanothin titanium layers on anodized alumina. *Surf. Coat. Tech.* **2016**, *307*, 610–621. [CrossRef]
32. Garskaite, E.; Gross, K.A.; Yang, S.W.; Yang, T.C.K.; Yang, J.C.; Kareiva, A. Effect of processing conditions on the crystallinity and structure of carbonated calcium hydroxyapatite (CHAp). *CrystEngComm* **2014**, *16*, 3950–3959. [CrossRef]
33. Mujahid, M.; Sarfraz, S.; Amin, S. On the formation of hydroxyapatite nano crystals prepared using cationic surfactant. *Mater. Res.* **2015**, *18*, 468–472. [CrossRef]
34. Natasha, A.N.; Sopyan, I.; Zuraid, A. Fourier transform infrared study on sol-gel derived manganese-doped hydroxyapatite. *Adv. Mater. Res.* **2008**, *47*, 1185–1188. [CrossRef]
35. Malakauskaite-Petruleviciene, M.; Stankeviciute, Z.; Niaura, G.; Garskaite, E.; Beganskiene, A.; Kareiva, A. Characterization of sol-gel processing of calcium phosphate thin films on silicon substrate by FTIR spectroscopy. *Vibr. Spectrosc.* **2016**, *85*, 16–21. [CrossRef]
36. Usinskas, P.; Stankeviciute, Z.; Niaura, G.; Maminskas, J.; Juodzbalys, G.; Kareiva, A. Sol-gel processing of calcium hydroxyapatite thin films on silicon nitride (Si_3N_4) substrate. *J. Sol Gel Sci. Technol.* **2017**, *83*, 268–274. [CrossRef]

37. Karampas, I.A.; Kontoyannis, C.G. Characterization of calcium phosphates mixtures. *Vibr. Spectrosc.* **2013**, *64*, 126–133. [CrossRef]
38. Sofronia, A.M.; Baies, R.; Anghel, E.M.; Marinescu, C.A.; Tanasescu, S. Thermal and structural characterization of synthetic and natural nanocrystalline hydroxyapatite. *Mater. Sci. Eng. C* **2014**, *43*, 153–163. [CrossRef] [PubMed]
39. Chernozem, R.V.; Surmeneva, M.A.; Krause, B.; Baumbach, T.; Ignatov, V.P.; Tyurin, A.I.; Loza, K.; Epple, M.; Surmenev, R.A. Hybrid biocomposites based on titania nanotubes and a hydroxyapatite coating deposited by RF-magnetron sputtering: Surface topography, structure, and mechanical properties. *Appl. Surf. Sci.* **2017**, *426*, 229–237. [CrossRef]
40. Ramesh, B.; Dillip, G.R.; Rambabu, B.; Joo, S.W.; Raju, B.D.P. Structural studies of a green-emitting terbium doped calcium zinc phosphate phosphor. *J. Molec. Struct.* **2018**, *1155*, 568–572. [CrossRef]
41. Huang, J.; Fan, X.; Xiong, D.; Li, J.; Zhu, H.; Huang, M. Characterization and one-step synthesis of hydroxyapatite-Ti(C,N)-TiO$_2$ composite coating by cathodic plasma electrolytic saturation and accompanying electrochemical deposition on titanium alloy. *Surf. Coat. Technol.* **2017**, *324*, 463–470. [CrossRef]
42. Brazda, L.; Rohanova, D.; Helebrant, A. Kinetics of dissolution of calcium phosphate (Ca–P) bioceramics. *Process. Appl. Ceram.* **2008**, *2*, 57–62. [CrossRef]
43. Wu, M.; Wang, T.; Wang, Q.; Huang, W. Preparation of bio-inspired polydopamine coating on hydrated tricalcium silicate substrate to accelerate hydroxyapatite mineralization. *Mater. Lett.* **2019**, *236*, 120–123. [CrossRef]
44. Xie, L.; Yang, Y.; Fu, Z.; Li, Y.; Shi, J.; Ma, D.; Liu, S.L.; Luo, D. Fe/Zn-modified tricalcium phosphate (TCP) biomaterials: Preparation and biological properties. *RSC Adv.* **2019**, *9*, 781–789. [CrossRef]
45. Chen, Z.; Zhai, J.; Wang, D.; Chen, C. Bioactivity of hydroxyapatite/wollastonite composite films deposited by pulsed laser. *Ceram. Int.* **2018**, *44*, 10204–10209. [CrossRef]
46. Shahrezaee, M.; Raz, M.; Shishehbor, S.; Moztarzadeh, F.; Baghbani, F.; Sadeghi, A.; Bajelani, K.; Tondnevis, F. Synthesis of magnesium doped amorphous calcium phosphate as a bioceramic for biomedical application: In vitro study. *Silicon* **2018**, *10*, 1171–1179. [CrossRef]
47. Heimann, R.B. Plasma-sprayed hydroxylapatite coatings as biocompatible intermediaries between inorganic implant surfaces and living tissue. *J. Therm. Spray Technol.* **2018**, *27*, 1212–1237. [CrossRef]
48. Roy, M.; Bandyopadhyay, A.; Bose, S. Induction plasma sprayed nano hydroxyapatite coatings on titanium for orthopaedic and dental implants. *Surf. Coat. Technol.* **2011**, *205*, 2785–2792. [CrossRef]
49. Xiao, G.Y.; Lü, Y.P.; Zhu, R.F.; Xu, W.H.; Jiao, Y. Fabrication of hydroxyapatite microspheres with poor crystallinity using a novel flame-drying method. *Transac. Nonfer. Metals Soc. China* **2012**, *22*, S169–S174. [CrossRef]
50. Zhu, H.; Guo, D.; Sun, L.; Li, H.; Hanaor, D.A.; Schmidt, F.; Xu, K. Nanostructural insights into the dissolution behavior of Sr-doped hydroxyapatite. *J. Eur. Ceram. Soc.* **2018**, *38*, 5554–5562. [CrossRef]

© 2019 by the authors. Licensee MDPI, Basel, Switzerland. This article is an open access article distributed under the terms and conditions of the Creative Commons Attribution (CC BY) license (http://creativecommons.org/licenses/by/4.0/).

Article

Deposition of Photocatalytic TiO$_2$ Coating by Modifying the Solidification Pathway in Plasma Spraying

Kui Wen [1,2], Min Liu [1,2], Xuezhang Liu [1,2,3,*], Chunming Deng [2] and Kesong Zhou [1,2,*]

1. School of Materials Science and Engineering, Central South University, Changsha 410083, China; wenkui@csu.edu.cn (K.W.); liumin_gz@163.net (M.L.)
2. National Engineering Laboratory for Modern Materials Surface Engineering Technology, The Key Lab of Guangdong for Modern Surface Engineering Technology, Guangdong Institute of New Materials, Guangzhou 510651, China; denghans@126.com
3. School of Materials and Mechanical Engineering, Jiangxi Science and Technology Normal University, Nanchang 330013, China
* Correspondence: xuezhang_liu@126.com (X.L.); kszhou2004@163.com (K.Z.)

Received: 21 September 2017; Accepted: 11 October 2017; Published: 13 October 2017

Abstract: The deposition of photocatalytic TiO$_2$ coatings with plasma spraying is attractive for large-scale applications due to its low cost and simplicity, but it is still a challenge to obtain a TiO$_2$ coating with high anatase content. The solidification pathway of inflight melted particles was investigated in the present paper, and TiO$_2$ coatings with enhanced photocatalytic activity were obtained without a significant loss of the microhardness. The coating microstructure, phase composition, and crystallite size were investigated by scanning electron microscopy (SEM) and X-ray diffraction (XRD). Photocatalytic performance was evaluated by decomposing an aqueous solution of methylene blue. Results showed that the anatase content in TiO$_2$ coating was augmented to 19.9% from 4%, and the time constant of the activity was increased to 0.0046 h^{-1} from 0.0017 h^{-1}.

Keywords: photocatalytic coatings; solidification; plasma spraying; TiO$_2$; microstructure

1. Introduction

Titanium dioxide (TiO$_2$) has been extensively investigated due to its high photocatalytic activity [1,2]. There are three normal crystal phases for TiO$_2$ material: brookite, anatase, and rutile. Under atmosphere pressure and temperature, rutile is the stable phase. Calcined at the temperature about 573 to 1073 K, anatase and brookite will irreversibly transform into rutile. However, the photocatalytic activity of rutile is lower than that of anatase due to an increased electron-hole recombination rate [3,4].

The deposition of TiO$_2$ coating well attached on the surface of a substrate can be more efficient and as compared to TiO$_2$ P25 powder or similar products, owing to the easy recovery of the photocatalyst from the water [5]. Several methods have been reported to prepare TiO$_2$ coatings, including vapor deposition [6,7], electrophoretic deposition [8,9], sol-gel [10], and thermal spraying [11]. In comparison to other methods, thermal spraying has several highlighted features to make it particularly attractive, which include flexibility and high efficiency.

Unfortunately, during the thermal spraying of a plasma jet with a temperature of 14,000 K, the powder feedstock is melted. These molten particles are then driven to deposit on a substrate. Compared with initial feedstock powders, the anatase transformation to rutile is clearly observed in sprayed coatings. Even given P25/20 TiO$_2$ granulated nanopowders as the starting powder, an anatase content of only 1.7%–5% were obtained in the coatings [12]. Then, retaining the metastable anatase phase in thermally sprayed TiO$_2$ coatings becomes a great challenge.

Bozorgtabar et al. [12] demonstrated that photocatalytic activity was strongly related to the process conditions of thermal spraying. Thus, Colmenares-Angulo et al. [13] adjusted spraying conditions to reduce the heat input, and expected to reduce the transformation of anatase TiO_2 to rutile. Ctibor et al. [14] reported that most of the reduced phases were formed in TiO_2 coatings. The stoichiometry was further found to be a function of process parameters. Zhang et al. [15] reported that oxygen vacancies at the surface remarkably prolonged the life of the photon-generated carrier. Thus, it significantly increased the activity. However, others have recently observed that oxygen vacancies caused a significant raise to the charge carrier recombination and resulted into a depressed photocatalytic activity [16].

Herein, modifying the solidification pathway of inflight melted particles was investigated by injecting distilled water into the plasma jet rather than adjusting the spray conditions, and photocatalytic TiO_2 coatings with enhanced activity were obtained. The elaborated coating microstructure was mainly characterized by SEM, while phase composition and crystallite size were investigated by XRD. Finally, the photocatalytic activity of the proposed TiO_2 coatings was evaluated by decomposing an aqueous solution of methylene blue.

2. Experimental

2.1. Atmospheric Plasma Spraying

An atmospheric plasma spraying system (MF-P1000, GTV, Luckenbach, Germany) was used to deposit TiO_2 coatings. A carrier gas with a flow rate of 3.5 L/min radially injected feedstock powder into the plasma jet, while distilled water was radially injected into the plasma jet through a nozzle with average diameter of 0.3 mm by a suspension feeder (GTV, Luckenbach, Germany). The powder injector was 8.0 mm apart from the gun outlet, while the solution injector was 15.0 mm away from the gun outlet. Three typical coatings were obtained with different feeding parameters. As seen in Table 1, the speed of the feed disc regulated the mass of the feedstock powder injected into the plasma jet, and the flow rate controlled the mass of distilled water injected into the jet.

Table 1. The details of plasma spraying parameters for TiO_2 coatings.

Spraying Parameters	Samples		
	TiO_2-12-0	TiO_2-12-30	TiO_2-20-30
Spraying current (A)	650	650	650
Spraying voltage (V)	69	69	69
Primary gas Ar (L/min)	40	40	40
Secondary gas H_2 (L/min)	10	10	10
Speed of feed disc (rpm)	1.2	1.2	2.0
Spraying distance (mm)	100	100	100
Flow rate (mL/min)	0	30	30

Micron-sized commercial TiO_2 powders (−45 + 15 μm, Sunspraying Science and Technology Co. Ltd., Beijing, China) were adopted as feedstock powders (Figure 1). Stainless steel plates (30 mm in diameter, STS316) were employed as substrates. Prior to the deposition, the substrates were cleaned with ethanol to remove contaminants, followed by blasting with Al_2O_3 abrasives.

2.2. Coating Characterization

Scanning electron microscopy (SEM, Nova-Nona-430, Thermo Fisher Scientific, Waltham, MA, USA) was utilized to investigate the microstructure of TiO_2 coatings. A UV-Vis spectrophotometer (HP8453, Agilent, Santa Clara, CA, USA) was used to record spectra in the wavelength range of 250 to 750 nm. Microhardness was obtained with a load of 100 g for 10 s using Vickers (2100B, Instron Tukon, Shanghai, China), and the value was averaged from five indents per specimen.

An X-ray diffractometer (XRD, D8-Advance, Bruker, Billerica, MA, USA) equipped with copper radiation X-ray was employed to assess the phase composition of the proposed coatings. The 2θ was acquired in the range of 20° to 90° with a 0.05° step size. The anatase content was simply determined by the use of the relationship given by Berger-Keller et al. [17]. Meanwhile, crystallite size was calculated using the Scherrer formula [18].

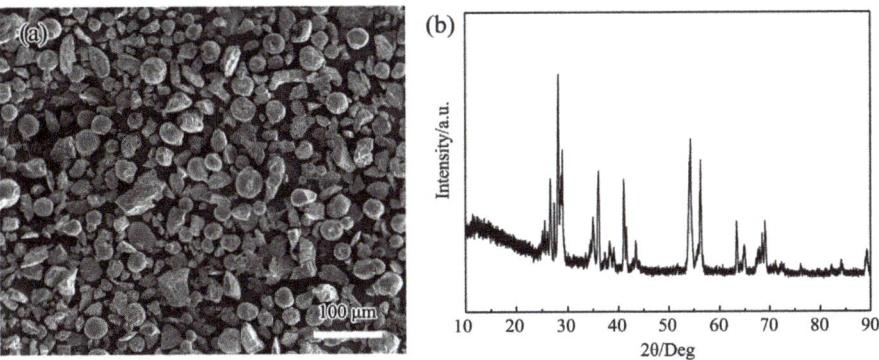

Figure 1. (a) SEM morphology and (b) XRD pattern of feedstock powders.

2.3. Photocatalytic Activity

The photocatalytic behavior of the TiO_2 coatings was assessed from decomposing methylene blue (MB) in a home-made setup. The coated samples were immersed in a glass reactor containing 50 mL of MB solution (5 ppm), and then illuminated with a UV-lamp (λ = 370 nm, I = 2.5 mW/cm^2). The concentration of MB was determined at certain intervals by measuring the solution absorbance with a UV-visible spectrometer at 664 nm wavelength, which was the maximum absorption peak of MB.

3. Results and Discussion

3.1. Crystal Structure and Phase Composition

The crystal structure and phase composition of as-deposited coatings were mainly analyzed by XRD. Figure 2 illustrates the patterns, where the dominant phase is rutile. The minor phase is anatase. It is interesting that there is some preferred orientation in XRD patterns. This may be resulted from the lamellar structure of as-deposited TiO_2 coatings stacked by flat granules, but it should be investigated further. Additionally, Ti_8O_{15} phases are observed in all of as-prepared samples, which is consistent with other works [13,19]. For TiO_2 coatings sprayed by thermal plasma, Magneli phases were widely reported [20]. This is mainly ascribed to the reduction of TiO_2, which results from the high temperature of the thermal plasma jet. There may be other reduced phases, but the analysis is inconclusive because the crystallite size is very small.

As shown in Figure 2, the peak intensity of 2θ = 25.25° (101) becomes stronger as distilled water is injected into the plasma jet. The XRD results illustrate that the total amount of anatase phase is increased by modifying the solidification pathway. Moreover, the anatase content and crystallite size are calculated by the Berger-Keller relationship and Scherrer formula, respectively. The results are demonstrated in Table 2.

At atmospheric pressure, anatase and brookite are two metastable phases. They can transform to rutile at the temperature range of 573–1073 K. However, from a thermodynamics view, the nucleation from the melt could be modified by the temperature and quenching rate [12]. Though anatase is a metastable phase, it preferentially solidifies with a high quenching rate due to its lower surface energy in comparison to rutile ($\gamma_{Anatase}$ = 0.38 J·m^{-2}, γ_{Rutli} = 0.93 J·m^{-2}) [11,21].

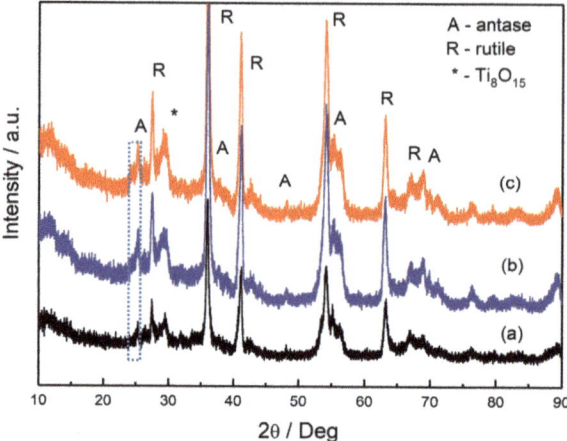

Figure 2. XRD patterns of TiO$_2$ coatings: (**a**) TiO$_2$-12-0, (**b**) TiO$_2$-12-30, and (**c**) TiO$_2$-20-30.

Table 2. Phase compositions and crystallites size of TiO$_2$ coatings.

Items	Samples		
	TiO$_2$-12-0	TiO$_2$-12-30	TiO$_2$-20-30
Anatase (Vol %)	4.0	11.0	19.8
Anatase crystallites average size (nm)	30.4	22.2	19.9
Rutile crystallites average size (nm)	73.0	40.2	27.0
Rate constant (h^{-1})	0.0017	0.0045	0.0046
R^2	0.9965	0.9966	0.9894

During the plasma spraying process, the plasma temperature is higher than 14,000 K. The speed can reach about 500 m/s [22]. TiO$_2$ particles are fully heated by the plasma jet. Consequently, these molten particles with high temperature deposit on the substrate. As the solidification temperature is close to the melting point of TiO$_2$, the particles are apt to nucleate into stable rutile. Therefore, the rutile phase (96%) in the coating is obtained, and the crystallite size is 73 nm.

However, when distilled water is injected into the plasma jet, the stream is fragmented into smaller drops and then evaporated. The evaporation of the droplet solvent can carry off a great deal of heat from melted particles, and those quenching particles with moderate temperature are deposited on the substrate. Thus, the particles solidify with the temperature lower than the melting point of TiO$_2$ and generate the anatase phase. As a result, the anatase content is 11% and 19.9% for TiO$_2$-12-30 and TiO$_2$-20-30, and the crystallite size is 22.21 and 19.9 nm.

3.2. Morphology Characterization

The photocatalytic performance of TiO$_2$ coatings is directly correlated to phase compositions and microstructural characteristics. Thus, the morphology and microstructure of TiO$_2$ coatings are investigated by SEM. The surface micrographs of TiO$_2$ coatings are shown in Figure 3. It can be observed that there is no significant discrepancy among these coated samples. They are all composed of fully molten areas and insufficient molten areas; the latter exhibit a rough surface. During the spraying process, the plasma jet of high enthalpy heats the injected particles and accelerates the melting to form solidified droplets (or splats) onto the substrate. Meanwhile, small voids are observed in coating surface. This is mainly ascribed to un-melted particles. As a result, the pores and roughness can increase the reaction area between the catalyzer and the solution during the degradation of the MB dye.

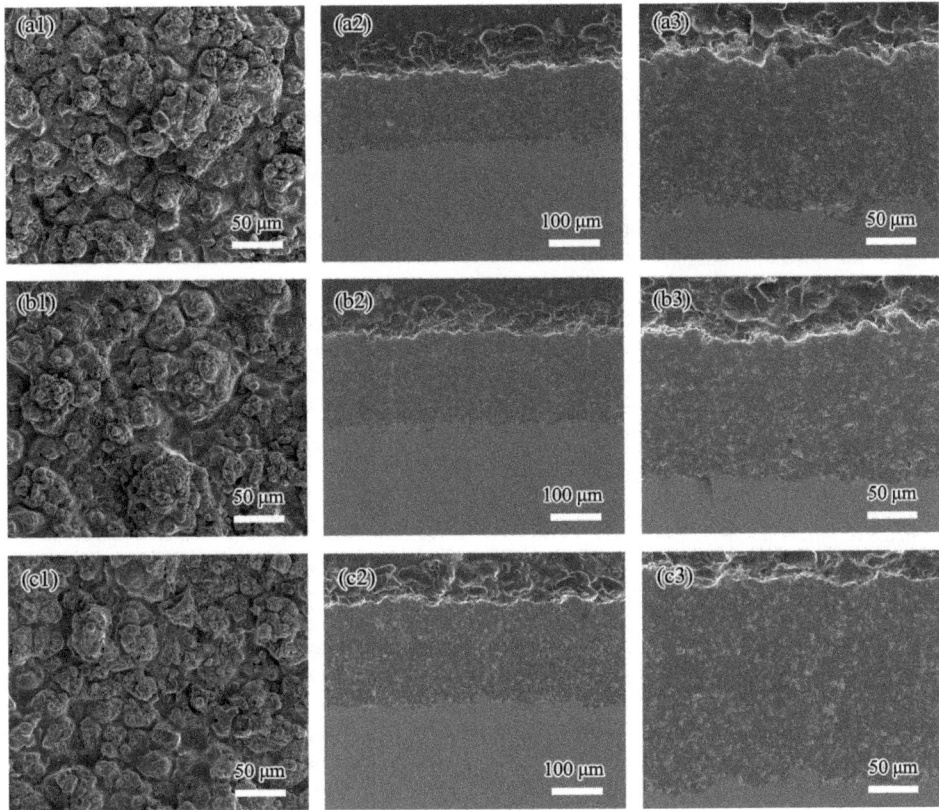

Figure 3. SEM images of TiO_2 coatings: (**a1–a3**) TiO_2-12-0, (**b1–b3**) TiO_2-12-30, and (**c1–c3**) TiO_2-20-30.

The corresponding cross-sections are illustrated in Figure 3 as well. Along with the morphology, those coatings all exhibit lamella structure. Meanwhile, the TiO_2 coatings are firmly bonded to the stainless steel substrates, and there is no crack at the interface. For the TiO_2-12-0 coating, some voids are observed, which may be resulted from the relaxation of the thermal stress of inter-laminar or incomplete contact between the plates and un-melted particles. When modifying the solidification pathway, the cross-section of the TiO_2-12-30 coating becomes fine and the porosity decreases obviously. On the other hand, the coating of TiO_2-20-30 presents a rough and porous cross-section that is caused by the lower melting point of the particles.

In addition, the thicknesses of TiO_2-12-0, TiO_2-12-30, and TiO_2-20-30 are about 150, 180, and 200 μm, respectively. The increased thicknesses of TiO_2-20-30 can be ascribed to the augmented speed of the feed disc (from 1.2 to 2.0 rpm). As the speed increases, more TiO_2 particles are injected into the plasma jet and take part in the coating deposition. However, it is interesting that the thickness of TiO_2-12-30 also becomes thicker when compared with TiO_2-12-0, because the speed of feed disc is 1.2 rpm. The results imply that distilled water injected into the plasma jet can not only modify the solidification pathway of inflight melted TiO_2 particles, but also can increase the deposition efficiency. Further study will be carried out to clarify this in future.

3.3. Microhardness

Functional coatings need sufficient mechanical properties and desired microstructures, while among the mechanical properties, the microhardness plays an important role on the long-term stability. Therefore, the influence of modifying the solidification pathway on the microhardness of TiO_2 coatings was reviewed.

The value of HV_{100} is 1148.27 for the TiO_2-12-0 coating. When adopting the modification, HV_{100} of the TiO_2-12-30 coating changes to 1137.8. This implies that there is no different discrepancy to the microhardness. With the modification, distilled water with a flow rate of 30 mL/min is radially injected into the plasma jet. The evaporation of water decreases the temperature of the plasma jet. However, the solution injector is 15.0 mm away from the gun outlet, which is located downstream of the powder injector, and thus the optimized injection does not violently affect the molten status of TiO_2 particles except those in the solidification pathway. Therefore, it does not cause a significant change to the microhardness.

However, the HV_{100} value slightly decreases to 1064.04 in the TiO_2-20-30 coating. The reason for this can be ascribed to the molten status of TiO_2 particles. As the speed of the feed disc increases from 1.2 to 2.0 rpm, more TiO_2 particles are injected into the plasma jet core. The temperature of the plasma jet already becomes lower with the modification of the solidification pathway, compared with TiO_2-12-0. Then, the heat energy shared by a single TiO_2 particle further decreases as more TiO_2 particles exist in the plasma jet, and some particles impact the substrate as un-melted or partially melted particles. As clarified in Figure 3(c2), this results in a rough and porous cross-section. Consequently, it slightly reduces the microhardness.

3.4. Photocatalytic Properties

The UV-Vis diffuse reflectance spectra of TiO_2 coatings were firstly recorded in the wavelength range of 250–750 nm. The results are illustrated in Figure 4. All coatings have an absorbance edge in the range of 390–400 nm, which is characteristic of semiconductor coatings. The absorbance edge is ascribed to the charge transfer from the valence band to the conduction band. For TiO_2 materials, the valence band is largely caused by $2p$ orbitals of the oxide anions, and the conduction band is constituted with $3d\ t_{2g}$ orbitals of the Ti^{4+} cations [23]. These charge carriers present high reactive activity and can act with different paths. One is the recombination of electron-hole pairs without taking part in the degradation. The other involves gathering at the surface, and reacting with the pollutant [24]. In comparison, all of the spectra show similar curve shapes without new spectrum phenomena. Although the solidification pathway is modified, no notable shift to longer or shorter wavelengths was observed in the TiO_2-12-30 and TiO_2-20-30 coatings. Perhaps the main reason for this is that the rutile phase still dominates in both coatings, though the anatase content increases.

The photocatalytic performances are shown in Figure 5. All coatings present different degrees of activity in decomposing methylene blue with the irritation of UV light. The coating of TiO_2-12-0 shows depressed photocatalytic activity. This is consistent with other works [12,13]. Without modifying the solidification pathway, the anatase content in this coating is as low as 4%. For the TiO_2-12-30 and TiO_2-20-30 coatings, they both exhibit a significant increase in photocatalytic activity. Obviously, the enhanced activity is ascribed to the augmented anatase content in both TiO_2 coatings because they present similar structure. As the TiO_2-20-30 coating has a higher anatase content than TiO_2-12-30, it presents higher activity.

Furthermore, the variation in MB concentration with irradiation time is coincident with Langmuir-Hinshelwood model. The kinetics equation can be depicted by $-\ln C/C_0 = kt$, where C and C_0 are the measured concentration of MB and the initial concentration, respectively, and k is the time constant of the activity (h^{-1}). By a first-order fitting of the exponential term to the time t, the values of k are obtained for all coatings and summarized in Table 2. As presented in Table 2, the k value for the TiO_2-12-30 or TiO_2-20-30 coating is higher than that of the TiO_2-12-0 coating, as the former is deposited with the modification of the solidification pathway. As clarified by the equation, the bigger value of k implies a higher photocatalytic activity.

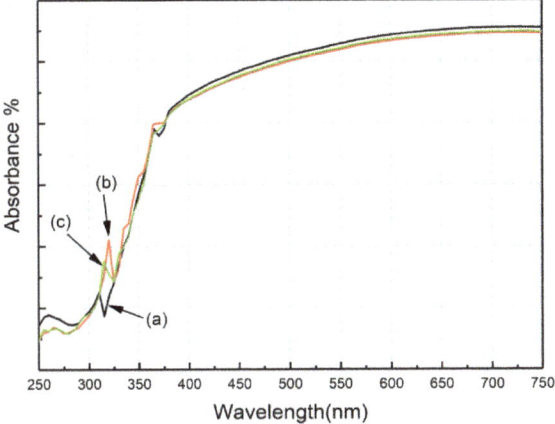

Figure 4. UV-Vis reflectance spectra of TiO_2 coatings: (**a**) TiO_2-12-0, (**b**) TiO_2-12-30, and (**c**) TiO_2-20-30.

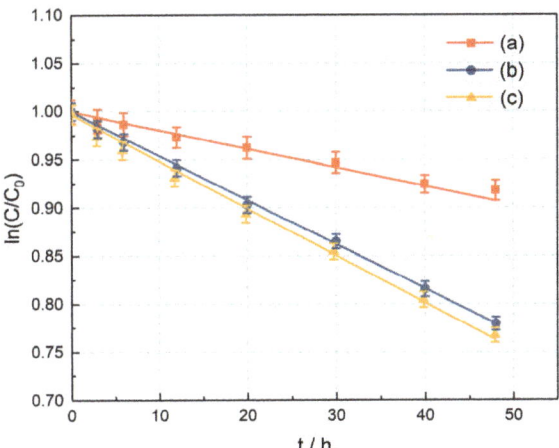

Figure 5. Photocatalytic activity in decomposing methylene blue of TiO_2 coatings: (**a**) TiO_2-12-0, (**b**) TiO_2-12-30, and (**c**) TiO_2-20-30.

4. Conclusions

Distilled water was injected into the plasma jet to modify the solidification pathway of inflight melted TiO_2 particles rather than adjusting the spray conditions. Results showed that the anatase phase was nucleated from melted TiO_2 particles under a high cooling rate due to its lower surface energy comparing to the rutile phase, and photocatalytic TiO_2 coatings with enhanced activity were obtained. Finally, the anatase content in TiO_2 coatings was augmented to 19.9% from 4%, and the time constant of the activity was increased to 0.0046 h^{-1} from 0.0017 h^{-1}. In addition, the modification of the solidification pathway did not remarkably cause the loss of coating microhardness.

Acknowledgments: This work has been financially supported by Guangdong Natural Science Foundation (2016A030312015), Guangdong Academy of Sciences (2017GDASCX-0202, 2017GDASCX-0843, 2016GDASPT-0206, 2016GDASPT-0317).

Author Contributions: Kui Wen, Xuezhang Liu and Kesong Zhou conceived and designed the experiments; Kui Wen and Xuezhang Liu performed the experiments; Kui Wen, Xuezhang Liu, Min Liu and Chunming Deng analyzed the data; Xuezhang Liu and Kui Wen wrote and revised the paper respectively.

Conflicts of Interest: The authors declare no conflict of interest.

References

1. Priyanka, K.P.; Revathy, V.R.; Rosmin, P.; Thrivedu, B.; Elsa, K.M.; Nimmymol, J.; Balakrishna, K.M.; Varghese, T. Influence of La doping on structural and optical properties of TiO_2 nanocrystals. *Mater. Charact.* **2016**, *113*, 144–151. [CrossRef]
2. Daram, P.; Banjongprasert, C.; Thongsuwan, W.; Jiansirisomboon, S. Microstructure and photocatalytic activities of thermal sprayed titanium dioxide/carbon nanotubes composite coatings. *Surf. Coat. Technol.* **2016**, *306*, 290–294. [CrossRef]
3. Zelekew, O.A.; Kuo, D.-H.; Yassin, J.M.; Ahmed, K.E.; Abdullah, H. Synthesis of efficient silica supported TiO_2/Ag_2O heterostructured catalyst with enhanced photocatalytic performance. *Appl. Surf. Sci.* **2017**, *410*, 454–463. [CrossRef]
4. Chung, L.; Chen, W.F.; Koshy, P.; Sorrell, C.C. Effect of Ce-doping on the photocatalytic performance of TiO_2 thin films. *Mater. Chem. Phys.* **2017**, *197*, 236–239. [CrossRef]
5. Cravanzola, S.; Jain, S.M.; Cesano, F.; Damin, A.; Scarano, D. Development of multifunctional TiO_2/MWCNT hybrid composite grafted on stainless-steel grating. *RSC Adv.* **2015**, *5*, 103255–103264. [CrossRef]
6. Quesada-Gonzalez, M.; Boscher, N.D.; Carmalt, C.J.; Parkin, I.P. Interstitial boron-doped TiO_2 thin films: The significant effect of boron on TiO_2 coatings grown by atmospheric pressure chemical vapor deposition. *ACS Appl. Mater. Interfaces* **2016**, *8*, 25024–25029. [CrossRef] [PubMed]
7. Taherniya, A.; Raoufi, D. The annealing temperature dependence of anatase TiO_2 thin films prepared by the electron-beam evaporation method. *Semicond. Sci. Technol.* **2016**, *31*, 125012. [CrossRef]
8. Chava, R.K.; Lee, W.M.; Oh, S.Y.; Jeong, K.U.; Yu, Y.T. Improvement in light harvesting and device performance of dye sensitized solar cells using electrophoretic deposited hollow TiO_2 NPs scattering layer. *Sol. Energy Mater. Sol. Cells* **2017**, *161*, 255–262. [CrossRef]
9. Liu, C.F.; Huang, C.P.; Hu, C.C.; Juang, Y.J. Photoelectrochemical degradation of dye wastewater on TiO_2-coated titanium electrode prepared by electrophoretic deposition. *Sep. Purif. Technol.* **2016**, *165*, 145–153. [CrossRef]
10. Anitha, V.S.; Lekshmy, S.S.; Joy, K. Effect of annealing on the structural, optical, electrical and photocatalytic activity of ZrO_2-TiO_2 nanocomposite thin films prepared by sol-gel dip coating technique. *J. Mater. Sci. Mater. Electron.* **2017**, *28*, 10541–10554. [CrossRef]
11. Mauer, G.; Guignard, A.; Vaßen, R. Plasma spraying of efficient photoactive TiO_2 coatings. *Surf. Coat. Technol.* **2013**, *220*, 40–43. [CrossRef]
12. Bozorgtabar, M.; Rahimipour, M.; Salehi, M.; Jafarpour, M. Structure and photocatalytic activity of TiO_2 coatings deposited by atmospheric plasma spraying. *Surf. Coat. Technol.* **2011**, *205*, S229–S231. [CrossRef]
13. Colmenares-Angulo, J.; Zhao, S.; Young, C.; Orlov, A. The effects of thermal spray technique and post-deposition treatment on the photocatalytic activity of TiO_2 coatings. *Surf. Coat. Technol.* **2009**, *204*, 423–427. [CrossRef]
14. Ctibor, P.; Seshadri, R.C.; Henych, J.; Nehasil, V.; Pala, Z.; Kotlan, J. Photocatalytic and electrochemical properties of single- and multi-layer sub-stoichiometric titanium oxide coatings prepared by atmospheric plasma spraying. *J. Adv. Ceram.* **2016**, *5*, 126–136. [CrossRef]
15. Zhang, J.; Zhao, Z.; Wang, X.; Yu, T.; Guan, J.; Yu, Z.; Li, Z.; Zou, Z. Increasing the oxygen vacancy density on the TiO_2 surface by La-doping for dye-sensitized solar cells. *J. Phys. Chem. C* **2010**, *114*, 18396–18400. [CrossRef]
16. Rajender, G.; Giri, P.K. Strain induced phase formation, microstructural evolution and bandgap narrowing in strained TiO_2 nanocrystals grown by ball milling. *J. Alloy. Compd.* **2016**, *676*, 591–600. [CrossRef]

17. Berger-Keller, N.; Bertrand, G.; Filiatre, C.; Meunier, C.; Coddet, C. Microstructure of plasma-sprayed titania coatings deposited from spray-dried powder. *Surf. Coat. Technol.* **2003**, *168*, 281–290. [CrossRef]
18. Chen, D.; Jordan, E.H.; Gell, M. Porous TiO$_2$ coating using the solution precursor plasma spray process. *Surf. Coat. Technol.* **2008**, *202*, 6113–6119. [CrossRef]
19. Shen, P.K.; He, C.; Chang, S.; Huang, X.; Tian, Z. Magnéli phase Ti8O15 nanowires as conductive carbon-free energy materials to enhance the electrochemical activity of palladium nanoparticles for direct ethanol oxidation. *J. Mater. Chem. A* **2015**, *3*, 14416–14423. [CrossRef]
20. Dosta, S.; Robotti, M.; Garcia-Segura, S.; Brillas, E.; Cano, I.G.; Guilemany, J.M. Influence of atmospheric plasma spraying on the solar photoelectro-catalytic properties of TiO$_2$ coatings. *Appl. Catal. B Environ.* **2016**, *189*, 151–159. [CrossRef]
21. Hanaor, D.A.H.; Sorrell, C.C. Review of the anatase to rutile phase transformation. *J. Mater. Sci.* **2011**, *46*, 855–874. [CrossRef]
22. Cizek, J.; Khor, K.A.; Dlouhy, I. In-flight temperature and velocity of powder particles of plasma-sprayed TiO$_2$. *J. Therm. Spray Technol.* **2013**, *22*, 1320–1327. [CrossRef]
23. Carneiro, J.O.; Azevedo, S.; Fernandes, F.; Freitas, E.; Pereira, M.; Tavares, C.J.; Lanceros-Méndez, S.; Teixeira, V. Synthesis of iron-doped TiO$_2$ nanoparticles by ball-milling process: The influence of process parameters on the structural, optical, magnetic, and photocatalytic properties. *J. Mater. Sci.* **2014**, *49*, 7476–7488. [CrossRef]
24. Toma, F.-L.; Berger, L.-M.; Shakhverdova, I.; Leupolt, B.; Potthoff, A.; Oelschlägel, K.; Meissner, T.; Gomez, J.A.I.; De Miguel, Y. Parameters influencing the photocatalytic activity of suspension-sprayed TiO$_2$ coatings. *J. Therm. Spray Technol.* **2014**, *23*, 1037–1053. [CrossRef]

© 2017 by the authors. Licensee MDPI, Basel, Switzerland. This article is an open access article distributed under the terms and conditions of the Creative Commons Attribution (CC BY) license (http://creativecommons.org/licenses/by/4.0/).

Article

Antibacterial Properties of Zn Doped Hydrophobic SiO$_2$ Coatings Produced by Sol-Gel Method

Bożena Pietrzyk *, Katarzyna Porębska, Witold Jakubowski and Sebastian Miszczak

Institute of Materials Science and Engineering, Faculty of Mechanical Engineering, Lodz University of Technology, Stefanowskiego Str. 1/15, 90-924 Lodz, Poland; katka.porebska@gmail.com (K.P.); witold.jakubowski@p.lodz.pl (W.J.); sebastian.miszczak@p.lodz.pl (S.M.)
* Correspondence: bozena.pietrzyk@p.lodz.pl; Tel.: +48-42-6313043

Received: 6 May 2019; Accepted: 29 May 2019; Published: 1 June 2019

Abstract: Bacteria existing on the surfaces of various materials can be both a source of infection and an obstacle to the proper functioning of structures. Increased resistance to colonization by microorganisms can be obtained by applying antibacterial coatings. This paper describes the influence of surface wettability and amount of antibacterial additive (Zn) on bacteria settlement on modified SiO$_2$-based coatings. The coatings were made by sol-gel method. The sols were prepared on the basis of tetraethoxysilane (TEOS), modified with methyltrimethoxysilane (MTMS), hexamethyldisilazane (HMDS) and the addition of zinc nitrate or zinc acetate. Roughness and surface wettability tests, as well as study of the chemical structure of the coatings were carried out. The antibacterial properties of the coatings were checked by examining their susceptibility to colonization by *Escherichia coli*. It was found that the addition of zinc compound reduced the susceptibility to colonization by *E. coli*, while in the studied range, roughness and hydrophobicity did not affect the level of bacteria adhesion to the coatings.

Keywords: SiO$_2$ coatings; sol-gel; Zn doping; antibacterial coatings; hydrophobic coatings; wettability

1. Introduction

The presence and spread of pathogenic microorganisms in our environment is still a present and growing problem. This problem is exacerbated by the development of microbial resistance to antibiotics and disinfectants. Among the many ways of spreading microorganisms, one of the most widespread is their transfer through tactile contact, in particular through all solid surfaces [1]. Microorganisms living on the touch surfaces can transfer onto the human body during interaction and contribute to the spread of infections [2]. One of the preconditions for the presence of microorganisms on the surfaces of materials is their ability to adhere, which allows the colonization of the surface and the development of a bacterial biofilm [3,4], that is, a source of potential infection associated with the use of materials [5,6]. In addition, microorganisms colonizing surfaces of materials may change their functionality and structure, and even cause degradation (biocorrosion phenomenon) [7].

There are different concepts and strategies for providing antibacterial properties of surfaces [8,9]. These properties, especially increased resistance to colonization and multiplication by bacteria, can be obtained by modifying the surface layer of the material [1,10–12] or by depositing antibacterial coatings with appropriate chemical and physical properties [13–16]. Particularly important, along with free surface energy and surface charge [3,4,17], are hydrophobic properties of the surface, preventing bacterial adhesion and precluding their multiplication. Very important is the degree of surface hydrophobicity [17] and the hydrophobic or hydrophilic properties of the cell walls of the bacteria themselves [18]. The more hydrophobic cells adhere more strongly to hydrophobic surfaces, while hydrophilic cells strongly adhere to hydrophilic surfaces [18]. Insufficient hydrophobicity of the surface, combined with the hydrophobicity of the bacteria, can increase the propensity of microorganisms to adhesion [17].

The solution, consisting in the production of a coating with antimicrobial properties, is a widely studied and frequently applied approach [13,14,19]. A particularly interesting method for producing such coatings can be the sol-gel method, which makes it possible to produce ceramic oxide coatings from the liquid phase in simple and convenient ways and at a relatively low temperature [20]. Colloidal solution, from which coatings are deposited in this method, allows the easy introduction of a wide spectrum of substances modifying the properties of the resulting thin oxide films, such as in terms of bactericidal activity [21–23]. In order to obtain anti-bacterial properties of such coatings, in addition to providing an appropriate surface wettability condition, it is possible to use the antibacterial properties of metal additives, such as Ag [24–26], Zn [21,27] or Cu [27,28], among others [29]. The impact of metals on bacteria, though widely described and proven [24,30–32], is very diverse and not fully understood, as it is most likely a combination of many different mechanisms [33,34].

The combination of the bactericidal effect of metal ions with the hydrophobicity of the surface of the coating can provide an antibacterial effect through the synergy of both phenomena.

In this work, studies of SiO_2 coatings prepared by sol-gel method, with additions of hydrophobizing compounds and zinc compounds, were carried out. The aim was to determine the impact of these additives on the structure and properties of SiO_2 coatings, in particular, on their antibacterial properties against *Escherichia coli*—understood as limiting the susceptibility of the coatings to colonization and survival of bacteria on their surfaces.

2. Materials and Methods

2.1. Preparation of Sols and Deposition of Coatings

A precursor for the preparation of sols was tetraethoxysilane (TEOS) and $Si(OC_2H_5)_4$ (98% Fluka). In this study, three types of sols were used: basic SiO_2 sol and two sols of SiO_2 modified with additives, increasing the hydrophobicity of coatings made from these sols. In order to increase hydrophobicity, the additives of methyltrimethoxysilane (MTMS), $CH_3Si(OCH_3)_3$ (98%, Aldrich) and hexamethyldisilazane (HMDS) $HN[Si(CH_3)_3]_2$ (98%, Fluka) were used.

The SiO_2 basic sol (sol S) was prepared by dissolving the precursor (TEOS) in anhydrous ethyl alcohol (EtOH) and adding 36% hydrochloric acid (HCl) as a catalyst. The molar ratios of TEOS/EtOH/HCl components were 1:20:0.6.

The sol labeled as M was obtained by adding to the sol S methyltrimethoxysilane (MTMS) so that the molar ratio of TEOS/MTMS was 1:0.64.

The third type of sol was obtained in two stages. In the first stage, the precursor (TEOS) was dissolved in anhydrous ethyl alcohol (EtOH) and 25% NH_4OH was added as the catalyst. The TEOS/EtOH/NH_4OH molar ratios were 1:20:0.1. Then, hexamethyldisilazane (HMDS) was added to the above mixture. The TEOS/HMDS molar ratio was 1:0.5. The second step was mixing the prepared liquid with the same volume of sol M (with the addition of MTMS). In the sol thus prepared, the MTMS/HMDS molar ratio was 1.28:1. This sol was labeled as MH.

After preparing the sols MH and M (hydrophobically modified) the amount of solvent (EtOH) was doubled to give the molar ratio TEOS/EtOH 1:40 to ensure a longer life (extended gelation time in the vessel) of these sols.

Zinc nitrate $Zn(NO_3)_2 \cdot 6H_2O$ (98%, Chempur) was used as additive for each type of prepared sol in a molar ratio of TEOS/Zn = 1:0.01, 1:0.05 and 1:0.1. After the addition of zinc nitrate, the sol was stirred continuously using a magnetic stirrer (Wigo, Piastów, Poland) for about 10 h at room temperature.

In addition, in order to compare the properties of the coatings doped with Zn derived from different chemical compounds, zinc acetate $Zn(CH_3COO)_2 \cdot 2H_2O$ (98%, Chempur), was used instead of zinc nitrate additive to the sol M. The molar ratio of TEOS/Zn was preserved and was 1:0.01, 1:0.05 and

1:0.1. After the addition of zinc acetate, the sol was stirred for 2 h at 60°C (to accelerate the dissolution of Zn compound), then 8 h at room temperature.

The priority of choosing the amount of additives was the balance between obtaining the properties of hydrophobic coatings and preserving the stability of the sol as well as the possibility of dissolving the appropriate amount of Zn compounds in the sol. The sols obtained in the manner described above remained stable (i.e., they did not show a tendency to gelate in the vessel or a distinct change in viscosity) for several months.

The labeling of the prepared sols and the chemical compounds used to prepare them are shown in Table 1.

Table 1. Labeling of prepared sols.

Sol Label	Precursor	Hydrophobizer	Zinc Compound	Molar Ratio Si/Zn
S	TEOS	–	–	–
M	TEOS	MTMS	–	–
MH	TEOS	MTMS + HMDS	–	–
SZn1	TEOS	–	Zinc nitrate	0.01
SZn5	TEOS	–	Zinc nitrate	0.05
SZn10	TEOS	–	Zinc nitrate	0.1
MZn1	TEOS	MTMS	Zinc nitrate	0.01
MZn5	TEOS	MTMS	Zinc nitrate	0.05
MZn10	TEOS	MTMS	Zinc nitrate	0.1
MHZn1	TEOS	MTMS+HMDS	Zinc nitrate	0.01
MHZn5	TEOS	MTMS+HMDS	Zinc nitrate	0.05
MHZn10	TEOS	MTMS+HMDS	Zinc nitrate	0.1
MaZn1	TEOS	MTMS	Zinc acetate	0.01
MaZn5	TEOS	MTMS	Zinc acetate	0.5
MaZn10	TEOS	MTMS	Zinc acetate	0.1

Prior to the deposition of coatings, the sols were aged at room temperature for 7 days. The coatings were applied to basic laboratory glass slides with a thickness of 1 mm, as well as on wafers of monocrystalline silicon with an orientation (100) and thickness of 0.3 mm. The surface of the substrates was about 2 cm^2. Before the deposition process, the substrates were washed in an ultrasonic washer (SONIC-3, Polsonic, Warszawa, Poland) in ethyl alcohol and dried with a stream of compressed air.

Coatings were deposited by dip-coating method using dip-coater (TLO 0.1, MTI Corporation, Richmond, CA, USA), by immersing the substrate in a sol and then withdrawing it at a constant rate of 0.2 mm/s. After applying the coating it was dried at room temperature.

2.2. Characterization of Coatings

The evaluation of coating morphology was carried out using optical microscopy (MO) (MM100, Optatech, Warszawa, Poland) and scanning electron microscopy (SEM) (S-3000M, Hitachi, Tokyo, Japan). Magnifications, 200× (MO) for coatings deposited on glass and silicon substrates and 1000× (SEM) for coatings deposited on a silicon substrates, were applied.

Surface topography of chosen coatings deposited on silicon substrate were measured using atomic force microscope (AFM) (MULTIMODE 5, Bruker Corporation, Billerica, MA, USA) working in tapping mode. All investigations were performed under ambient conditions. Image acquisition was performed with the use of Nanoscope 7.3 software and further image processing was done using Nanoscope Analysis 1.5 software (Bruker Corporation). For each sample, the area of 10×10 µm^2 was scanned. From topography images, commonly used roughness parameters were determined (average values taken from 512 surface profiles).

The thickness of the coatings deposited on silicon substrates was determined using X-ray reflectometry by measuring the intensity of the X-ray beam reflected from the surface of the sample. The tests were carried out using an X-ray diffractometer (EMPYREAN, Malvern Panalytical, Worcestershire,

UK) using characteristic CoKα (1,7903 A) radiation. Measurements were carried out in the parallel beam geometry in the omega–2theta angle range of 0.1 to 3 degrees. The value of the critical incidence angle was 0.19–0.21 degrees.

The contact angle, defined as the angle formed between the surface on which the drop of water has been deposited and the surface tangent to the drop at the point of contact with the surface, was determined using the drop shape analyzer (DSA100, Kruss GmBH, Hamburg, Germany). The measurements were carried out for coatings deposited on glass substrates.

The chemical structure of the coatings was studied using an Fourier-transform infrared (FTIR) spectrometer (Nicolet iS50, Thermo Fisher Scientific, Waltham, MA, USA) in the range of 4000–400 cm^{-1}. The study was carried out in transmission mode, recording the absorbance of IR radiation passing through coatings deposited on silicon.

2.3. Antibacterial Properties

2.3.1. Bacterial Colonization

The susceptibility to bacterial colonization of *Escherichia coli* (strain DH5α) was examined on coatings deposited on glass substrates. Directly before the test, the samples were sterilized in water vapor at 121 °C for 20 minutes using a Prestige Medical autoclave. Samples were placed each into a separate flask and immersed in the media containing NaCl (1%), bactopeptone (1%) and yeast extract (0.5%), pH = 7.0. The medium was supplemented with a small number (2×10^3) of *E. coli* cells. The samples were incubated for 24 h at 37 °C. After incubation, sample surfaces were extensively washed with deionized water and gently dried. Next, the solution of fluorescent dyes for the visualization of *E. coli* cells was applied to the surface of sample [35].

2.3.2. Visualization of *E. coli* Cells at the Sample Surface

E. coli cells were observed using fluorescence microscope inspection after the application of bis-benzidine and propidium iodide, making the visualization of both live and dead cells possible.

Each surface was robed with the dyes by applying 10 μL of stock solution of each dyes (100 μg/mL). The dyes were allowed to penetrate the cells and to bind to dsDNA. This process was carried out for 5 minutes at 28 °C in the dark. Finally, bacterial cells present on the sample surface were detected with the usage of the fluorescence microscope (Olympus GX71) and photos were taken with a CCD camera (DC73). Results for six randomly selected separate areas were inspected for each sample. Image acquisition was carried out using the analySIS DOCU software while counting bacteria using the ImageJ software with plugin "cell counter".

The numbers of killed and surviving bacteria cells were determined for each coating, uncoated glass and uncoated stainless steel 316L, which was used as a reference in each experiment. The number of observed bacteria in each case was related to the control and was presented as a percentage of the control, i.e., the percentage of the number of bacteria on the steel control substrate. At the same time, the level of toxicity of coatings for bacterial cells was demonstrated by calculating the percentage of living cells in relation to all present cells on the tested surface—thanks to the use of live/dead fluorescence staining.

The obtained results were calculated as the mean ± standard deviation of the test. The presented results contain averaged values of test results for each of the type of coatings.

The tests were carried out for at least four coatings of each type—one or two coatings from at least three series. The series of coatings refers to coatings of the same type produced in a separate experiment. The series of coatings were produced every few weeks under the same laboratory conditions.

3. Results and Discussion

3.1. Characteristics of Coatings

3.1.1. Microscopic Study

During microscopic examination, it was found that the produced coatings were homogeneous, smooth and without cracks and discontinuities. Only in the case of MH coatings with the addition of zinc nitrate, a small number of fine precipitations (visible in the small, limited areas) were observed.

3.1.2. Thickness of Coatings

Thicknesses were determined for coatings made from S, M and MH sols and their counterparts with the addition of zinc compound in the highest amount used (Zn10). The coatings made from unmodified sol had a thickness of about 64 nm (S—67 nm; SZn10—61 nm). The thickness of the coatings obtained from M and MH modified sols was lower—about 22nm (M—21 nm, MZn10—27 nm, MH—24 nm, MHZn10—18 nm). This is the result of a lower concentration of precursors in these sols.

3.1.3. Topography and Roughness of Coatings

The topography and surface roughness examinations were carried out for coatings made from S, M, MH and SZn10, MZn10, MHZn10 sols, produced on silicon substrates. The results of the roughness measurements are presented in Figure 1.

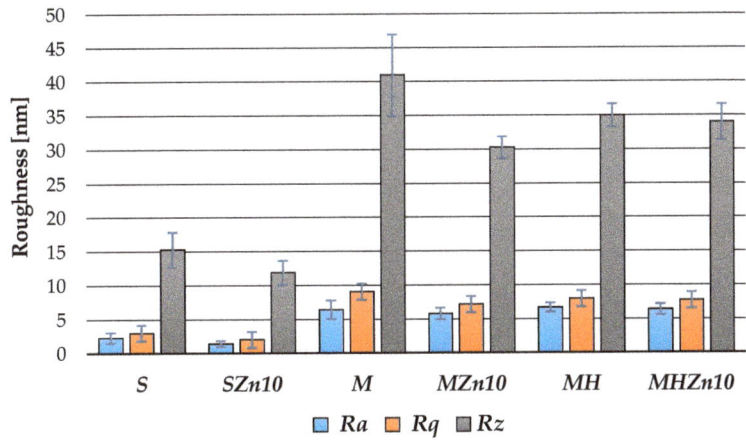

Figure 1. Roughness of coatings with and without Zn additive.

It was found that the modification of coatings using MTMS (M sol) as well as MTMS and HMDS (MH sol) raised the surface roughness compared with unmodified (S) coatings. On the other hand, there was no noticeable change in roughness of coatings with the addition of Zn compared with their Zn-free counterparts. However, it should be noted that despite roughness differences, Ra, Rq and Rz parameters for all of the coatings were very low: Ra = 1.4–6.7 nm; Rq = 2.0–9.1 nm; Rz = 12–41 nm.

The surface topography of S and SZn10, as well as MH and MHZn10 coatings, is shown in Figure 2. Similarly, as in the case of roughness parameters, there was a pronounced difference between coatings made from S and MH sols, while surface topographies of Zn doped and corresponding non-doped coatings (S vs. SZn10 and MH vs. MHZn10) were similar.

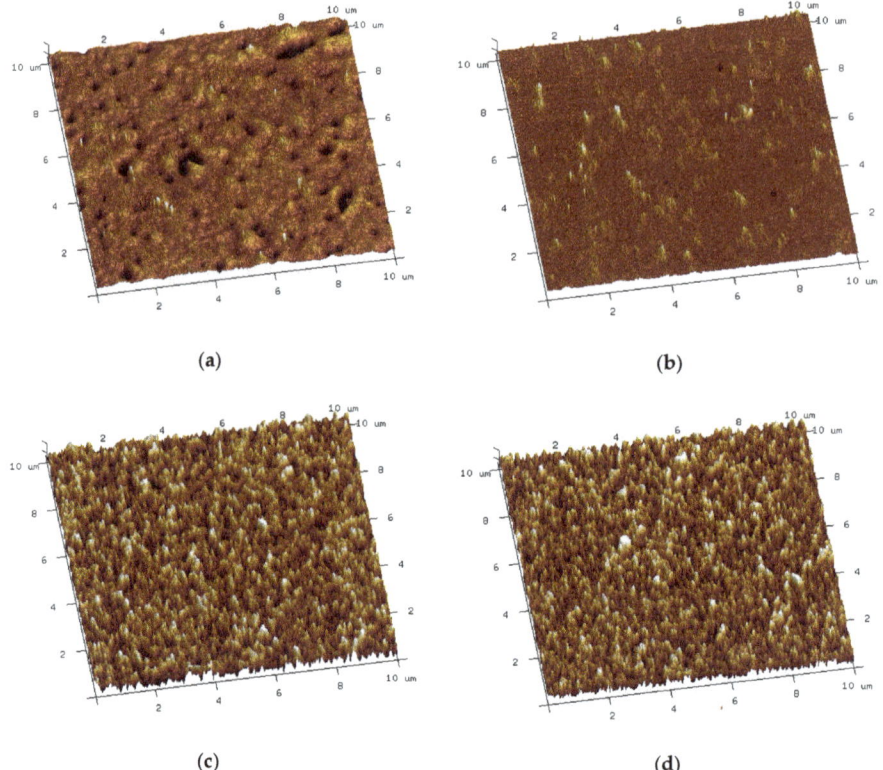

Figure 2. Topography of coatings: (**a**) SiO$_2$ (sol S) coating; (**b**) SiO$_2$+Zn (sol SZn10) coating; (**c**) Modified SiO$_2$ coating (sol MH); (**d**) Modified SiO$_2$+Zn coating (sol MHZn10).

3.1.4. Wettability of Coatings

In order to evaluate the results of coatings modifications with hydrophobic additives (MTMS and HMDS), the measurements of surface contact angles were performed. The results are shown in Figure 3.

The contact angles of coatings deposited from unmodified S sols were between 40° and 46° and were higher than the contact angle of the surface of untreated glass (30°). The modification of sols using hydrophobizers had a significant effect on the level of surface wettability. For coatings prepared from M sols, the contact angle was about 90°–95°. The highest level of hydrophobicity was obtained for coatings prepared from MH sol and was 102°. Figure 3 shows that the Zn additive does not significantly affect the wettability of coatings—differences of contact angles for the same type of sol with various Zn content are very small.

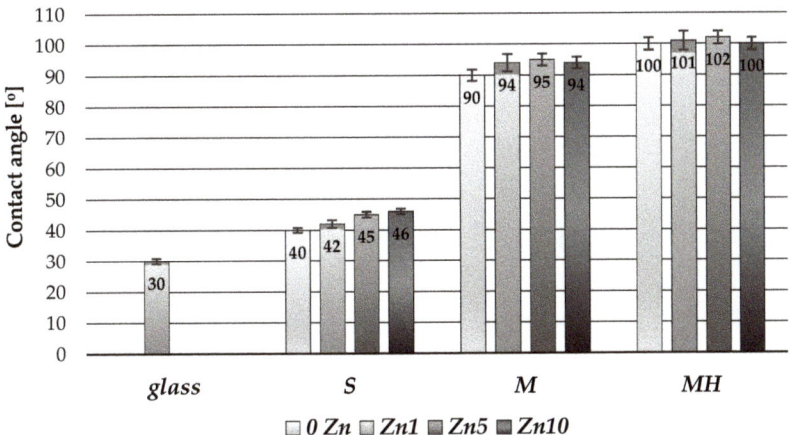

Figure 3. Wettability of coatings.

3.1.5. Chemical Structure of Coatings

The chemical structure of the coatings was analyzed using FTIR infrared spectroscopy for all types of coatings without the addition of Zn (S, M, MH) and with the addition of zinc nitrate in the highest applied amount (SZn10, MZn10, MHZn10). Exemplary FTIR spectra of coatings are presented in Figures 4–6.

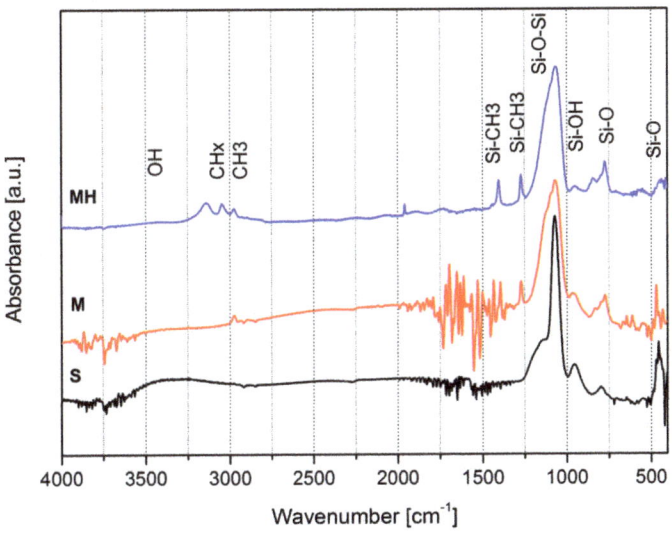

Figure 4. FTIR spectra of coatings obtained from sols S, M and MH.

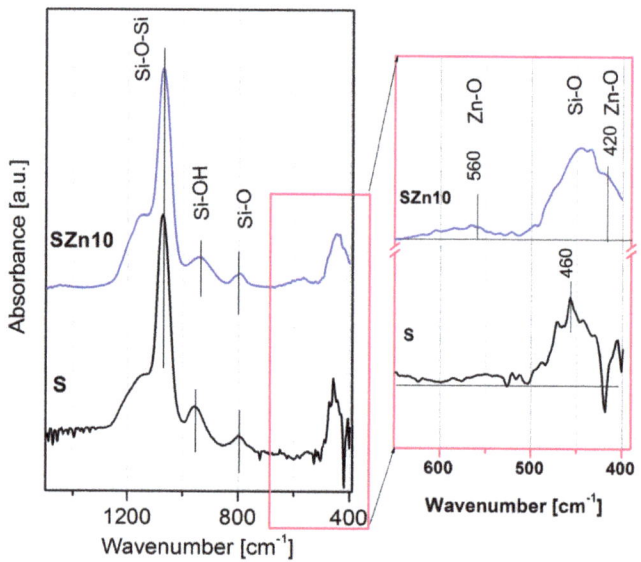

Figure 5. FTIR spectra of S and SZn10 coatings.

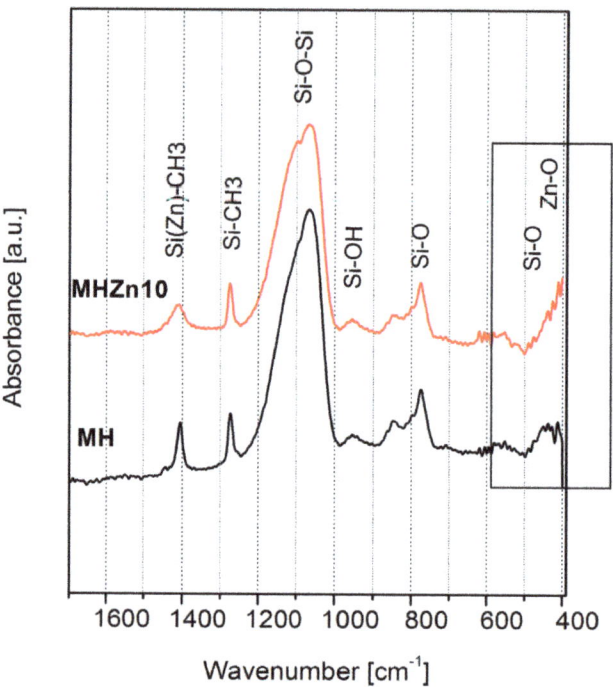

Figure 6. FTIR spectra of MH and MHZn10 coatings.

In the spectra of coatings S, M and MH shown in Figure 4, absorption bands typical for SiO_2 were observed: 460, 800, 1095 cm^{-1}, respectively. In the range of 3000–3500 cm^{-1}, in the spectrum of the S coating, a broad band typical of hydroxyl bonds was visible. In the spectra of coatings M and MH, absorption bands resulting from the presence of methyl groups –CH_3 for 1274 and 2970 cm^{-1} were also visible in the coating structure. Moreover, in the spectrum of the coating MH, bands were also observed from these groups of about 1400 cm^{-1} and in the range of 2970–3100 cm^{-1}. In the spectra of M and MH coatings, the amount of hydroxyl groups (ranging from 3000–3500 cm^{-1}) was reduced as the amount of methyl groups increased. The presence of methyl groups in M and MH coatings was responsible for lowering their wettability (increase of contact angle) [36,37].

Changes in the chemical structure caused by Zn doping of the SiO_2 coating are shown in Figure 5. They were visible in the range of 400–600 cm^{-1}, characteristic for Zn–O bonds. They consisted in the appearance of an additional band of 420 cm^{-1} and a weak band of 560 cm^{-1} in the spectrum of the coating with the addition of Zn compared with the spectrum of the undoped coating.

In the spectra of hydrophobized coatings M and MH, the changes caused by the addition of zinc nitrate were analogous to those obtained from the S sol. These changes are visible within the marked fragment in Figure 6 showing the FTIR spectra of the MH and MHZn10 coatings. Additionally, in the FTIR spectrum of MHZn10 coating, the widening of the 1415 cm^{-1} band was observed, which may indicate the replacement of the Si atom by Zn in the Si–CH_3 connections [38]. The analysis of the spectra shown in Figures 5 and 6 demonstrate that the zinc atoms were bound in the coating by a chemical bond with oxygen (Zn–O) as well as in the Si(Zn)–CH_3 bonds.

3.2. Antibacterial Properties of the Coatings

The results of testing the susceptibility of the surface on microbial colonization and survival of the bacteria on the surfaces of the coatings deposited on glass substrates are shown in Figures 7 and 8. The results of bacterial adhesion testing are presented as the number of bacteria on the tested surface with respect to the number of bacteria on the surface of the steel control sample (% of control). The survival rate of the bacteria are presented as a percentage share of the bacteria living among all the bacteria present on the surface.

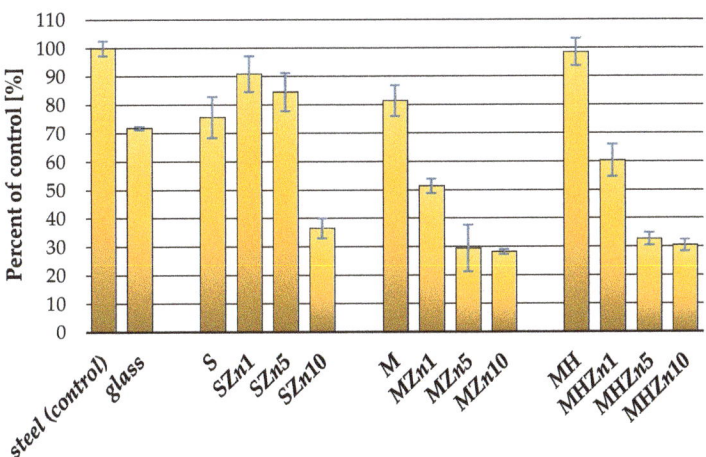

Figure 7. The colonization of bacteria on tested surfaces.

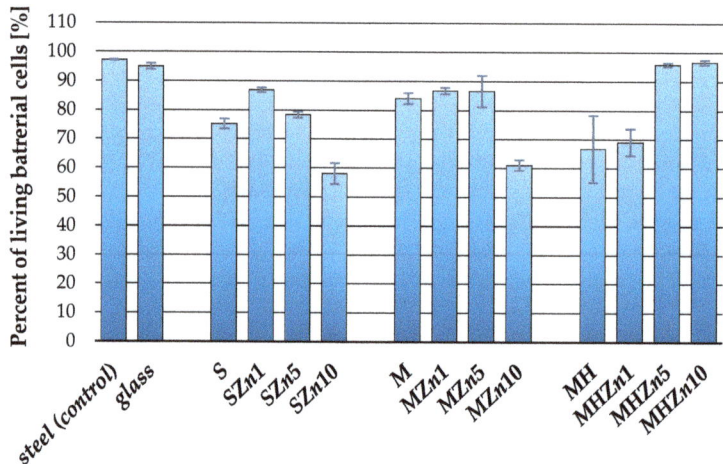

Figure 8. Survivability of bacteria on tested surfaces.

Based on the results shown in Figure 7, it can be concluded that the addition of zinc reduces the degree of surface colonization by *E. coli* bacteria, reducing the amount of bacteria colonizing the surfaces of the coatings compared with those without the addition of Zn. The effect of reducing the number of bacteria adhered to the surface was observed on all types of coatings with the addition of zinc (SZn, MZn and MHZn). For each type of coating, the number of bacteria bound to the surface decreased with the increase in the amount of added zinc.

The best antibacterial effects were obtained for coatings with the maximum amount of used additive Zn10. For all types of coatings doped with zinc Zn10, the number of bacteria was about 30% of the number of bacteria adhered to the surface of the steel control sample.

At the same time, there was no relationship between the survivability of bacteria on the tested surfaces and the type of coating (Figure 8). It can only be concluded that the number of living bacteria *E. coli* on the surface of the coatings was lower than on substrates without coating (steel, glass).

The percentage of living bacteria on the coatings was 60%–85% of the control, with the exception of coatings MHZn5 and MHZn10, where it was comparable to the control surface. However, the conducted studies do not show a clear correlation between the reduction in bacterial survivability and the type of hydrophobic modification or the content of zinc.

In order to determine the influence of zinc addition on coating properties, the properties of coatings doped with various zinc compounds were compared: nitrate $Zn(NO_3)_2$ and acetate $Zn(CH_3CO_2)_2$. Results of colonization and bacterial survivability studies on M type coatings doped by different Zn compounds are presented in Figure 9.

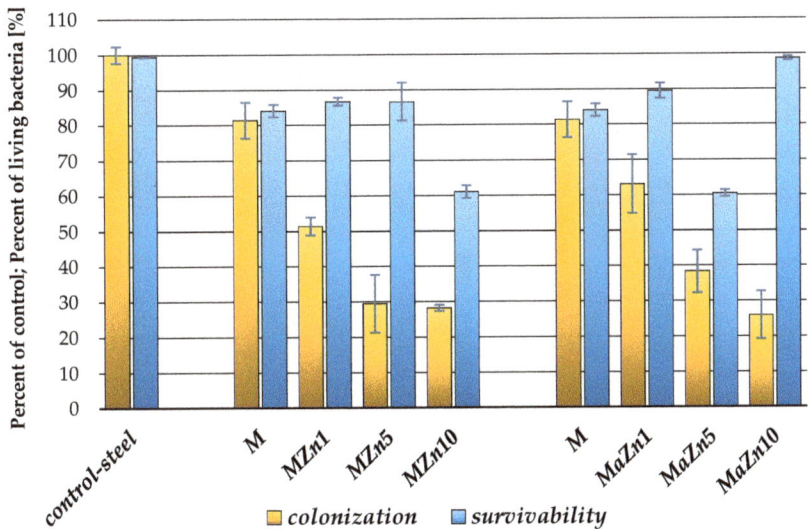

Figure 9. Comparison of antibacterial properties of M type coatings doped by different Zn compounds: nitrate (MZn1, MZn5, MZn10) and acetate (MaZn1, MaZn5, MaZn10).

The antibacterial properties of the tested coatings were similar regardless of the type of compound used to introduce Zn into the coating. For both types of Zn compound, with the increase in its content, the susceptibility to colonization of *E. coli* decreased to less than 30% relative to the control surface (stainless steel), while the survivability of the bacteria did not show a clear tendency with the change of the Zn additive amount.

4. Summary and Conclusions

This paper presents research on topography, surface wettability, chemical structure and level of colonization by *E. coli* of SiO_2 coatings modified with hydrophobizing additives and zinc compounds.

Obtained coatings were homogeneous and smooth (without significant defects or discontinuities) and were not damaged (cracks or detachments from the substrate) after the sterilization processes. No correlation was observed between the coating's roughness and its antibacterial properties. Changes in roughness as a result of the performed modifications were too small (Ra = 1.4–6.7 nm, Rq = 2.0–9.1 nm) to have had a significant effect on bacterial behavior [39].

Coatings showed differences in microscopic surface topography and surface wettability as a result of their modification with hydrophobizers (MTMS and HMDS). Changes in wettability of unmodified and MTMS/HMDS modified coatings resulted from the introduction of –CH_3 (methyl) groups into the coating structure. It was found that MTMS/HMDS-modified coatings showed an increase in hydrophobicity, but the extent of this change was not sufficient enough to significantly reduce the adhesion of *E. coli*. This coincides with literature reports in which the effect of the smaller contact angles in the hydrophobic range on the ability to settle the bacteria is not clear [40,41] and only the achievement of very high wetting angles (above 150°) provides a pronounced resistance to colonization [17].

A factor that had a significant impact on the antibacterial properties of the coatings was the additive of zinc. With the increase of Zn content, the susceptibility of the surface to colonization by *E. coli* decreased regardless of the type of compound used to introduce Zn into the coating. Investigations of the chemical structure of the coatings showed that the Zn atoms (introduced into the silica sol in the form of salts) were incorporated into the coating structure by chemical bonds with oxygen (Zn–O) and alkyl groups, which ensured its antibacterial properties.

A summary of data on wettability and colonization of *E. coli* on the surfaces of all types of coatings without Zn additive and with the highest concentration of doped Zn is shown in Figure 10.

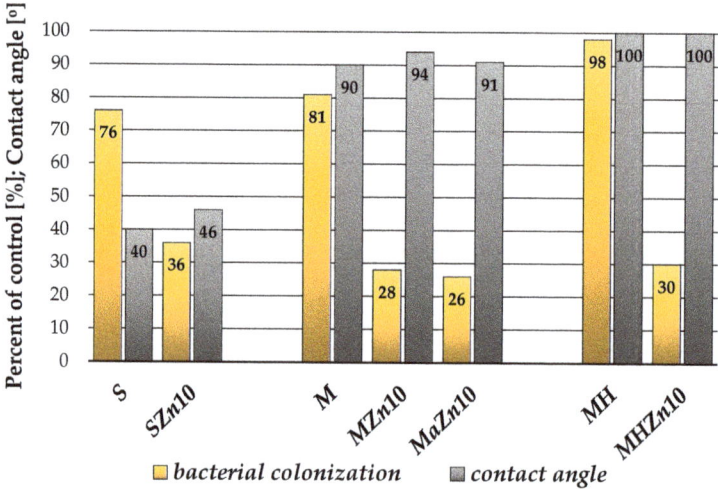

Figure 10. Comparison of mean wettability of the coatings and their colonization by *E. coli*.

Another studied feature of the coatings was the survival rate of the adhered bacteria, which can be treated as a measure of surface toxicity against microorganisms. The conducted research showed no statistically significant effect of any of the studied factors on the survival of *E. coli* bacteria already settled on the surface. Figure 8 shows that the share of living bacteria among the bacteria inhabiting the surface of the coating was slightly lower than that for the control surfaces, but no trend can be pointed out in this respect. This result, however, indicates a beneficial mechanism of antibacterial activity of Zn doped coatings, which reduce the number of bacteria colonizing the surface more than only contributing to their death. In the absence of the reduction of colonization intensity on the surface, dead bacteria would create a layer isolating newly settled cells from the coating containing an antibacterial agent, which would promote the formation of a bacterial biofilm [3,42]. Prevention of bacterial biofilm formation, observed in the case of Zn doped coatings, is a key feature of antibacterial touch surfaces [1,6,43].

The conducted research allows us to conclude that examined coatings show antimicrobial activity by limiting their colonization by bacteria (*E. coli*). The obtained effect is a result of the use of Zn additive in the coatings, not due to the hydrophobic properties of their surface (Figure 10). Therefore, the expected synergy of hydrophobicity and Zn doping in terms of antibacterial activity was not found.

The simple and convenient way of coatings produced using the sol-gel method, with a low thickness and manufacturing process which takes place entirely at room temperature, enables their potential application as antibacterial coatings on many types of substrates.

Author Contributions: P.B.: Concept, Methodology, Investigations, Results Analysis, Writing, Editing and Supervising the Manuscript; P.K.: Methodology, Investigations, Data and Results Analysis; J.W.: Biological Investigations, Data Analysis, Discussion; M.S.: Results Analysis, Discussion, Writing and Editing the Manuscript.

Funding: This research received no external funding.

Acknowledgments: The authors acknowledge Anna Sobczyk-Guzenda for recording of FTIR spectra and Łukasz Kołodziejczyk for obtaining AFM images.

Conflicts of Interest: The authors declare no conflict of interest.

References

1. Villapún, V.M.; Dover, L.G.; Cross, A.; González, S. Antibacterial metallic touch surfaces. *Materials* **2016**, *9*, 736. [CrossRef]
2. Page, K.; Wilson, M.; Parkin, I.P. Antimicrobial surfaces and their potential in reducing the role of the inanimate environment in the incidence of hospital-acquired infections. *J. Mater. Chem.* **2009**, *19*, 3819–3831. [CrossRef]
3. Carniello, V.; Peterson, B.W.; van der Mei, H.C.; Busscher, H.J. Physico-chemistry from initial bacterial adhesion to surface-programmed biofilm growth. *Adv. Colloid Interface Sci.* **2018**, *261*, 1–14. [CrossRef]
4. Song, F.; Koo, H.; Ren, D. Effects of material properties on bacterial adhesion and biofilm formation. *J. Dent. Res.* **2015**, *94*, 1027–1034. [CrossRef] [PubMed]
5. Ramasamy, M.; Lee, J. Recent nanotechnology approaches for prevention and treatment of biofilm-associated infections on medical devices. *BioMed Res. Int.* **2016**, *2016*, 1851242. [CrossRef] [PubMed]
6. Jamal, M.; Ahmad, W.; Andleeb, S.; Jalil, F.; Imran, M.; Nawaz, M.A.; Hussain, T.; Ali, M.; Rafiq, M.; Kamil, M.A. Bacterial biofilm and associated infections. *J. Chin. Med. Assoc.* **2018**, *81*, 7–11. [CrossRef]
7. Kip, N.; van Veen, J.A. The dual role of microbes in corrosion. *ISME J.* **2014**, *9*, 542–551. [CrossRef] [PubMed]
8. Campoccia, D.; Montanaro, L.; Arciola, C.R. A review of the biomaterials technologies for infection-resistant surfaces. *Biomaterials* **2013**, *34*, 8533–8554. [CrossRef] [PubMed]
9. Sun, D.; Babar Shahzad, M.; Li, M.; Wang, G.; Xu, D. Antimicrobial materials with medical applications. *Mater. Technol.* **2015**, *30*, B90–B95. [CrossRef]
10. Elbourne, A.; Crawford, R.J.; Ivanova, E.P. Nano-structured antimicrobial surfaces: From nature to synthetic analogues. *J. Colloid Interface Sci.* **2017**, *508*, 603–616. [CrossRef] [PubMed]
11. Henriques, P.C.; Borges, I.; Pinto, A.M.; Magalhães, F.D.; Gonçalves, I.C. Fabrication and antimicrobial performance of surfaces integrating graphene-based materials. *Carbon* **2018**, *132*, 709–732. [CrossRef]
12. Champagne, V.; Sundberg, K.; Helfritch, D. Kinetically deposited copper antimicrobial surfaces. *Coatings* **2019**, *9*, 257. [CrossRef]
13. Swartjes, J.J.T.M.; Sharma, P.K.; van Kooten, T.G.; van der Mei, H.C.; Mahmoudi, M.; Busscher, H.J.; Rochford, E.T.J. Current developments in antimicrobial surface coatings for biomedical applications. *Curr. Med. Chem.* **2015**, *22*, 2116–2129. [CrossRef] [PubMed]
14. Cloutier, M.; Mantovani, D.; Rosei, F. Antibacterial coatings: Challenges, perspectives, and opportunities. *Trends Biotechnol.* **2015**, *33*, 637–652. [CrossRef] [PubMed]
15. Adlhart, C.; Verran, J.; Azevedo, N.F.; Olmez, H.; Keinänen-Toivola, M.M.; Gouveia, I.; Melo, L.F.; Crijns, F. Surface modifications for antimicrobial effects in the healthcare setting: a critical overview. *J. Hosp. Infection* **2018**, *99*, 239–249. [CrossRef]
16. Li, R.; Jin, Z.T.; Liu, Z.; Liu, L. Antimicrobial double-layer coating prepared from pure or doped-titanium dioxide and binders. *Coatings* **2018**, *8*, 41. [CrossRef]
17. Zhang, X.; Wang, L.; Levänen, E. Superhydrophobic surfaces for the reduction of bacterial adhesion. *RSC Adv.* **2013**, *3*, 12003–12020. [CrossRef]
18. Krasowska, A.; Sigler, K. How microorganisms use hydrophobicity and what does this mean for human needs? *Front. Cell. Infect. Microbiol.* **2014**, *4*, 112. [CrossRef] [PubMed]
19. Salwiczek, M.; Qu, Y.; Gardiner, J.; Strugnell, R.A.; Lithgow, T.; McLean, K.M.; Thissen, H. Emerging rules for effective antimicrobial coatings. *Trends Biotechnol.* **2014**, *32*, 82–90. [CrossRef] [PubMed]
20. Levy, D.; Zayat, M. *The Sol-Gel Handbook: Synthesis, Characterization and Applications*, 1st ed.; Wiley-VCH Verlag GmbH & Co. KGaA: Weindheim, Germany, 2015.
21. Jaiswal, S.; McHale, P.; Duffy, B. Preparation and rapid analysis of antibacterial silver, copper and zinc doped sol-gel surfaces. *Colloids Surf. B Biointerfaces* **2012**, *94*, 170–176. [CrossRef]
22. Owens, G.J.; Singh, R.K.; Foroutan, F.; Alqaysi, M.; Han, C.-M.; Mahapatra, C.; Kim, H.-W.; Knowles, J.C. Sol-gel based materials for biomedical applications. *Prog. Mater. Sci.* **2016**, *77*, 1–79. [CrossRef]
23. Khokhlova, M.; Dykas, M.; Krishnan-Kutty, V.; Patra, A.; Venkatesan, T.; Prellier, W. Oxide thin films as bioactive coatings. *J. Phys. Condens. Matt.* **2018**, *31*, 033001. [CrossRef]
24. Chernousova, S.; Epple, M. Silver as antibacterial agent: Ion, nanoparticle, and metal. *Angew. Chem. Int. Ed.* **2013**, *52*, 1636–1653. [CrossRef] [PubMed]

25. Jung, W.K.; Koo, H.C.; Kim, K.W.; Shin, S.; Kim, S.H.; Park, Y.H. Antibacterial activity and mechanism of action of the silver ion in staphylococcus aureus and *Escherichia coli*. *Appl. Environ. Microbiol.* **2008**, *74*, 2171. [CrossRef] [PubMed]
26. Jeon, H.-J.; Yi, S.-C.; Oh, S.-G. Preparation and antibacterial effects of Ag-SiO2 thin films by sol-gel method. *Biomaterials* **2003**, *24*, 4921–4928. [CrossRef]
27. Jäger, E.; Schmidt, J.; Pfuch, A.; Spange, S.; Beier, O.; Jäger, N.; Jantschner, O.; Daniel, R.; Mitterer, C. Antibacterial silicon oxide thin films doped with zinc and copper grown by atmospheric pressure plasma chemical vapor deposition. *Nanomaterials* **2019**, *9*, 255. [CrossRef]
28. Trapalis, C.C.; Kokkoris, M.; Perdikakis, G.; Kordas, G. Study of antibacterial composite Cu/SiO2 thin coatings. *J. Sol-Gel Sci. Technol.* **2003**, *26*, 1213–1218. [CrossRef]
29. Turner, R.J. Metal-based antimicrobial strategies. *Microb. Biotechnol.* **2017**, *10*, 1062–1065. [CrossRef] [PubMed]
30. Kumar, V.V.; Anthony, S.P. Antimicrobial studies of metal and metal oxide nanoparticles. In *Surface Chemistry of Nanobiomaterials*; Grumezescu, A.M., Ed.; William Andrew Publishing: Norwich, NY, USA, 2016; pp. 265–300.
31. Vincent, M.; Hartemann, P.; Engels-Deutsch, M. Antimicrobial applications of copper. *Int. J. Hyg. Environ. Health* **2016**, *219*, 585–591. [CrossRef] [PubMed]
32. Pasquet, J.; Chevalier, Y.; Pelletier, J.; Couval, E.; Bouvier, D.; Bolzinger, M.-A. The contribution of zinc ions to the antimicrobial activity of zinc oxide. *Colloids Surf. A Physicochem. Eng. Aspects* **2014**, *457*, 263–274. [CrossRef]
33. Lemire, J.A.; Harrison, J.J.; Turner, R.J. Antimicrobial activity of metals: mechanisms, molecular targets and applications. *Nat. Rev. Microbiol.* **2013**, *11*, 371. [CrossRef] [PubMed]
34. Mittapally, S.; Taranum, R.; Parveen, S. Metal ions as antibacterial agents. *JDDT* **2018**, *8*, 411–419. [CrossRef]
35. Jakubowski, W.; Bartosz, G.; Niedzielski, P.; Szymanski, W.; Walkowiak, B. Nanocrystalline diamond surface is resistant to bacterial colonization. *Diam. Relat. Mater.* **2004**, *13*, 1761–1763. [CrossRef]
36. Cai, S.; Zhang, Y.; Zhang, H.; Yan, H.; Lv, H.; Jiang, B. Sol-gel preparation of hydrophobic silica antireflective coatings with low refractive index by base/acid two-step catalysis. *ACS Appl. Mater. Interfaces* **2014**, *6*, 11470–11475. [CrossRef]
37. Philipavičius, J.; Kazadojev, I.; Beganskienė, A.; Melninkaitis, A.; Sirutkaitis, V.; Kareiva, A. Hydrophobic antireflective silica coatings via sol-gel process. *Mater. Sci.* **2008**, *14*, 283–287.
38. Long, D.A. Infrared and Raman Characteristic Group Frequencies. Tables and charts George Socrates John Wiley and Sons, Ltd, Chichester, Third Edition, 2001. *J. Raman Spectrosc.* **2004**, *35*, 905.
39. Preedy, E.; Perni, S.; Nipič, D.; Bohinc, K.; Prokopovich, P. Surface roughness mediated adhesion forces between borosilicate glass and Gram-Positive Bacteria. *Langmuir* **2014**, *30*, 9466–9476. [CrossRef]
40. Tang, L.; Pillai, S.; Revsbech, N.P.; Schramm, A.; Bischoff, C.; Meyer, R.L. Biofilm retention on surfaces with variable roughness and hydrophobicity. *Biofouling* **2011**, *27*, 111–121. [CrossRef]
41. Bazaka, K.; Crawford, R.J.; Ivanova, E.P. Do bacteria differentiate between degrees of nanoscale surface roughness? *Biotechnol. J.* **2011**, *6*, 1103–1114. [CrossRef]
42. Garrett, T.R.; Bhakoo, M.; Zhang, Z. Bacterial adhesion and biofilms on surfaces. *Prog. Nat. Sci.* **2008**, *18*, 1049–1056. [CrossRef]
43. Bjarnsholt, T. The role of bacterial biofilms in chronic infections. *APMIS* **2013**, *121*, 1–58. [CrossRef] [PubMed]

© 2019 by the authors. Licensee MDPI, Basel, Switzerland. This article is an open access article distributed under the terms and conditions of the Creative Commons Attribution (CC BY) license (http://creativecommons.org/licenses/by/4.0/).

Article

Influence of Silicon-Modified Al Powders (SiO₂@Al) on Anti-oxidation Performance of Al₂O₃-SiO₂ Ceramic Coating for Carbon Steel at High Temperature

Bo Yu [1,2], Guoyan Fu [1,3], Yanbin Cui [1], Xiaomeng Zhang [1], Yubo Tu [1,2], Yingchao Du [1,2], Gaohong Zuo [1,4], Shufeng Ye [1] and Lianqi Wei [1,*]

1. State Key Laboratory of Multiphase Complex System, Institute of Process Engineering, Chinese Academy of Sciences, PO Box 353, Beijing 100190, China; yubo@ipe.ac.cn (B.Y.); gyfu@ipe.ac.cn (G.F.); ybcui@ipe.ac.cn (Y.C.); xmzhang@ipe.ac.cn (X.Z.); ybtu@ipe.ac.cn (Y.T.); ycdu@ipe.ac.cn (Y.D.); yglu@ipe.ac.cn (G.Z.); sfye@ipe.ac.cn (S.Y.)
2. Department of Chemical Engineering, University of Chinese Academy of Sciences, No. 19(A) Yuquan Road, Beijing 100049, China
3. China ENFI Engineering Corporation, No. 12 Fuxing Road, Beijing 100038, China
4. Department of Textile and Material Engineering, Dalian Polytechnic University, No. 1 Light Industry Court, Ganjingzi District, Dalian 116034, China
* Correspondence: lqwei@ipe.ac.cn; Tel.: +86-10-8254-4899

Received: 24 January 2019; Accepted: 26 February 2019; Published: 4 March 2019

Abstract: In this paper, silicon-modified Al powders (SiO₂@Al) were prepared by tetraethyl orthosilicate (TEOS) hydrolysis under alkaline conditions. Using SiO₂@Al as additives, a new Al₂O₃-SiO₂ ceramic coating (ASMA) was formed on carbon steel to prevent carbon steel from oxidization at 1250 °C for 120 min. Compared with the Al₂O₃-SiO₂ ceramic coating without additive (AS), ASMA showed a remarkably better anti-oxidation performance, especially during the temperature-rise period. According to the characterization conducted by TG-DTA, XRD and SEM-EDS, it was found that the metallic Al in ASMA melted at 660 °C and reacted with SiO₂ on its surface, which generated local high temperature and accelerated the sintering of ceramic raw materials. The mullite and hercynite formed in ASMA also played a major role for enhancing the anti-oxidation performance of ceramic coating.

Keywords: ceramic coating; anti-oxidation; SiO₂@Al additive; carbon steel

1. Introduction

Carbon steel always involves a slab reheating and hot rolling process to obtain thin carbon steel. During the reheating process, slabs suffer from serious oxidation over 1250 °C, which results in weight loss of slabs and energy waste [1–3]. Moreover, oxidation could potentially lead to serious problems for the surface quality of carbon steel, such as decarburization and micro-defects [4,5]. Different methods have been taken to protect slabs from oxidization during the reheating process. The most effective countermeasure for the reduction of the oxidation intensity is to isolate slabs from oxidants. Therefore, reheating in reducing atmosphere, such as vacuum or inert gas atmosphere, is once thought to be an ideal solution. However, the rigid operation conditions and expensive equipment limit its industrial application [6–8].

Protective coating, an effective and economical method, is considered as an alternative solution to prevent the slabs from oxidization at high temperature [9,10]. Ceramic coating (MgO, Al₂O₃, SiO₂, CoO, ZnO, ZrO₂ etc.) usually has a high melting point and excellent chemical stability at high temperature [11,12]. By using ceramic coating, a protective layer can be formed on carbon steel surface at temperatures ranging from 1100 to 1300 °C so that the internal diffusion of oxygen and external diffusion of iron ions are slowed down [11,13–16].

Among the ceramic coatings, Al_2O_3-SiO_2 coating (AS) is widely used due to its high anti-oxidation performance and low cost [17]. Silicates with low melting points are always added into Al_2O_3-SiO_2 coating to accelerate the sintering under low temperature. However, a eutectic crystal of Fe–2FeO·SiO_2 is formed over 1170 °C, which is harmful to the anti-oxidation performance of the protective coating at high temperature [18,19]. In order to accelerate the sintering of Al_2O_3-SiO_2 coating under low temperature and avoid the formation of Fe–2FeO·SiO_2 over 1170 °C, silicates with low melting point should be replaced by other additives [20,21]. Aluminite powder is an alternative additive because it can melt at 660 °C and react with SiO_2 [22]. The exothermic reaction between Al and SiO_2 leads to local high temperature and thus the sintering of ceramic materials is accelerated. Sodium silicate and colloidal silica are widely used as binder agents in ceramic coating preparation. Functional colloidal sol would form under the hydrolysis of sodium silicate or metasilicic acid and be balanced with the free OH^-. If Al powders are directly added into ceramic coating slurry containing alkaline binder such as sodium silicate and colloidal silica, a exothermic reaction ($2Al + 2H_2O + 2OH^- \rightarrow 2AlO_2^- + 3H_2$) will take place and break the balance of hydrolysis within the sodium silicate solution ($SiO_3^{2-} + 2H_2O \rightarrow H_2SiO_3 + 2OH^-$). It was the consumption of OH^- that would accelerate the hydrolysis and enhance the viscosity of the slurry. Viscous slurry is hard to disperse by spraying due to its poor fluidity. In addition, the hydrogen produced by the reaction is flammable and it is hazardous in the reheating workshop. Therefore, Al powders should be modified to be stable with the coating slurry.

In this paper, silicon-modified Al powders (SiO_2@Al) were prepared by tetraethyl orthosilicate (TEOS) hydrolysis under alkaline conditions. A new Al_2O_3-SiO_2 ceramic coating (ASMA) was prepared with SiO_2@Al as additives to prevent carbon steel from oxidization. The anti-oxidation performance of ASMA was investigated by a heating process from room temperature to 1250 °C with a rate of 10 °C/min and maintained for 120 min at 1250 °C. A protective mechanism of ASMA was also clarified. Some achievements in this paper were not only applied to anti-oxidation of carbon steel at high temperature but also provided new ideas for the design of low-temperature sintering ceramic.

2. Experiment Procedure

2.1. Preparation of Carbon Steel Sample

Carbon steel, J55 (C 0.28 wt %, Si 0.27 wt %, Mn 1.35 wt % and Fe balance), with a specimen size of $55 \times 50 \times 5$ mm^3 was used for high temperature treatment in muffle furnace. A carbon steel specimen ($25 \times 24 \times 5$ mm^3) was cut from cold-rolled plates for continuous thermo-balance investigation. The specimens were cleaned with alcohol in an ultrasonic bath and dried in an oven.

2.2. Preparation of Modified Al Powders (SiO_2@Al)

Al powders with the size of 45 μm, tetraethyl orthosilicate (TEOS), ethanol, ammonia water, and deionized water were used for the preparation of SiO_2@Al. A mixture of 10 g Al powders and 250 mL ethanol was stirred in a beaker at 50 °C for 1 h. The mixture was blended with 15 mL TEOS under agitation, which was diluted with 150 mL alcohol. After that, 20 mL ammonia water diluted with 25 mL deionized water was dropped into the mixture by a peristaltic pump at 2 mL/min as the catalyst for TEOS hydrolysis. Then, the beaker was kept in a water bath at 50 °C for 8 h. The obtained SiO_2@Al was separated from the resulting mixture by vacuum filtration, followed by washing several times with ethanol. Then SiO_2@Al was dried at 60 °C under vacuum.

The synthesis of SiO_2@Al through TEOS hydrolysis under alkaline condition is illustrated in Figure 1. The SEM-EDS results in Figure 2 indicated that Al powders were successfully encapsulated by SiO_2 and thus $SiO_2@Al$ was obtained. The size of SiO_2 encapsulating on the surface of Al powders was about 100 nm.

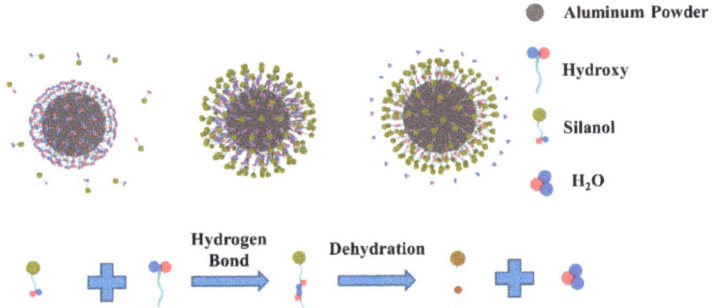

Figure 1. The schematic illustration of $SiO_2@Al$ synthesis.

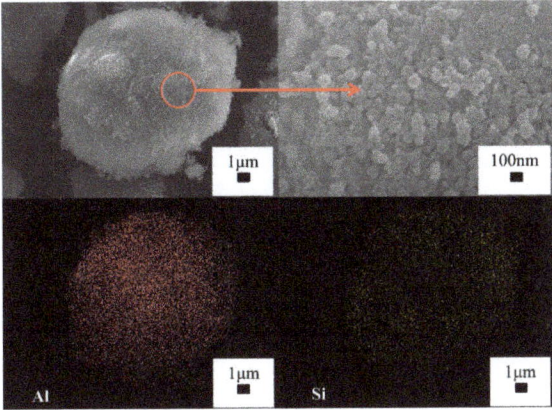

Figure 2. SEM and EDS images of $SiO_2@Al$.

2.3. Preparation of Coating Slurry

As shown in Table 1, ceramic coating slurry with different amount of $SiO_2@Al$ (0, 5%, 10%, 15%, 20%, 25%) were prepared to determine an appropriate proportion of $SiO_2@Al$ additives. The mixture was ball-milled with water for 4 h until the particle size was <45 μm. The milling ball was ZrO_2 and the diameter of ZrO_2 balls was 6–10 mm. During the ball-milled process, $SiO_2@Al$ stably existed in the slurry without any hydrogen releasing. The prepared slurry was coated on carbon steel by spraying gun, which was connected to an air compressor with pressure of 8 bar. The thickness of the coating was 0.4 mm.

Table 1. The composition of the ceramic coating (wt %).

No.	Al_2O_3	Sodium Silicate Solution (40 wt %)	$SiO_2@Al$	Surface Active Agent Sodium Polyacrylate	H_2O
1	70.5	23.5	0	4	2
2	66	21	5	4	4
3	62	18.5	10	4	5.5
4	58	16	15	4	7
5	54	13.5	20	4	8.5
6	50	11	25	4	10

2.4. Evaluation of Anti-oxidation Performance

To evaluate the anti-oxidation performance of ceramic coating, the weight changes of heated carbon steel samples were investigated. The carbon steel samples were heated in a muffle furnace from room temperature to different temperatures (1050, 1100, 1150, 1200, and 1250 °C) and maintained for 120 min. The anti-oxidation ability E, which was related to the weight loss of carbon steel oxidation, was calculated by Equations (1) and (2). Two samples were used in each test and the average value of samples was adopted.

$$\text{Steel yield } \alpha = \frac{m_2}{m_1} \times 100\% \tag{1}$$

$$\text{Anti} - \text{oxidation ability } E = \frac{\alpha_{\text{coated}} - \alpha_{\text{bare}}}{1 - \alpha_{\text{bare}}} \times 100\% \tag{2}$$

where, m_1 and m_2 are the weights of samples before and after high temperature treatment without scale. α_{coated} and α_{bare} are the yields of the coated and bare samples after high temperature treatment.

To evaluate the non-isothermal kinetic of carbon steel oxidation, the correlation between oxidation reaction rate and heating temperature was investigated. The samples were heated at a rate of 10 °C/min to certain temperature (1050, 1100, 1150, 1200, and 1250 °C) and maintained for 5 min. The reaction rate (v) was calculated by Equations (3) and (4).

$$\text{Weight loss per unit area } \Delta m = M_2 - M_1 \tag{3}$$

$$\text{Reaction rate } v = \frac{\Delta m}{\Delta t} \tag{4}$$

where, Δm is the weight loss per unit area of the sample during the holding stage at certain temperature. Δt is the time that the samples were maintained at certain temperature.

To evaluate the anti-oxidation ability of ceramic coating at a certain temperature, the isothermal kinetic was also carried out. The isothermal kinetic was conducted by a continuous thermos-balance (RZ, Luoyang Precondar, Luoyang, China) at a heating rate of 10 °C/min to 1250 °C and maintained for 120 min. The weight change of the sample was calculated by Equation (5).

$$\Delta \omega = \omega_i - \omega_0 \tag{5}$$

Here, ω_0 is the weight change per unit area of the sample heated to certain temperature and ω_i is that of the sample heated and maintained at certain temperature.

2.5. Characterization

The morphology of the ceramic coating and carbon steel substrate was characterized by scanning electron microscopy (SEM; JSM-6700F, JEOL, Tokyo, Japan) equipped with an energy dispersive X-ray spectroscopy (EDS, NORAN, Thermo Fisher, Waltham, MA, USA). The phase-transition occurred at different temperature was characterized by X-ray diffraction (XRD; X'Pert Pro, Philips, Amsterdam, The Netherlands) operated from 5° to 90° for 5.25 min. The weight change and thermal change during heating process were detected by TG-DTA (TG-DTA; STA449, Netzsch, Nuremberg, Germany). The TG-DTA test was operated from room temperature to 1250 °C at a heating rate of 10 °C/min. The test was conducted under flowing air.

3. Results and Discussion

3.1. Performance of Coating

The anti-oxidation performances of the coatings with different proportion of SiO_2@Al were investigated by the weight loss during the heating process in muffle furnace. As shown in Figure 3, when temperatures ranged from 1050 to 1150 °C, a higher proportion of SiO_2@Al led to a better anti-oxidation performance. However, when the proportion exceeded 10%, the enhancement of the anti-oxidation performance was not obvious. When temperatures ranged from 1200 to 1250 °C, the anti-oxidation performance of coating improved with the increasing amount of SiO$_2$@Al until 10% and reduced with further SiO$_2$@Al addition, especially at 1250 °C. Therefore, in view of the anti-oxidation performance and the cost, the ceramic coating with 10% of SiO$_2$@Al was chosen as the appropriate one to conduct the following investigations.

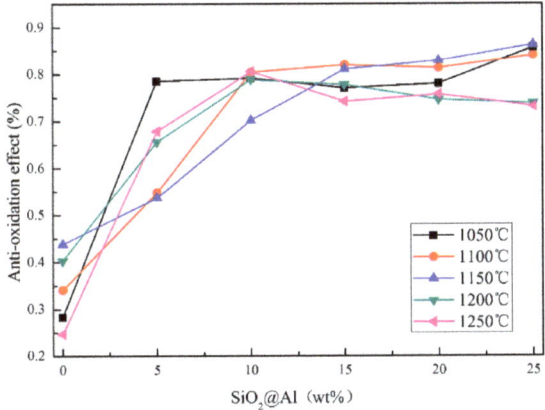

Figure 3. Anti-oxidation performances of the coatings with different proportion of SiO$_2$@Al.

The non-isothermal kinetic for sample protected by ASMA, sample protected by AS, and bare sample are shown in Figure 4. The apparent activation energy, (E_a), was calculated by the Arrhenius equation. As shown in Equation (6), if the concentration of reactant (B) was constant, the reaction rate (v) would have a positive correlation with the reaction rate constants (k), which is the function of temperature described by Equation (7) (Arrhenius equation).

$$v = B \times k \tag{6}$$

$$k = A \times e^{(E_a/RT)} \tag{7}$$

The relationship between reaction rate (v) and apparent activation energy (E_a) could be deduced with Equations (6) and (7) by a logarithmic operation. The ultimate equation was described as Equations (8) and (9).

$$\ln k = -\frac{E_a}{RT} + \ln A \tag{8}$$

$$\ln v = -\frac{E_a}{RT} + \ln A + \ln B \tag{9}$$

where, A and B were constant. R was 8.314 J·mol^{-1}·K^{-1}.

Therefore, the dependent variable (ln$v$) had a linear relationship with the variable (1/T). As described in Equations (3) and (4), the reaction rate (v) could be calculated by the weight loss per unit area and the duration time of samples heated to a certain temperature. Based on the kinetic results at different temperatures as shown in Figure 4, a linear relationship was observed between ln$v$ and 1/T. The slope of each fitting was equal to ($-E_a$/R) as described in Equation (8). It was a remarkable fact that the result of the sample protected by AS was not able to be fitted by a single linear equation because there was an inflection point at about 1150 °C. Therefore, the kinetic of the sample protected by AS should be divided into two parts (AS1 and AS2) by 1150 °C.

As shown in Figure 4, the apparent activation energy (E_a) of bare sample was 120.11 kJ/mol, while that of the sample protected by ASMA was 326.50 kJ/mol. The apparent activation energy (E_a)

of the sample protected by AS between 1025 and 1150 °C was 106.14 kJ/mol, which was approximate to the bare sample. The apparent activation energy (E_a) of the sample protected by AS between 1150 and 1250 °C was 224.05 kJ/mol, which was between that of the sample protected by ASMA and the bare sample. Generally, higher apparent activation energy resulted in a better anti-oxidation performance of coating. Therefore, ASMA exhibited an excellent anti-oxidation performance between 1025 and 1250 °C, while AS only worked above 1150 °C. Additionally, the anti-oxidation performance of ASMA was much better than AS due to its larger apparent activation energy (E_a). To further prove the enhancement of anti-oxidation of ASMA, the non-isothermal experiment was conducted by a continuous thermo-balance. As shown in Figure 5, compared with AS, ASMA showed an obviously better anti-oxidation performance during the temperature-rise period.

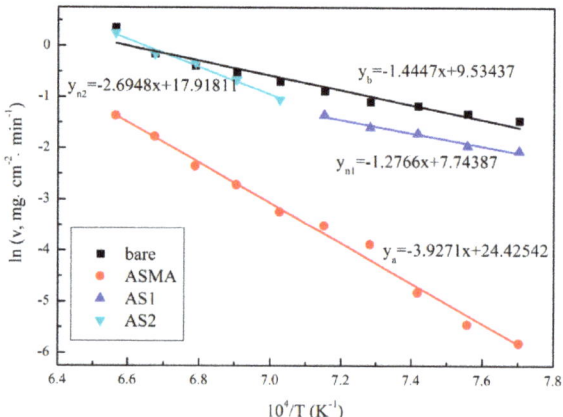

Figure 4. Results of non-isothermal kinetics.

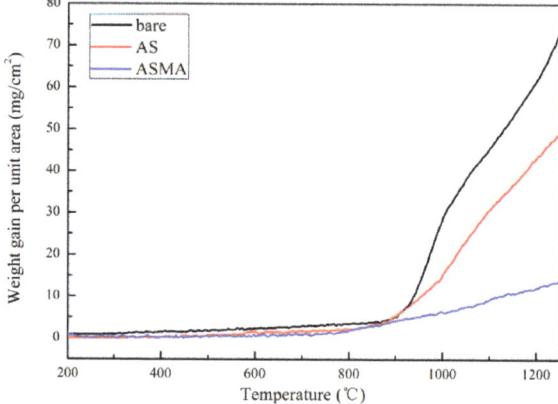

Figure 5. Results of continuous thermo-balance from 200 to 1250 °C with a rate of 10 °C/min.

Since the serious oxidation of carbon steel always occurred over 1250 °C, the isothermal kinetic was conducted at 1250 °C. It was widely accepted that the oxidation of carbon steel followed the parabolic law [3,23,24], which meant that the oxidation process was controlled by the diffusion. The reaction constant could be calculated by Equation (10) [2].

$$(\Delta \omega)^2 = H + k_p \times t_i \tag{10}$$

Here, t_i (s) is the duration of oxidation, $\Delta\omega$ is the weight gain per unit area (mg/cm^2), H is a constant, and k_p (mg$^2 \cdot$cm$^{-4} \cdot$s^{-1}) is the reaction rate constant.

In this study, the linear relationship between $\Delta\omega$ and $t_i^{1/2}$ was used to evaluate k_p [2]. As shown in Figure 6, the reaction rate constant (k_p) of the bare sample, the sample protected by AS, and the sample protected by ASMA was 2.09, 0.096, and 0.046 mg$^2 \cdot$cm$^{-4} \cdot$s^{-1}, respectively. It can be concluded that ASMA possessed a better anti-oxidation performance than AS due to its smaller reaction rate constant at 1250 °C.

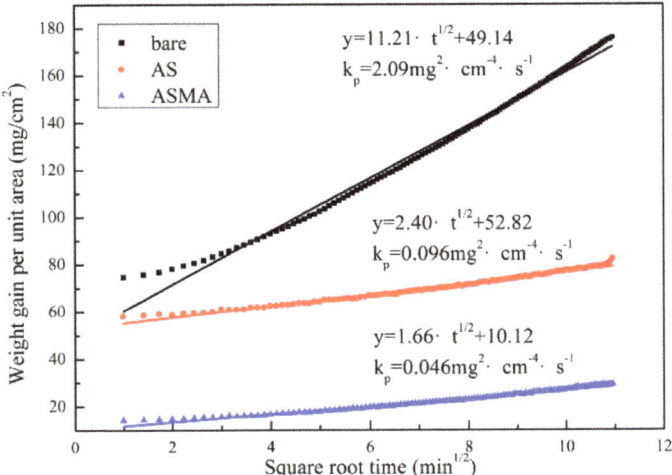

Figure 6. Results of isothermal kinetics.

3.2. Morphology of Coating

The microstructure was an intuitive evidence for evaluating the compactness of coating which controlled the inner diffusion rate of oxygen. Therefore, anti-oxidation performance of coating was better with a more compact structure. The outer-surface microstructures of sample protected by the ASMA and sample protected by the AS were investigated after the thermal treatments of different temperature (900, 1150, and 1250 °C). As described in Figure 7, ASMA formed a compact structure at 900 °C and the compactness improved with the increasing of temperature until 1250 °C. The amount of liquid phase increased with the rising temperature, the porosity reduced and thus the sintering of coating was enhanced at the same time. However, the sample protected by AS possessed a porous structure composed of isolated particles at 900 °C, while the compactness improved at 1150 °C. The above results suggested that SiO$_2$@Al could accelerate the sintering of the ceramic coating and enhance the compactness.

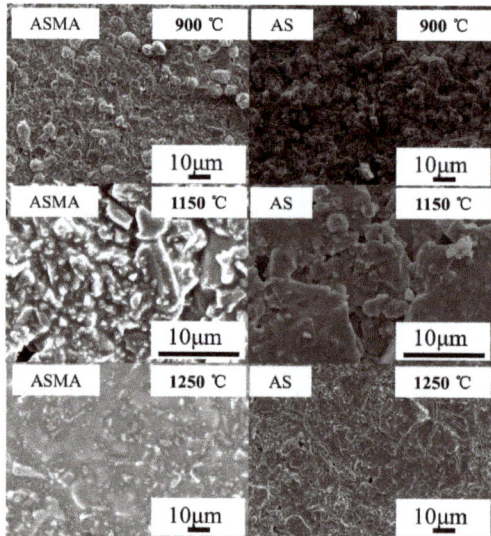

Figure 7. SEM images of outer-surfaces belonging to sample protected by ASMA and sample protected by AS at different temperatures.

3.3. Protection Mechanism

Figure 8 showed the TG-DTA data of the ASMA and AS slurry. As shown in Figure 8a, there was an endothermic peak at about 662.85 °C, an exothermic peak at 977.7 °C, and an exothermic peak at 1153.91 °C for ASMA slurry, while there was only a gradual exothermic trend during heating process for AS slurry. In order to interpret the possible reactions during the heating process, the phase transitions were investigated by XRD method at 500, 800, 1100, and 1250 °C, which were near the endothermic or exothermic temperature. Compared with Figure 9a,b, it was found that no specific phase formed between 500 and 800 °C, thus the endothermic peak at 662.82 °C could be attributed to the melting of Al powders in $SiO_2@Al$. Phases of Si and $Al_{3.21}Si_{0.47}$ were newly formed within ASMA when temperature increased from 800 to 1100 °C, which indicated that an exothermic reaction occurred as described in Equation (11) and it gave rise to the exothermic peak at 977.7 °C [25]. Compared with Figure 9c,d, it was remarkable that mullite and cristobalite formed within ASMA when temperature ranged from 1100 to 1250 °C. Therefore, the exothermic peak at 1153.91 °C might result from the exothermic reaction expressed as Equations (12) and (13) [26,27]. In contrast, there was no new phase formed within AS during the whole heating process except for cristobalite. To sum up, the Al powders in $SiO_2@Al$ melted at 662.85 °C and reacted with SiO_2 on its surface, which generated local high temperature and the sintering of ceramic raw materials, thus contributing to its excellent anti-oxidation performance, which was consistent with the non-isothermal kinetics. Besides, mullite and cristobalite formed at 1150 °C could further enhance the coating compactness and improve the anti-oxidation performance above 1150 °C [19]. However, ceramic raw materials just sintered above 1150 °C within AS, and thus showed an unobvious anti-oxidation effect below this temperature.

$$4Al + 3SiO_2 \rightarrow 3Si + 2Al_2O_3 \tag{11}$$

$$SiO_2 \text{ (amorphous)} \rightarrow SiO_2 \text{ (cristobalite)} \tag{12}$$

$$3Al_2O_3 + 2SiO_2 \rightarrow 3Al_2O_3 \cdot 2SiO_2 \tag{13}$$

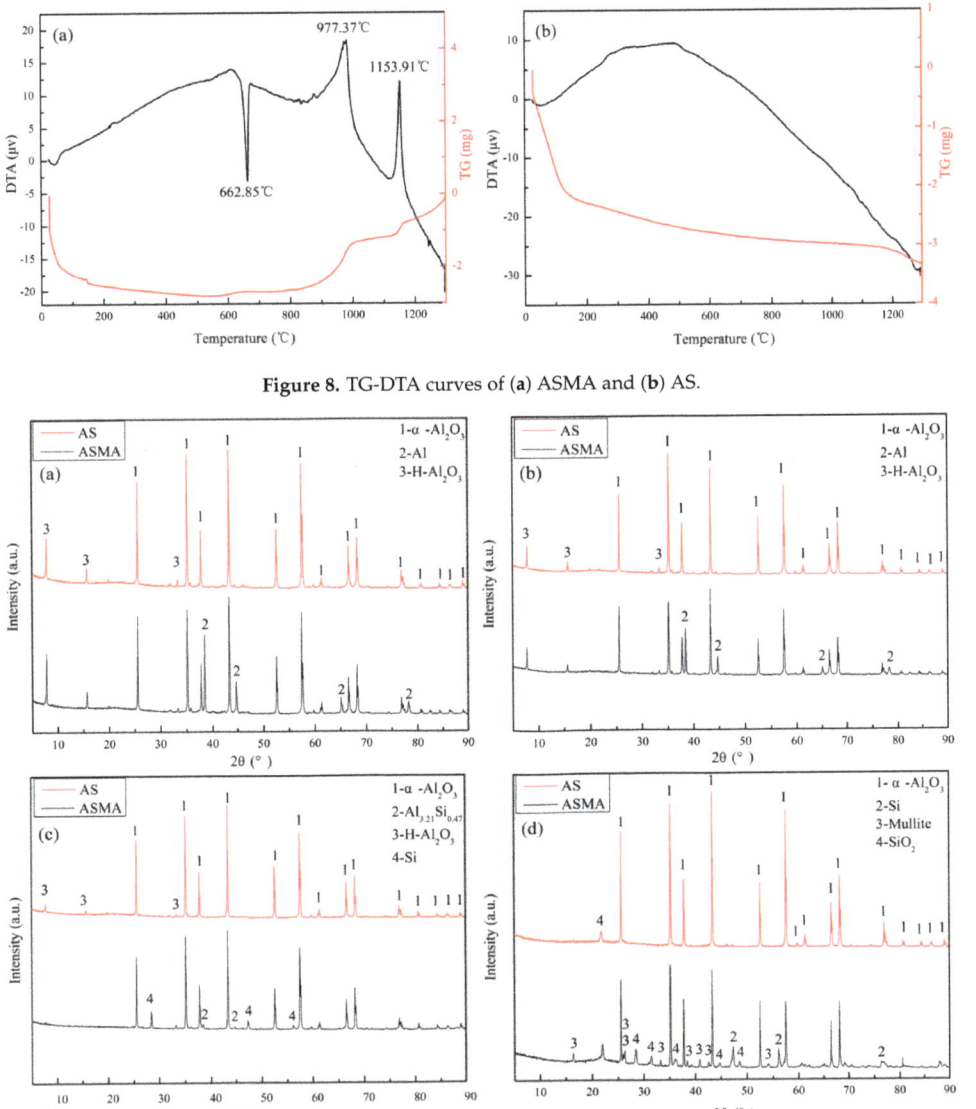

Figure 8. TG-DTA curves of (**a**) ASMA and (**b**) AS.

Figure 9. Comparisons of XRD result of the sample protected by ASMA and AS slurry at dffierent temperatures: (**a**) 500 °C, (**b**) 800 °C, (**c**) 1100 °C, (**d**) 1250 °C.

According to the SEM-EDS results of coating cross-sections exhibited in Figure 10, both the coatings were able to prevent the carbon steel substrate from oxidization at 1250 °C. Compared with Figure 11a,b, ASMA was composed of a hercynite layer, a mullite layer, and a ceramic layer with high aluminum content, while AS consisted of a Fe_3O_4 layer, an olivine layer composed of FeO and $2FeO \cdot SiO_2$, a ceramic layer composed of Al_2O_3 and cristobalite. Equations (14) or (15) possibly occurred to form the hercynite ($Al_2O_3 \cdot FeO$) within ASMA. As reported in the literature [28], the structure of hercynite could be expressed as $(Fe^{2+}_{1-x}Al^{3+}_x)_{tet}(Al^{3+}_{2-x}Fe^{2+}_x)_{oct}O_4$ where $0 < x < 0.25$, while that of Fe_3O_4 could be expressed as $(Fe^{3+})_{tet}(Fe^{3+}Fe^{2+})_{oct}O_4$ [29]. Referring to Freer and O'Reilly [30], the diffusion activation energy of Fe^{2+} at octahedral site and tetrahedral site were 0.2(±0.05) eV and 0.71

($\pm$0.09) eV, which indicated that the diffusion of iron ions within hercynite was slower than Fe_3O_4 due to the lower fraction of Fe^{2+} occupied the octahedral site. Therefore, hercynite also played a significant role for enhancing the anti-oxidation ability of ASMA. However, hercynite was easily oxidized as shown in Equation (16) under oxidation condition. As mentioned in Section 3.1, the compactness of ASMA was greatly improved with the addition of SiO_2@Al, which effectively prohibited the inner diffusion of oxygen. As a result, the hercynite layer could be stable within the ceramic coating. The result agreed well with the finding in isothermal kinetic experiment.

$$3Al_2O_3 \cdot 2SiO_2 + 3FeO \rightarrow 3Al_2O_3 \cdot FeO + 6SiO_2 \quad (14)$$

$$Al_2O_3 + FeO \rightarrow Al_2O_3 \cdot FeO \quad (15)$$

$$4Al_2O_3 \cdot FeO + O_2 \rightarrow 4Al_2O_3 + 2Fe_2O_3 \quad (16)$$

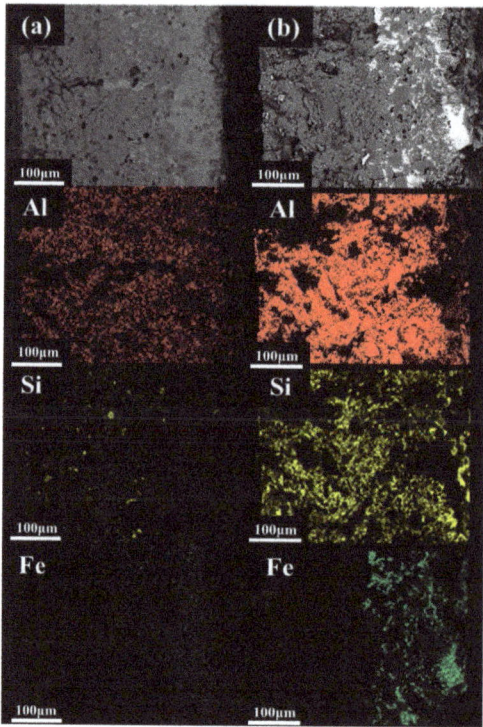

Figure 10. SEM-EDS images of cross-sections belonging to (**a**) the sample protected by ASMA and (**b**) the sample protected by AS at 1250 °C.

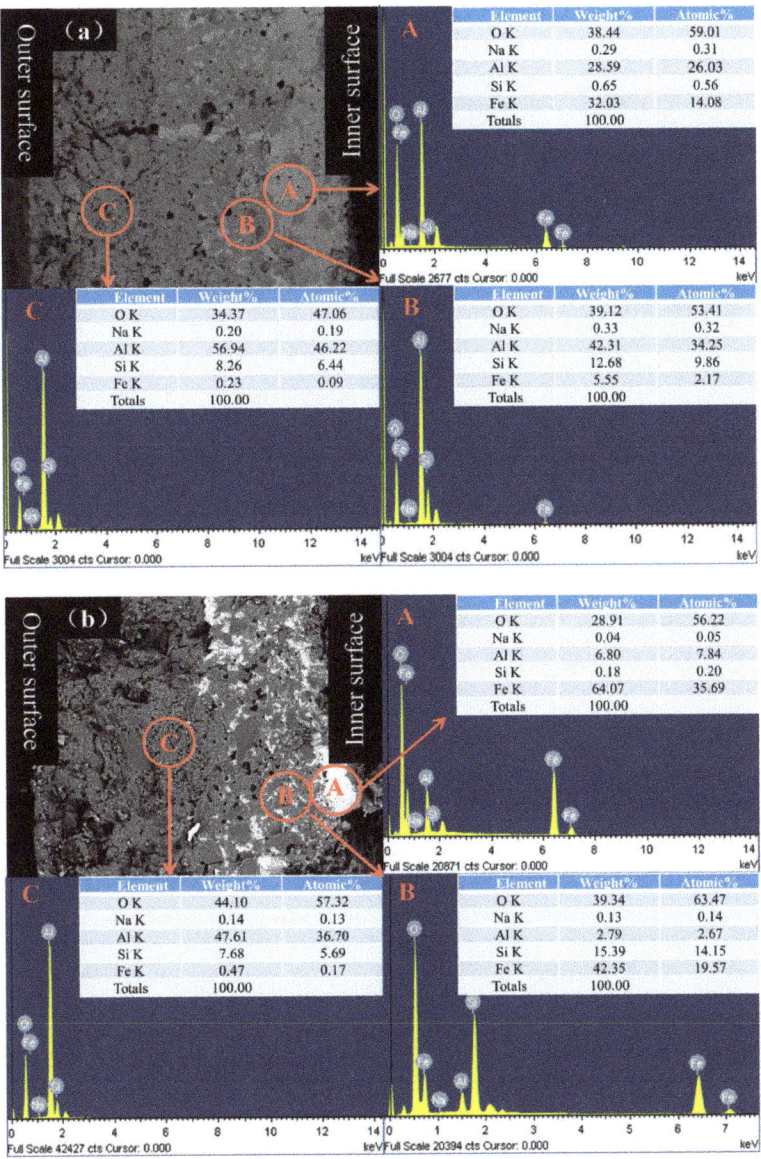

Figure 11. SEM images with results of EDS measurements performed on selected areas of cross-sections belonging to (**a**) sample protected by ASMA and (**b**) sample protected by AS at 1250 °C.

4. Conclusions

In present work, modified Al powders (SiO$_2$@Al) were prepared by the hydrolysis of tetraethyl orthosilicate (TEOS) under alkaline conditions. A new ceramic coating was prepared with the SiO$_2$@Al as additives to prevent the carbon steel from oxidization at 1250 °C. The anti-oxidation performance and a probable protective mechanism of the coating were clarified.

Some results were obtained as follows:

- The Al powders were successfully encapsulated by SiO_2 and the SiO_2@Al stably existed in the as-prepared alkaline coating slurry. The appropriate proportion of SiO_2@Al in the coating was 10%.
- The metallic Al in ASMA melted at 662.85 °C. What generated local high temperature and accelerated the sintering of ceramic raw materials was the exothermal reaction between the melted Al and SiO_2 ($4Al + 3SiO_2 \rightarrow 3Si + 2Al_2O_3$). The well-sintered structure decreased the inner diffusion rate of oxygen and enhanced the anti-oxidation effect of ASMA during the temperature-rise period.
- The hercynite layer formed within ASMA played a significant role for slowing down the external diffusion of iron ions due to the lower fraction of Fe^{2+} occupied the octahedral site, which facilitated the improvement of anti-oxidation ability at 1250 °C.

Author Contributions: Conceptualization, B.Y. and L.W.; Formal analysis, B.Y., G.Z. and L.W.; Funding acquisition, S.Y.; Investigation, B.Y., X.Z., Y.D. and L.W.; Methodology, B.Y., G.F. and L.W.; Software, G.Z.; Supervision, Y.C., Y.D. and S.Y.; Validation, B.Y. and Y.T.; Writing–original draft, B.Y.; Writing–review & editing, B.Y., Y.C. and L.W.

Funding: This research was funded by the National Strategic Priority Research Program 2017 (No. YFC0703205).

Conflicts of Interest: The authors declare no conflict of interest.

References

1. Frolenkov, K.Y.; Frolenkova, L.Y.; Shadrin, I.F. High-temperature oxidation of low-alloyed steel under glass coatings. *Prot. Met. Phys. Chem. Surf.* **2010**, *46*, 103–109. [CrossRef]
2. Fu, G.; Wei, L.; Zhang, X.; Cui, Y.; Wang, Y.; Yu, B.; Lv, C.; Ye, S. A MgO-SiO_2-Al_2O_3-ZnO ceramic-glass coating to improve the anti-oxidation of carbon steel at high temperature. *ISIJ Int.* **2018**, *58*, 929–935. [CrossRef]
3. Hu, X.J.; Zhang, B.M.; Chen, S.H.; Fang, F.; Jiang, J.Q. Oxide scale growth on high carbon steel at high temperatures. *J. Iron Steel Res. Int.* **2013**, *20*, 47–52. [CrossRef]
4. Liu, Y.; Zhang, W.; Tong, Q.; Sun, Q.; Chao, Y.; Yang, W.; Li, T.; Li, L. Effects of chemical composition on decarburization layer depth of high carbon steels in 2% oxygen atmosphere. *Heat Treat. Met.* **2017**, *42*, 143–148.
5. Heo, N.H.; Lee, J.K. Grain boundary segregation of phosphorus and intergranular surface cracking accompanied by decarburization in plain carbon steels. *ISIJ Int.* **2011**, *51*, 673–678. [CrossRef]
6. Torkar, M.; Glogovac, B. Diminution of scaling by the application of a protective coating. *J. Mater. Process. Technol.* **1996**, *58*, 217–222. [CrossRef]
7. Munther, P.; Lenard, J. The effect of scaling on interfacial friction in hot rolling of steels. *J. Mater. Process. Technol.* **1999**, *88*, 105–113. [CrossRef]
8. Mao, Z.L.; Yang, X.J.; Zhu, S.L.; Cui, Z.D.; Lu, Y. Pack cementation processing parameters for SiC coatings on C/C for optimum tribological properties. *Surf. Coat. Technol.* **2014**, *254*, 54–60. [CrossRef]
9. Fu, G.; Wei, L.; Shan, X.; Zhang, X.; Ding, J.; Lv, C.; Liu, Y.; Ye, S. Influence of a Cr_2O_3 glass coating on enhancing the oxidation resistance of 20MnSiNb structural steel. *Surf. Coat. Technol.* **2016**, *294*, 8–14. [CrossRef]
10. Fu, G.Y.; Wei, L.Q.; Zhang, X.M.; Cui, Y.B.; Lv, C.C.; Ding, J.; Yu, B.; Ye, S.F. A high-silicon anti-oxidation coating for carbon steel at high temperature. *Surf. Coat. Technol.* **2017**, *310*, 166–172. [CrossRef]
11. Wei, L.; Peng, L.; Ye, S.; Xie, Y.; Chen, Y. Preparation and properties of anti-oxidation inorganic nano-coating for low carbon steel at an elevated temperature. *J. Wuhan Univ. Technol.* **2006**, *21*, 48–52.
12. Wang, D. Formation and property of ceramic layer on a low-carbon steel. *Int. J. Process. Sci. Charact. Appl. Adv. Mater.* **2007**, 134–138.
13. Zhou, X.; Ye, S.F.; Xu, H.W.; Liu, P.; Wang, X.J.; Wei, L.Q. Influence of ceramic coating of MgO on oxidation behavior and descaling ability of low alloy steel. *Surf. Coat. Technol.* **2012**, *206*, 3619–3625. [CrossRef]
14. Zhou, X.; Wei, L.Q.; Liu, P.; Wang, X.J.; Ye, S.F.; Chen, Y.F. Preparation and characterization of high temperature protective ceramic coating for plain carbon steel. *Chin. J. Process Eng.* **2010**, *10*, 167–172.

15. Li, G.; Jie, X.; He, L. High temperature oxidation behavior of (Ti,Al) C ceramic coatings on carbon steel prepared by electrical discharge coating in kerosene. *Adv. Mat. Res.* **2011**, *189–193*, 186–192. [CrossRef]
16. Nguyen, M.D.; Bang, J.W.; Kim, Y.H.; Bin, A.S.; Hwang, K.H.; Pham, V.H.; Kwon, W.T. Slurry spray coating of carbon steel for use in oxidizing and humid environments. *Ceram. Int.* **2018**, *44*, 8306–8313. [CrossRef]
17. Wang, C.; Chen, S. The high-temperature oxidation behavior of hot-dipping Al–Si coating on low carbon steel. *Surf. Coat. Technol.* **2006**, *200*, 6601–6605. [CrossRef]
18. Yang, C.H.; Lin, S.N.; Chen, C.H.; Tsai, W.T. Effects of temperature and straining on the oxidation behavior of electrical steels. *Oxid. Met.* **2009**, *72*, 145–157. [CrossRef]
19. Ren, Q.; He, X.; Wu, X. Effects of nano-additives on the structure and properties of 99% alumina ceramics. *Rare Met. Mater. Eng.* **2008**, *37*, 452–454.
20. Yin, J.; Wang, X.; Zhang, Y.; Zhao, C.; Wang, X.; Gao, H.; Yang, J. Effect of sintering aids on high purity alumina ceramic densification process. *Chin. Rare Earths* **2014**, *35*, 16–20.
21. Chen, R.Y.; Yuen, W.Y.D. The effects of steel composition on the oxidation kinetics, scale structure, and scale-steel interface adherence of low and ultra-low carbon steels. *Mater. Sci. Forum* **2006**, *522–523*, 451–460. [CrossRef]
22. Le, M.T.; Kim, C.; Lee, J. Effect of silica addition on ceramic layer in centrifugal-thermit reaction. *Mater. Trans.* **2008**, *49*, 1410–1414. [CrossRef]
23. Jiang, Z. Research on TG-DSC of carbon steel oxidation kinetics. *Hot Work. Technol.* **2015**, *44*, 89–91.
24. Nagelberg, A.S. Observations on the role of Mg and Si in the directed oxidation of Al-Mg-Si alloys. *J. Mater. Res.* **1992**, *7*, 265–268. [CrossRef]
25. Aksay, I.A.; Pask, J. A stable and metastable equilibria in system SiO_2-Al_2O_3. *J. Am. Ceram. Soc.* **1975**, *58*, 507–512. [CrossRef]
26. Zhu, Z.; Wei, Z.; Sun, W.; Hou, J.; He, B.; Dong, Y. Cost-effective utilization of mineral-based raw materials for preparation of porous mullite ceramic membranes via in-situ reaction method. *Appl. Clay Sci.* **2016**, *120*, 135–141. [CrossRef]
27. Odashima, H.; Kitayama, M. Oxidation-inhibition mechanism and performance of a new protective coating for slab reheating of 3% Si-steel. *ISIJ Int.* **1990**, *30*, 255–264. [CrossRef]
28. Dong, Y.; Lu, H.; Cui, J.; Yan, D.; Yin, F.; Li, D. Mechanical characteristics of $FeAl_2O_4$ and $AlFe_2O_4$ spinel phases in coatings—A study combining experimental evaluation and first-principles calculations. *Ceram. Int.* **2017**, *43*, 16094–16100. [CrossRef]
29. Hazen, R.M.; Jeanloz, R. Wüstite ($Fe_{1-x}O$): A review of its defect structure and physical-properties. *Rev. Geophys.* **1984**, *22*, 37–46. [CrossRef]
30. Freer, R.; Oreilly, W. The diffusion of Fe^{2+} ions in spinels with relevance to the process of maghemitization. *Miner. Mag.* **1980**, *43*, 889–899. [CrossRef]

© 2019 by the authors. Licensee MDPI, Basel, Switzerland. This article is an open access article distributed under the terms and conditions of the Creative Commons Attribution (CC BY) license (http://creativecommons.org/licenses/by/4.0/).

Article

Ti(C, N) as Barrier Coatings

Katarzyna Banaszek [1] and Leszek Klimek [2],*

[1] Department of General Dentistry, Medical University of Lodz, Pomorska Str. 251, 92-213 Lodz, Poland
[2] Institute of Materials Science and Engineering, Lodz University of Technology, Stefanowskiego Str. 1/15, 90-924 Lodz, Poland
* Correspondence: leszek.klimek@p.lodz.pl

Received: 22 April 2019; Accepted: 3 July 2019; Published: 8 July 2019

Abstract: Metals and their alloys are materials that have long been used in stomatological prosthetics and orthodontics. The side effects of their application include reactions of the body such as allergies. Their source can be corrosion products as well as metal ions released in the corrosion process, which penetrate the surrounding tissue. In order to prevent the harming effect of metal alloys, intensive research has been performed to purify metal prosthetic restorations by way of modifying their surface. The study presents the investigation results of Ti(C, N)-type coatings applied to alloy Ni–Cr by means of the magnetronic method. Five coatings differing in the nitrogen and carbon content were investigated. The studies included the determination of the coatings' chemical composition, construction, as well as the amount of ions released into the environment: distilled water, 0.9% NaCl and artificial saliva. The performed investigations showed that, in reference to an alloy without a coating, each coating constitutes a barrier reducing the amount of ions transferred into the examined solutions. So, Ti(C, N)-type coatings can be considered for biomedical applications as protective coatings of non-precious metal alloys.

Keywords: Ni–Cr alloy; Ti(C, N) coatings; ion release

1. Introduction

Metals and their alloys are materials that have been used in stomatological prosthetics and orthodontics for a long time now. The side effects of their use can be some reactions of the body, e.g., allergies. Their source can be the corrosion products as well as the metal ions released in the corrosion process, which penetrate the surrounding tissue. Metals and basic metal alloys under the conditions of the oral cavity do not exhibit a hazardous effect as long as they are corrosion resistant. However, the exceptionally strong corrosive properties of the biological environment make all the basic metal alloys unable to resist corrosion [1,2]. Often, prosthetic restorations require the use of several types of metal alloys, which additionally contribute to the creation of electrogalvanic elements and intensifies the corrosion speed [3]. The process of corrosive destruction is accompanied by the penetration of hazardous alloys, corrosion products, and/or metal ions into the environment [4]. The metal ions, released during the corrosion process from the metal restoration constructions, are transferred into the digestive system and accumulate in the stomach, liver, spleen, bones, and mucous membrane. The metal ions released from metallic implants accumulate in the surrounding tissues, often reaching very high concentrations. After exceeding the critical concentration, toxic as well as allergic reactions take place [5]. The degree of the harmful effect of prosthetic metal constructions depends mainly on the amount of released corrosion products and/or metal ions (being in direct relation with the restorations' proneness to corrosion, as well as their size and the conditions in which they exist, e.g., the presence of another metal in the oral cavity, the pH, etc.) and also on the degree of their harmful (toxic) effect [6]. One should remember, however, that the hazardous effect of prosthetic restorations is also a result of the cooperation of many other factors, such as the galvanic currents or mechanical poling.

The experimental studies demonstrated a hazardous effect of the basic metal alloys [7–10]. The results of the clinical tests proved also a harmful effect of prosthetic constructions made of the basic metal alloys [11]. In stomatology, usually, we observe local toxicity of metals and their alloys. It is thought that, among many metal alloys used in clinical applications, NiCr and CoCr alloys characterize in the highest degree of hazardous effect [4,6,12–14]. This is also confirmed by clinical observations of gums in contact with crowns made of NiCr alloys, compared to precious metal alloys [15].

More and more often, one can observe allergic reactions to metal alloys used in stomatological prosthetics and orthodontics. Hypersensitivity to metals causes inflammatory reactions [16], in the case of patients both with skeletal prostheses and those made of CoCr alloys, as well as crowns made of CrNi steels, NiCr alloys, and alloys based on palladium and orthodontic devices [11,17]. So, the clinical application of metal alloys in prosthetics poses the risk of the occurrence of an allergic reaction.

Despite the corrosion and the harmful effect of metal alloys, they continue to be applied for long-term construction elements of prosthetic restorations [18], being the most frequently used despite the fact that new "metal-free" technologies are being introduced [19,20], together with full-ceramic restoration materials [21] and constructions based on zirconium oxide [22,23]. In order to prevent the hazardous effect of metal alloys, intensive research has been performed aiming at the development of fully biocompatible materials, i.e., such that undergo both a mechanical and functional integration with the tissue. These studies are concentrated on:

- The development of metal alloys with a higher corrosion resistance;
- The purification of metal prosthetic restorations through their surface modification.

In recent years, more and more often, layers applied by various methods, e.g., chemical vapor deposition (CVD), Physical vapor deposition (PVD), and sol-gel, have been used for that purpose. Among the many coatings obtained by these methods, the most frequently applied ones include metal carbides, oxides and nitrides [24–29]. A special attention should be paid to titanium carbides and nitrides. This mainly results from their high strength and corrosion resistance. Research works have been performed aiming at modifying the technology of obtaining nitride layers in order to improve their properties, which depend, among other things, on the TiN/Ti_2N ratio in the layer [30]. Another research direction is obtaining layers made of titanium carbonitride Ti(C, N). As it was shown by the preliminary investigations [31–34], Ti(C, N) layers demonstrate a better corrosion resistance and wear resistance, and they significantly reduce the amount of the released metal ions and so, they can potentially be used as coatings for prosthetic and orthodontic metal restorations.

The tests performed on Ti(C, N) coatings applied on Ni–Cr alloys showed that they exhibit the proper mechanical and physicochemical properties, and also they are much less toxic than Ni–Cr alloys [34–37]. The toxicity reduction is probably connected with the barrier effect of the coating, consisting in a reduction of the amount of substrate ions being transferred into the environment. So, the aim of the study became a comparison of the amount of ions released into the solutions in samples with Ti(C, N) coatings with different carbon and nitrogen contents with samples without a coating.

2. Materials and Methods

The test material was constituted by samples made of NiCr Heraenium NA by Heraeus Kulzer (Hanau, Germany), alloy in the form of cylinders, 8 mm in diameter and 10 mm high, for the chemical composition and microscopic tests (Figure 1a), as well as bars cast from the same alloy, 2 mm in diameter and 45 mm long – for the ion release studies (Figure 1b). The initial composition of the alloy determined by the X-ray fluorescent analysis method with the use of an SRS300 spectrometer by SIEMENS (Munich, Germany) has been given in Table 1.

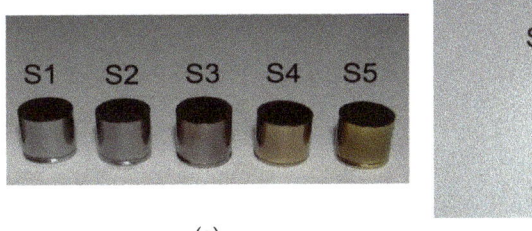 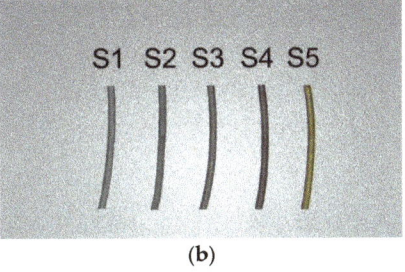

(a) (b)

Figure 1. Test samples: (**a**) microscopic and chemical composition tests; (**b**) ion release tests.

Table 1. Chemical composition of tested alloy.

Element Percentage wt.%						
Cr	Mo	Si	Fe	Co	Mn	Ni
24.79	8.89	1.57	1.33	0.17	0.12	residue

The disks were divided into five groups, according to the content of carbon and nitrogen. The coatings were deposited by the method of magnetron sputtering. The samples were cleaned in a detergent, i.e., acetone, and next, after they were placed in the chamber, they underwent ionic cleaning. The ionic cleaning was carried out with the use of an ionic gun applying argon as the working gas. The pressure in the chamber was 0.0021 Pa. The samples prepared in this way were coated with layers by means of the magnetron sputtering method. In order to improve the adhesion of the Ti(C, N) layers, first, an adhesive sublayer made of pure titanium was applied for the time of 120 s, with the argon pressure of 0.24 Pa and the following parameters of the magnetron's work: 3 kW/around, 4.5 A. After 2 min, a reactive gas was slowly released: nitrogen, acetylene or a mixture of the two. The deposition time of the main coating was the same for all the processes and equalled 7200 s. The constant voltage polarization during the deposition was −100 V. During the deposition, the samples were moved above the target's surface in a swinging motion, in order to homogenize the thickness. In each case, the pressure of the main process equalled 0.27 Pa. The type of reactive gases and their flows have been given in Table 2. These were the only varying parameters of the processes.

Table 2. Reactive gas flow.

Gas	Flow Unit	Samples				
		S1	S2	S3	S4	S5
N_2	[sccm]	0	4	8	12	16
C_2H_2		8	6	4	2	0

After the deposition, the vacuum chamber with the samples was cooled down and only then the charge was removed.

The obtained samples underwent chemical composition tests performed on the coatings as well as microscopic observations made on cross-sections. The assessment of the chemical composition of the coatings was performed by means of an optical emission spectrometer with a glow discharge spectrometer (GDS) 850 A by LECO (St. Joseph, MI, USA), and with an alternating current lamp RF (with radio frequency) with a 4 mm diameter anode. The microscopic tests on the samples' cross-sections were performed in a scanning electron microscope S-3000N (HITACHI, Tokyo, Japan). The obtained images of the coatings have been presented in Figure 2.

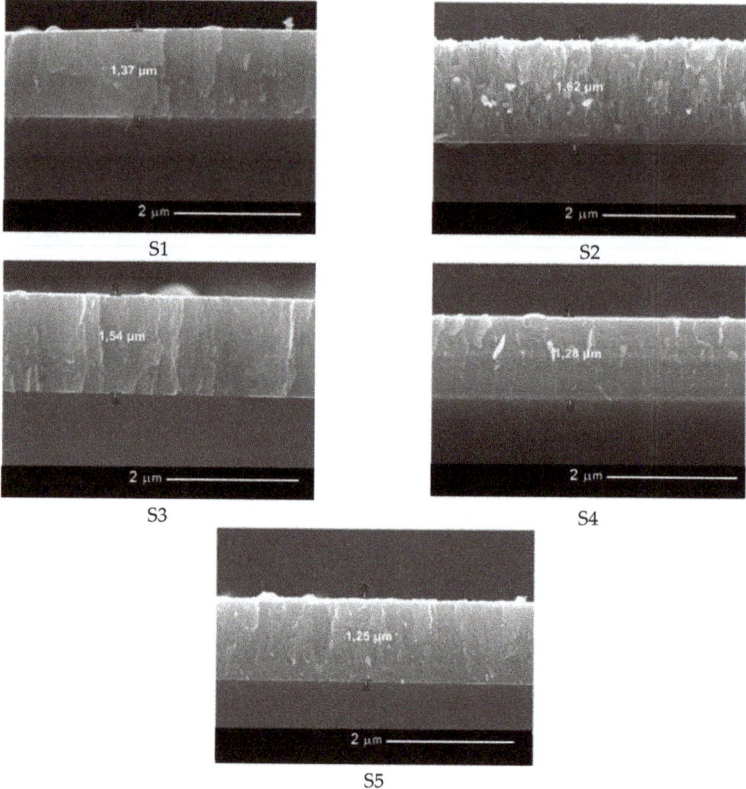

Figure 2. Cross-sections of particular coatings.

The ion release tests were carried out in the following environments:

- Distilled water;
- Physiological saline solution (0.9% NaCl in H_2O);
- Artificial saliva according to FusayamaMayer (2 dm^3 distilled water, 0.8 g NaCl, 0.8 g KCl, 1.59 g $CaCl_2 \cdot 2H_2O$, 1.56 g $NaH_2PO_4 \cdot 2H_2O$, 0.01 g $Na_2S \cdot 9H_2O$ and 2 g urea) [38].

The metal bars, 2.0 mm in diameter and 45 mm long, made of alloy NiCr, with Ti(C, N)-type coatings as well as without any, were placed in polystyrene vessels, volume 30 mL. In each vessel, a mixing element was placed. Each container was filled with 10.0 mL of the liquid. The analyzed systems were incubated at 36.7 ± 0.2 °C. In each environment, the tests were carried out for 10, 30 and 90 days. After the completion of the given cycle, three 3 mL samples were collected from each vessel and an analysis by means of an inductively coupled plasma optical emission spectrometer ICP OES Optima 8300 by PERKIN ELMER (Waltham, MA, USA) was performed. The content of the following ions was examined: Ni, Cr, and Mo. The spectrometer calibration (determination of the calibration curves) was performed by means of standard water solutions with the concentration of the determined elements of 0.5, 1.0, 5.0, and 10 mg/L. The measurement of the released ions was made directly on solutions without dilution or mineralization of the sample. The concentrations of the examined elements were calculated in reference to the prepared analytical curves. The element concentration in each sample was measured twice. Two analytical lines were applied for each measurement, as shown in Table 3. The application of different analytical lines was aimed at identifying the possible spectral interferences.

The presented results constitute an average of three measurements from two analytical lines. For each measurement, about 1.5 mL solution was used.

Table 3. Analytical lines used for the measurement.

Element	Analytical Lines [nm]	
Ni	232.003	341.476
Cr	267.716	357.869

The mass X_j of the ions released since the beginning of the process was calculated from the following formula:

$$X_j = C_j \times V_c \tag{1}$$

where:

- X_j—mass of the ions released from the sample [mg];
- C_j—spectrometrically determined ion concentration [mg/dm^3];
- V_c—volume of the liquid in which the samples were submerged [dm^3].

The mass of the ions released from the sample per surface unit, was calculated from the following formula:

$$M_j = X_j/S_p \tag{2}$$

where:

- M_j—mass of the released ions per surface unit [mg/mm^2];
- X_j—mass of the ions released from the sample [mg];
- S_p—sample surface in contact with the liquid [mm^2].

The obtained results concerning the amount of the released ions were subjected to a statistical analysis by means of the non-parametrical Mann Witney U test [39]. Statistical analyses were conducted using PQStat statistical software (version 1.6.4.122).

3. Test Results

Chemical composition of deposited Ti(C, N) layers is shown in Table 4.

Table 4. Chemical composition of tested coatings.

Coating	Element Percentage at.%			Element Percentage wt.%		
	Ti	C	N	Ti	C	N
S1	53.50	48.50	0.00	80.18	19.82	0.00
S2	52.91	33.91	13.80	79.51	13.90	6.60
S3	51.94	28.22	19.84	78.76	11.67	9.57
S4	47.78	20.05	32.17	75.26	8.61	16.12
S5	56.79	0.00	53.21	79.78	0.00	20.22

The effect of the performed process was obtaining coatings with the thickness from 1.3 to 1.6 µm and a diversified chemical composition, from pure titanium carbide with the content of about 48 at.% C, to pure titanium nitride with the content of about 53 at.% N. Additionally, three intermediate layers with the contents of about 34 at.% C and about 14 at.% N, about 28 at.% C and about 20 at.% N, about 20 at.% C and 32 at.% N were obtained. The total content of carbon and nitrogen in the coatings varied within the scope of 47% to 52%.

Figure 2 shows the microscopic images of the samples' cross-sections together with the particular coatings and their thicknesses (SE contrast).

The microscopic observations made it possible to evaluate the coatings' thickness, which equalled 1.25 to 1.62 μm. The coatings were air-tight, uniformly distributed and without defects. On the fractures of the coatings, one can see their columnar construction, where, besides some small differences in thickness, a difference in the crystallites' shape can also be observed. The coatings from groups S2, S3, S4 seem to have a more fine-grained structure.

Figures 3–8 show the amount of released nickel and chromium ions for the particular samples referred to a sample's surface unit.

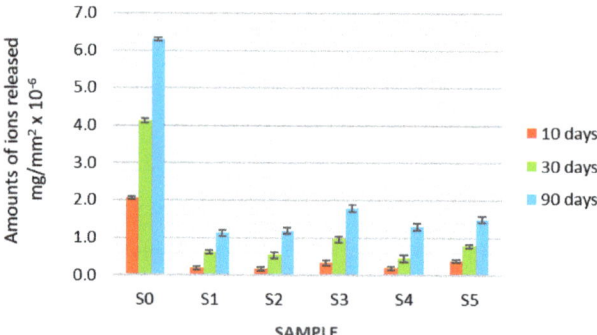

Figure 3. Amounts of ions Ni released into distilled water for the particular coatings after 10, 30, and 90 days.

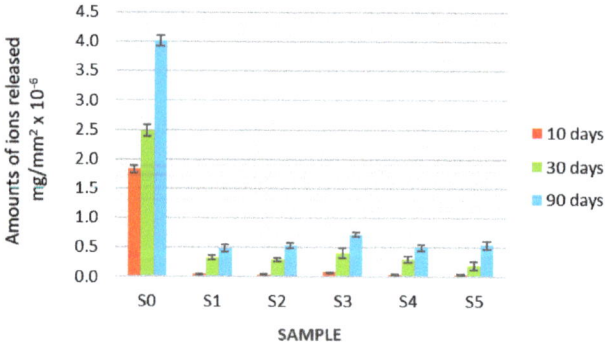

Figure 4. Amounts of ions Cr released into distilled water for the particular coatings after 10, 30, and 90 days.

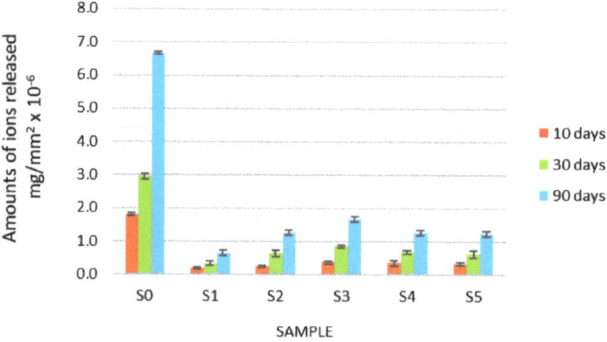

Figure 5. Amounts of Ni ions released into 0.9% solution NaCl for the particular coatings after 10, 30, and 90 days.

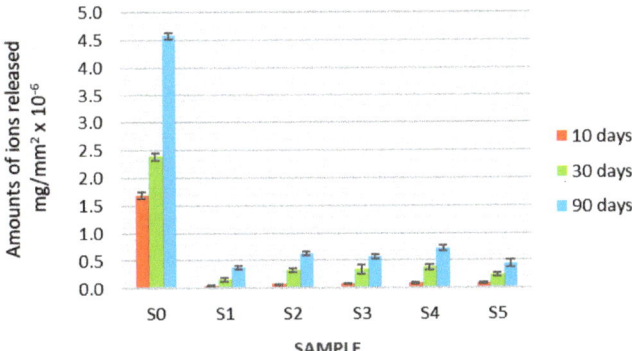

Figure 6. Amounts of Cr ions released into 0.9% solution NaCl for the particular coatings after 10, 30, and 90 days.

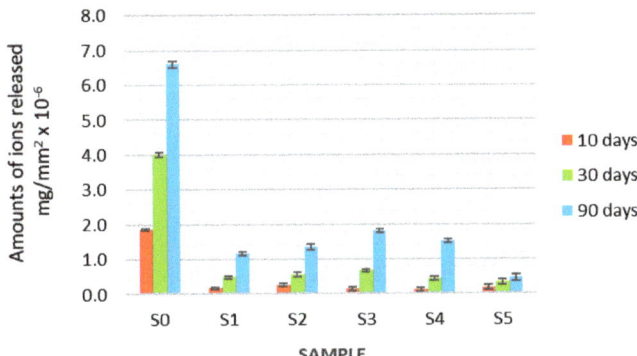

Figure 7. Amounts of Ni ions released into artificial saliva for the particular coatings after 10, 30, and 90 days.

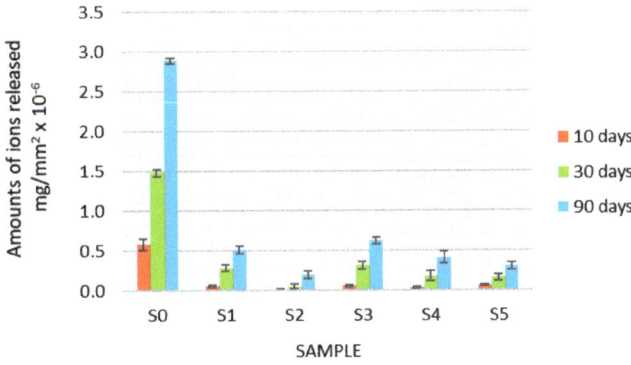

Figure 8. Amounts of Cr ions released into 0 artificial saliva for the particular coatings after 10, 30, and 90 days.

3.1. Examinations of the Amount of Released Ions

In the case of the control group S0, for each examined environment, an increase of the amount of released ions in respect of the results obtained after 10 days was observed—after 30 days about a double increase, after 90 days about a triple increase. In the case of the coated samples, the dynamics of the ion release was as follows.

3.2. Distilled Water Environment

In the environment of distilled water, after 10 days, in the samples covered with the studied coatings, a decrease of the released ions was recorded in reference to the uncoated alloy Ni–Cr (group S0)—from 5 to 12 times lower for the nickel ions and about a dozen times lower for the chromium ions. The lowest amount of ions was released from sample S4 (Ni ions—about 10× less, Cr—about 60× less). In turn, in reference to sample S0, the largest amount of ions was released from the sample with coating S3 (Ni ions—about 6× less, Cr—about 26× less).

After 30 days, in distilled water, about 6 to 10 times less nickel ions and about 3 to 8 times less chromium ions were recorded in reference to the non-coated alloy Ni–Cr (group S0). The lowest amount of ions was released from the samples in group S4, both in the case of nickel and chromium. The largest amount was recorded for group S3 and S5 (TiN), i.e., about 4.5 and about 6 times less nickel ions, and 6 and 3 times less chromium ions, respectively, in comparison with group S0.

In the environment of distilled water, after 90 days, in reference to the non-coated alloy Ni–Cr (group S0), the lowest amount of ions was released from the samples in group S1 and S4 (about 6 and about 5 times less for the Ni ions, respectively, and about 8 times less chromium ions, in comparison to the samples of pure alloy Ni–Cr (S0)). In the case of the remaining coatings, the decrease was about 3–4 times for the nickel ions and about 6–8 times for the chromium ions. The largest amount of nickel ions was released from samples S2 and S3, i.e., about 3 and 3.5 times less, whereas for chromium—6 and 8 times less, in respect of the non-coated alloy in group S0.

3.3. 0.9% NaCl Environment

In the 0.9% NaCl environment, after 10 days, in the samples covered with the tested coatings, a significant decrease of the released ion amount in reference to the non-coated alloy Ni–Cr (group S0) was recorded—from 4 to 10 times lower for the nickel ions and over a dozen times lower for the chromium ions. The lowest amount of ions was released from samples S1 (Ni ions—about 10× less, Cr—about 40× less). In turn, in reference to sample S0, the largest amount of ions was released from sample S3 (Ni ions—about 5× less, Cr—about 20× less). The amounts of released Cr ions in samples S2, S4 and S5 were at a similar level to those in the case of S3.

After 30 days, in the 0.9% NaCl environment, about 4 to about 9 times less nickel ions and about 6 to about 15 times less chromium ions was recorded in reference to the non-coated alloy Ni–Cr (group S0). The lowest amount of nickel and chromium ions was released from S1 (nearly 9 times and 15 times less, respectively). The largest amount of nickel ions was released from groups S3, S2, S4 (about 4 times less than in the non-coated group). With regard to the chromium ions, the largest amount was released in group S4—about 4 times less compared to group S0. In group S5, the amount of the released nickel ions was slightly more advantageous than in groups S3, S4 and S2, while in the case of the released chromium ions in group S5, the result was much better than in groups S2, S3, and S4, i.e., 10 times less compared to group S0.

In the 0.9% NaCl environment, after 90 days, in reference to the non-coated alloy Ni–Cr (group S0), the lowest amount of ions was released from the samples in group S1 (about 10 times less for the Ni ions and about 12 times less for the chromium ions in reference to the samples of pure alloy Ni–Cr). In the case of the remaining coatings, the decrease equalled about 4 times for the nickel ions and about 6–10 times for the chromium ions. The amount of the released nickel ions in groups S2, S4 and S5 was at a similar level, i.e., about 5 times less, and in S3—about 4 times less. Chromium was released in the largest amounts from sample S4, i.e., 6× less in reference to the non-coated alloy in group S0. In the remaining groups, i.e., S2, S3, and S5, the results were similar and the amount of the released chromium ions decreased about 7, 8 and 10 times, respectively, in reference to the alloy without a coating.

3.4. Artificial Saliva Environment

In the environment of artificial saliva, after 10 days, in the samples covered with the examined coatings, a significant decrease of the released ions was recorded in reference to the non-coated alloy Ni–Cr (group S0), which was from 5 to 11 times lower for the nickel ions and from 9 to 45 times lower for the chromium alloys. In reference to group S0, the Ni ion release was as follows: 11 times in group S1, 7.5 times in group S2 and about 5 times in groups S3, S4 and S5. The lowest amount of chromium, in reference to the samples without coatings, was released in group S2, i.e., 45 times less, whereas in group S4—over 14 times less. In groups S3 and S5, about 10 times less and in S5—9 times less Cr ions were recorded.

After 30 days, in the artificial saliva environment, from about 6 to about 12 times less nickel ions and from 5 to about 30 times less chromium ions were recorded in reference to the non-coated alloy Ni–Cr (group S0). The lowest amount of nickel ions was released from the samples in group S5 (12 times less compared to the non-coated alloy), whereas the lowest amount of chromium was released from sample S2 (about 30 times less in reference to the non-coated alloy). The highest amount of Ni and Cr ions was released from the samples in group S3, where 6 and 4.5 times less ions was released compared to the non-coated alloy. In groups S2, S1 and S4, 7×–9× less nickel ions were released than in the group without coating. With regard to the chromium ions in groups S4 and S5, 8–9 times less released ions were recorded, while in group S1—about 5 times less (close to that in group S3) compared to group S0.

In the artificial saliva environment, after 90 days, the results for the released Ni and Cr ions in the examined groups were diversified. The amounts of released Ni ions were from 3.5 to 14.5 times lower in respect of the non-coated alloy Ni–Cr, whereas in the case of the chromium ions—from 4.5 to 14 times lower. The least nickel ions was released from the samples in group S5 (about 14.5 times less), and the least chromium ions in group S2—14 times less. The highest amount of released ions was observed in group S3—3 times less nickel ions and 4 times less chromium ions in respect of the group without coatings.

4. Discussion

The amounts of the released nickel and chromium ions are different, which, among other reasons, results from the content of these elements in the alloy (about twice as much nickel as chromium). However, the ratio of the released nickel ions to the chromium ions is about 3 to 4 times higher. So, one should presume that the examined coatings constitute a better barrier for chromium than for the nickel. This can result from the difference of the size of the nickel Ni^{2+} and chromium Cr^{3+} ions, which equals about 20%.

The performed tests showed that, in the case of the Ti(C, N)-type coating, the amounts of Ni and Cr ions released into each of the tested environments was much lower than in the case of the alloy without a coating. In the case of each environment, the statistical analysis of the obtained results demonstrated a statistically significant decrease in the amount of released ions in the coated samples in respect of the samples without a coating. No statistically significant differences in the amount of released ions were observed between the samples covered with the particular coatings. The lack of significant differences in the penetration of the ions should not be surprising; the coatings have a similar construction and corrosion resistance. However, a more thorough analysis of the results makes it possible to notice a certain tendency showing that, in most cases, the amount of ions penetrating the coatings from groups S2, S3, and S4 is a little higher than that of the ions penetrating the coatings from groups S1 and S5, and this difference results from their thickness, as coatings S2 and S3 are the thickest and so, they should constitute the best barrier. This can be explained in two ways. First, these coatings are more fine-grained. Their fine-grained structure can be demonstrated by the results of diffraction tests included in the study [31] While the results presented there refer to the phase identification of coatings, as well as their texture and stresses, in the analysis of the width of the reflexes from phases Ti(C, N) (samples S2, S3, S4), we observe their slight widening in respect of phases TiC (S1) and TiN

(S5), which can prove their more fine-grained structure. A fine-grained structure significantly increases the amount of grain boundaries, and these, as we know, are the areas of facilitated diffusion. So, nickel and chromium from the alloy can penetrate the coating more easily. Also, one should take into account the fact that these processes take place at ambient temperatures, where other diffusion mechanisms are hindered. Another reason for such a behaviour of the coatings can be their crystallographic structure. Carbide TiC has a similar structure of the crystal lattice as that of nitride TiN, and they differ only in the parameters of the lattice. In the carbonitrides being formed, the carbon atoms are successively replaced by the nitrogen atoms (the latter are placed in the position of the former, without a change in the crystal lattice). As carbon atoms are bigger than nitrogen atoms, this change causes a certain relaxation of the lattice, which can create a situation of facilitated diffusion.

In the analysis of the amount of ions released into the particular examined solutions, no statistically significant differences were established, yet certain tendencies could be observed. The highest amount of ions released into the environment of a 0.9% NaCl solution results from the aggressiveness of the chlorine ions. Chloride ions have a clear electronegative character and they favour the metals' passing from the atomic state into the ionic state. This is especially visible in the alloy without a coating. The amounts of the ions released from the pure metal are slightly higher, both in the environment of the 0.9% NaCl solution and artificial saliva, as both solutions contain chlorine ions. The slightly lower amount of ions released into distilled water can be explained by the diffusion into a non-saturated environment. The lowest amount of ions from the coated samples were released into the solution of artificial saliva, which might be explained by the fact that the latter contains phosphates and sulfides, which can form low-soluble phosphates and sulfides with the substrate metals, and they, in turn, can block the discontinuity of the structure and the porosity of the coating, thus hindering the passing of the ions.

Covering metal alloy elements with coatings can significantly reduce the cytotoxicity of these elements, thus providing the possibility of their use in contact with various tissues of the human body. The investigations performed at many scientific centres have confirmed the positive results of the use of different coatings. In a study concerning the application of diamond like carbon (DLC) coatings on metal elements, the authors demonstrated this in short-term and long-term investigations in vivo performed on rat males, breed WISTAR. The modified samples were implanted under the skin for 24 h, 1, 4 and 26 weeks. After that time, the animals were put down and the tissues surrounding the modified elements and the spleen were collected. The evaluation of two routine dyed pieces of tissue and spleen showed that: 1) the carbon coating protects the tissue surrounding the metal implants against metal ion penetration, 2) no immunological reaction of the tissue or the spleen occurs, whereas the non-coated elements cause a strong reaction, 3) the statistically important differences (pb0.05) refer to the number and type of the cell elements and the degree of reaction in respect of the modified and non-modified elements [40]. Also, studies of similar coatings were performed, where the obtained amounts of released metal ions were correlated with the estimated biological response of the cell viability test (osteoblast cell line Saos-2) and the bacterial colonization test (strain Escherichia coli DH5α). The results showed that depositing hybrid DLC-type coatings by means of the magnetron sputtering radio frequency plasma assisted chemical vapour deposition (RF PACVD/MS) method makes it possible to obtain an air-tight coating, which prevents diffusion of harmful elements from the metallic substrate [41].

A reduced toxicity level is also exhibited by nitride and carbide coatings. Jang et al. studied biological Ti samples covered with TiN and 3-1-2,5-diphenyl tetrazolium bromide test (MTT) coatings, among other things, by examining their vitality by means of the MTT test after 8 days of being placed in the medium. The test showed a higher survivability of the cells from the group coated with TiN (by 125%) and TiAlN (by 117%) compared to the pure Ti group (100%) [42]. In the studies of Bramm et al., titanium was coated with titanium carbide TiC in the pulsed laser deposition (PLD) technology and evaluated in vitro in 3 cell lines as well as in vivo in respect of the integration with the bone, in tests performed on animals. The homogeneous TiC coatings covering titanium had a positive effect on the bones forming the cells, both in vitro and in vivo. According to the authors, these effects result from many factors, including both the chemistry of the coating and its morphology, with the consideration of the proper roughness. The TiC coatings simultaneously increase the implant's hardness and protect titanium from the aggressive attacks of the body fluids and tissues. The authors conclude that the use of TiC coatings on titanium improves the biocompatibility; also, with the appropriate roughness, it stimulates the proliferation of osteoblasts, their adhesion and diversity, as well as improves the osteointegration of the implant [43].

The biological tests performed on Ni–Cr alloys covered with the presented coatings also showed an improvement of the biocompatibility of the coated alloy in reference to the non-coated alloy. The evaluation by means of the MTT viability test performed on human microvascular endothelial cells coming from the skin—HMEC-1—showed that, in the case of the samples incubated for 24 and 96 h, we can observe statistically significant differences (with the value $p < 0.05$) between the non-coated samples and all the other samples coated with Ti (C, N) [36]. Also, the investigations with the use of fibroblasts proved that no Ti(C, N) coating affects the activity of the oxidoreductive enzymes, ether in the case of the direct cell culture or the application of extracts of the tested materials [37]. Such a behaviour of the Ti(C, N)-type coatings has also been observed by other researchers. Serro et al. [44] compared the cytotoxicity of 48-h extracts from implants coated with carbonitride Ti(C, N) (containing 21.6% C) and nitride TiN on mouse fibroblasts. Examining the morphology and biological activity and comparing them with the cells bred in a normal medium, they did not observe a cytotoxic effect of Ti (C, N) and TiN coatings on those cells. In the studies concerning a TiN coating on a nickel alloy [45], it has been established that it does not cause a cytotoxic irritation; a significant increase of the gum fibroblast on the surfaces of nickel alloys coated with TiN has been recorded, as opposed to the surface of the alloys without a TiN coating.

It can be inferred from the performed studies that there are no significant differences in the amounts of the ions which have passed through the particular coatings. Also, their biological properties are similar [36,37]. So, it is impossible to point to the best one. During clinical proceedings, their properties should rather be considered with regard to a particular application. As it was demonstrated in earlier studies, they exhibit big differences in hardness, modulus of elasticity and shear resistance [34].

The positive effect of the coatings is caused by various factors. Undoubtedly, a decrease in the harmful ions has basic importance. Ti(C, N)-type coatings additionally prevent oxidation, which, as is suggested by Toshifumi, could have a significant effect on the surface wettability [46]. This parameter is important, e.g., during the cells' adhesion to the substrate. The wettability measurements of the presented coatings demonstrated their higher hydrophilicity than alloy Ni–Cr [34]. Also, Li confirmed that NiTi coated with TiN exhibits a much higher hydrophilicity than in the case when no coating is applied. Additionally, he pointed to a significant role of the surface roughness [47], which can be especially important during the osteointegration of the implants with the bone [48].

Correlating the results of the amounts of the ions released from alloys with biological responses, we can see that the presence of the coatings causing a reduction of the amount of the ions released into the environment clearly favours a reduction of the alloys' cytotoxicity.

5. Conclusions

- All the examined Ti(C, N)-type coatings significantly limit the amount of the metal ions released into the environment.
- Ti(C, N)-type coatings can be considered for biomedical applications as protective coatings for non-precious metal alloys.
- Because there are no statistically significant differences in the amounts of the ions released by the particular coatings, during the selection of the latter, their other properties should be taken into consideration.

Author Contributions: Conceptualization, K.B. and L.K.; Methodology, L.K.; Validation, L.K. and K.B.; Formal Analysis, L.K.; Investigation, L.K.; Resources, L.K.; Data Curation, K.B.; Writing—Original Draft Preparation, K.B.; Writing—Review and Editing, L.K.; Visualization, K.B.; Supervision, L.K.; Funding Acquisition, K.B. and L.K.

Funding: This research received no external funding.

Conflicts of Interest: The authors declare no conflict of interest. We declare that the research is disclosed all conflicts interest statement, or explicitly state that there are none.

References

1. Raap, U.; Stiesch, M.; Reh, H.; Kapp, A.; Werfel, T. Investigation of contact allergy to dental metals in 206 patients. *Contact Dermat.* **2009**, *60*, 339–343. [CrossRef] [PubMed]
2. Beck, K.A.; Sarantopoulos, D.M.; Kawashima, I.; Berzins, D.W. Elemental release from CoCr and NiCr alloys containing palladium. *J. Prosthodont.* **2012**, *21*, 88–93. [CrossRef] [PubMed]
3. Galo, R.; Ribeiro, R.F.; Rodrigues, R.C.S.; Rocha, L.A.; Mattos, M.D.G.C.D. Effects of chemical composition on the corrosion of dental alloys. *Braz. Dent. J.* **2012**, *23*, 141–148. [CrossRef] [PubMed]
4. Małkiewicz, K.; Sztogryn, M.; Mikulewicz, M.; Wielgus, A.; Kamiński, J.; Wierzchon, T. Comparative assessment of the corrosion process of orthodontic archwires made of stainless steel, titanium–molybdenum and nickel–titanium alloys. *Arch. Civ. Mech. Eng.* **2018**, *18*, 941–947. [CrossRef]
5. Grimsdottir, M.; Hensten-Pettersen, A. Cytotoxic and antibacterial effects of orthodontic appliances. *Eur. J. Oral Sci.* **1993**, *101*, 229–231. [CrossRef]
6. Kansu, G.; Aydin, A.K. Evaluation of the biocompatibility of various dental alloys: Part I–Toxic potentials. *Eur. J. Prosthodont. Restor. Dent.* **1996**, *4*, 129–136.
7. Howie, D.W.; Rogers, S.D.; McGee, M.A.; Haynes, D.R. Biologic effects of cobalt chrome in cell and animal models. *Clin. Orthop. Relat. Res.* **1996**, *329*, S217–S232. [CrossRef]
8. Wataha, J.C.; Hanks, C.T.; Sun, Z. In vitro reaction of macrophages to metal ions from dental biomaterials. *Dent. Mater.* **1995**, *11*, 239–245. [CrossRef]
9. Wataha, J.C.; Malcolm, C.T.; Hanks, C.T. Correlation between cytotoxicity and the elements released by dental casting alloys. *Int. J. Prosthodont.* **1995**, *8*, 9–14.
10. Cicciù, M.; Fiorillo, L.; Herford, A.S.; Crimi, S.; Bianchi, A.; D'Amico, C.; Laino, L.; Cervino, G. Bioactive titanium surfaces: Interactions of eukaryotic and prokaryotic cells of nano devices applied to dental practice. *Biomedicines* **2019**, *7*, 12. [CrossRef]
11. Bruce, G.J.; Hall, W.B. Nickel hypersensitivity-related periodontitis. *Compend. Contin. Educ. Dent.* **1995**, *16*, 180–184.
12. Setcos, J.C.; Babaei-Mahani, A.; Di Silvio, L.; Mjör, I.A.; Wilson, N.H. The safety of nickel containing dental alloys. *Dent. Mater.* **2006**, *22*, 1163–1168. [CrossRef] [PubMed]
13. Faccioni, F.; Franceschetti, P.; Cerpelloni, M.; Fracasso, M.E. In vivo study on metal release from fixed orthodontic appliances and DNA damage in oral mucosal cells. *Am. J. Orthod. Dentofac. Orthop.* **2003**, *124*, 687–693. [CrossRef]
14. Noda, M.; Wataha, J.C.; Lockwood, P.E.; Volkmann, K.R.; Kaga, M.; Sano, H. Low-dose, long-term exposures of dental material components alter human monocyte metabolism. *J. Biomed. Mater. Res.* **2002**, *62*, 237–243. [CrossRef] [PubMed]
15. Wataha, J.C.; Lewis, J.B.; Lockwood, P.E.; Rakich, D.R. Effect of dental metal ions on glutathione levels in THP-1 human monocytes. *J. Oral Rehabil.* **2000**, *27*, 508–516. [CrossRef] [PubMed]
16. Könönen, M.; Rintanen, J.; Waltimo, A.; Kempainen, P. Titanium framework removable partial denture used for patient allergic to other metals: A clinical report and literature review. *J. Prosth. Dent.* **1995**, *73*, 4–7. [CrossRef]
17. De Silva, B.D.; Doherty, V.R. Nickel allergy from orthodontic appliances. *Contact Dermat.* **2000**, *42*, 102–103.
18. Napankangas, R.; Raustia, A. An 18-year retrospective analysis of treatment outcomes with metal-ceramic fixed partial dentures. *Int. J. Prosthodont.* **2011**, *24*, 314–319.
19. Miyazaki, T.; Hotta, Y. CAD/CAM systems available for the fabrication of crown and bridge restorations. *Aust. Dent. J.* **2011**, *56*, 97–106. [CrossRef]

20. Drago, C.; Howell, K. Concepts for designing and fabricating metal implant frameworks for hybrid implant prostheses. *J. Prosthodont.* **2012**, *21*, 413–424. [CrossRef]
21. von Steyern Vult, P. All-ceramic fixed partial dentures. Studies on aluminum oxide- and zirconium dioxide-based ceramic systems. *Swed. Dent. J. Suppl.* **2005**, *173*, 1–69.
22. Heintze, S.D.; Rousson, V. Survival of zirconia- and metal-supported fixed dental prostheses: A systematic review. *Int. J. Prosthodont.* **2010**, *23*, 493–502. [PubMed]
23. Almazdi, A.A.; Khajah, H.M.; Monaco, E.A., Jr.; Kim, H. Applying microwave technology to sintering dental zirconia. *J. Prosthet. Dent.* **2012**, *108*, 304–309. [CrossRef]
24. Pawlak, R.; Tomczyk, M.; Walczak, M. The favorable and unfavorable effects of oxide and intermetallic phases in conductive materials using laser micro technologies. *Mater. Sci. Eng. B* **2012**, *177*, 1273–1280. [CrossRef]
25. Krzak-Roś, J.; Filipiak, J.; Pezowicz, C.; Baszczuk, A.; Miller, M.; Kowalski, M.; Będziński, R. The effect of substrate roughness on the surface structure of TiO_2, SiO_2, and doped thin films prepared by the sol-gel method. *Acta Bioeng. Biomech.* **2009**, *11*, 21–29. [PubMed]
26. Szymanowski, H.; Sobczyk, A.; Gazicki-Lipman, M.; Jakubowski, W.; Klimek, L. Plasma enhanced CVD deposition of titanium oxide for biomedical applications. *Surf. Coat. Technol.* **2005**, *200*, 1036–1040. [CrossRef]
27. Rylska, D.; Klimek, L. Microstructure and corrosion characteristic of prosthetic dental alloys coated by TiN. *Acta Metall. Slovaca* **2004**, *10*, 938–942.
28. Wang, G.; Zreiqat, H. Functional coatings or films for hard-tissue applications. *Materials* **2010**, *3*, 3994–4050. [CrossRef]
29. Pietrzyk, B.; Miszczak, S.; Szymanowski, H.; Kucharski, D. Plasma enhanced aerosol-gel deposition of Al_2O_3 coatings. *J. Eur. Ceram. Soc.* **2013**, *33*, 2341–2346. [CrossRef]
30. Januszewicz, B.; Klimek, L. Investigation of TiCN coatings on steel substrates deposited by means of low pressure cathode ARC technique. *Acta Metall. Slovaca* **2004**, *10*, 926–929.
31. Banaszek, K.; Januszewicz, B.; Wołowiec, E.; Klimek, L. Complex XRD and XRF characterization of TiN-TiCN-TiC surface coatings for medical applications. *Solid State Phenom.* **2015**, *225*, 159–168. [CrossRef]
32. Sáenz de Viteri, V.; Barandika, M.G.; de Gopegui, U.R.; Bayón, R.; Zubizarreta, C.; Fernández, X.; Igartua, A.; Agullo-Rueda, F. Characterization of Ti–C–N coatings deposited on Ti6Al4V for biomedical applications. *J. Inorg. Biochem.* **2012**, *117*, 359–366. [CrossRef] [PubMed]
33. Sáenz deViteri, V.; Barandika, M.G.; Bayon, R.; Fernandeza, X.; Ciarsolo, I.; Igartua, A.; Tanoira, R.P.; Moreno, J.E.; Peremarch, C.P.J. Development of Ti–C–N coatings with improved tribological behavior and antibacterial properties. *J. Mech. Behav. Biomed. Mater.* **2016**, *55*, 75–86. [CrossRef] [PubMed]
34. Banaszek, K.; Pietnicki, K.; Klimek, L. Effect of carbon and nitrogen content in Ti(C, N) coatings on selected mechanical properties. *Met. Form* **2015**, *26*, 33–45.
35. Banaszek, K.; Klimek, L. Wettability and surface free energy of Ti(C, N) coatings on nickel-based casting prosthetic alloys. *Arch. Foundry Eng.* **2015**, *15*, 11–16. [CrossRef]
36. Banaszek, K.; Wiktorowska-Owczarek, A.; Kowalczyk, E.; Klimek, L. Possibilities of applying Ti (C, N) coatings on prosthetic elements–research with the use of human endothelial cells. *Acta Bioeng. Biomech.* **2016**, *18*, 119–126.
37. Banaszek, K.; Klimek, L.; Zgorzynska, E.; Swarzynska, A.; Walczewska, A. Cytotoxicity of titanium carbonitride coatings for prostodontic alloys with different amounts of carbon and nitrogen. *Biomed. Mater.* **2018**, *13*, 045003. [CrossRef]
38. Dalard, F.; Morgon, L.; Grosgogeat, B.; Schiff, N.; Lissac, M. Corrosion resistance of three orthodontic brackets: A comparative study of three fluoride mouthwashes. *Eur. J. Orthod.* **2005**, *27*, 541–549.
39. Berenson, M.L.; Levine, D.M.; Rindskopf, D. *Applied Statistics. A First Course*; Prentice Hall: Englewood Cliffs, NJ, USA, 1988.
40. Bociaga, D.; Jakubowski, W.; Komorowski, P.; Sobczyk-Guzenda, A.; Jedrzejczak, A.; Batory, D.; Olejnik, A. Surface characterization and biological evaluation of silver-incorporated DLC coatings fabricated by hybrid RF PACVD/MS method. *Mater. Sci. Eng. C* **2016**, *63*, 462–474. [CrossRef]
41. Bociaga, D.; Mitura, K. Biomedical effect of tissue contact with metallic material used for body piercing modified by DLC coatings. *Diam. Relat. Mater.* **2008**, *17*, 1410–1415. [CrossRef]
42. Hak Won, J.; Hyo-Jin, L.; Jung-Yun, H.; Kyo-Han, K.; Tae-Yub, K. Surface characteristics and osteoblast cell response on TiN- and TiAlN-coated Ti implant. *Biomed. Eng. Lett.* **2011**, *1*, 99–107.

43. Brama, M.; Rhodes, N.; Hunt, J.; Ricci, A.; Teghil, R.; Migliaccio, S.; Della Rocca, C.; Leccisotti, S.; Lioi, A.; Scandurra, M.; et al. Effect of titanium carbide coating on the osseointegration response in vitro and in vivo. *Biomaterials* **2007**, *28*, 595–608. [CrossRef] [PubMed]
44. Serro, A.; Completo, C.; Colaco, R.; Dos Santos, F.; Da Silva, C.L.; Cabral, J.M.; Araujo, H.; Pires, E.; Saramago, B. A comparative study of titanium nitrides, TiN, TiNbN and TiCN, as coatings for biomedical applications. *Surf. Coat. Technol.* **2009**, *203*, 3701–3707. [CrossRef]
45. Chien, C.C.; Liu, K.T.; Duh, J.G.; Chang, K.W.; Chung, K.H. Effect of nitride film coatings on cell compatibility. *Dent. Mater.* **2008**, *24*, 986–993. [CrossRef] [PubMed]
46. Toshifumi, S. CVD-titanium carbonitride coatings as corrosion-preventing barriers for steel in acid-brine steam at 200 °C. *Mater. Lett.* **1999**, *38*, 227–234.
47. Zhao, L.; Hong, Y.; Yang, D.; Lv, X.; Xi, T.; Zhang, D.; Hong, Y.; Yuan, J. The underlying biological mechanisms of biocompatibility differences between bare and TiN-coated NiTi alloys. *Biomed. Mater.* **2011**, *6*, 025012.
48. Annunziata, M.; Oliva, A.; Basile, M.A.; Giordano, M.; Mazzola, N.; Rizzo, A.; Lanza, A.; Guida, L. The effects of titanium nitride-coating on the topographic and biological features of TPS implant surfaces. *J. Dent.* **2011**, *39*, 720–728. [CrossRef]

 © 2019 by the authors. Licensee MDPI, Basel, Switzerland. This article is an open access article distributed under the terms and conditions of the Creative Commons Attribution (CC BY) license (http://creativecommons.org/licenses/by/4.0/).

Article

Microstructure Observation and Nanoindentation Size Effect Characterization for Micron-/Nano-Grain TBCs

Haiyan Liu [1], Yueguang Wei [1,*], Lihong Liang [2,*], Yingbiao Wang [3], Jingru Song [3], Hao Long [1] and Yanwei Liu [1]

1. Department of Mechanics and Engineering Science, College of Engineering, Peking University, Beijing 100871, China; 1706387228@pku.edu.cn (H.L.); mtlonghao@pku.edu.cn (H.L.); 1901111600@pku.edu.cn (Y.L.)
2. College of Mechanical and Electrical Engineering, Beijing University of Chemical Technology, Beijing 100029, China
3. LNM, Institute of Mechanics, Chinese Academy of Sciences, Beijing 100190, China; wangyingbiao123@163.com (Y.W.); songjingru@lnm.imech.ac.cn (J.S.)
* Correspondence: weiyg@pku.edu.cn (Y.W.); lianglh@mail.buct.edu.cn (L.L.); Tel.: +86-10-6275-7389 (Y.W.)

Received: 21 February 2020; Accepted: 31 March 2020; Published: 2 April 2020

Abstract: Microstructure observation and mechanical properties characterization for micron-/nano-grain thermal barrier coatings were investigated in this article. Scanning electron microscope images demonstrated that both micron-grain coating and nano-grain coating had micrometer-sized columnar grain structures; while the nano-grain coating had the initial nanostructures of the agglomerated powders reserved by the unmelted particles. The mechanical properties (hardness and modulus) of micron-/nano-grain coatings were characterized by using nanoindentation tests. The measurements indicated that the nano-grain coating possessed larger hardness and modulus than the micron-grain coating; which was related to the microstructure of coatings. Nanoindentation tests showed that the measured hardness increased strongly with the indent depth decreasing; which was frequently referred to as the size effect. The nanoindentation size effect of hardness for micron-/nano-grain coatings was effectively described by using the trans-scale mechanics theory. The modeling predictions were consistent with experimental measurements; keeping a reasonable selection of the material parameters.

Keywords: micron-/nano-grain coatings; microstructure; nanoindentation size effect; trans-scale mechanics theory

1. Introduction

Thermal barrier coatings (TBCs) with low thermal conductivity provide excellent thermal protection and have been widely used to protect hot-components in aero-engines from high-temperature environment [1–3]. The thermal and mechanical properties of TBCs have a significant effect on the lifetime of TBCs. Therefore, the measurements for thermal and mechanical properties are essential. In addition, the thermal and mechanical properties of materials are closely related to their microstructures, so it is worth investigating microstructures of TBCs.

Studies have shown that improvement in physical, thermal, and mechanical properties of new materials can be attributed to improvement of microstructure when compared with conventional materials [4]. For plasma sprayed coatings, particle size of the feedstock powders [5–8] and spray parameters [9–11] have a great effect on the microstructure (such as porosity, micro-crack content, and grain size) of the coatings. Liang et al. reported that the feedstock powders used for depositing nanostructured coating (particle size of agglomerated powders ranged in 30–50 μm, the size of nanoparticles inside the agglomerated powders varied from 50 to 80 nm) was smaller than the powders used for depositing conventional coating (particle size ranged in 60–100 μm), which contributed to the improvement of microstructure (such as reduction of grain size and contents of pores and micro-cracks) of the nanostructured coating [6]. Bai et al. investigated the effect of spray parameters on the particles temperature and velocity which determined the melting state of the particles, and found that the spray distance had the greatest impact [9]. Their results showed that the content of unmelted nanoparticles decreased from 12% to 6% when the temperature and velocity of in-flight particles increased by 10% and 14%, respectively.

Recently, a great deal of research with respect to the effect of microstructure on thermal properties such as thermal shock (oxidation) resistances [6–8,12,13] and thermal insulation [9,14], and mechanical properties such as hardness and modulus [5,14–22] have been done. Wang et al. found that the nanostructured zirconia coating possessed a better thermal shock resistance than that of the conventional zirconia coating [7]. The enhanced toughness and decreased porosity and micro-cracks in the nanostructured coating contribute to improvement of thermal shock resistance, which is consistent with what Liang reported [6]. Fan et al. discovered the small grain size had a positive influence on thermal fatigue life, and the columnar crystal microstructure had a great effect on thermal shock resistance of TBCs [13]. Wu et al. found that the reduction of porosity resulted in increase of thermal diffusivity, suggesting a significant degradation in the thermal barrier effect [14]. Some researchers have studied the effect of microstructure on hardness and modulus at high temperature [14–18]. They all discovered that the hardness and modulus of TBCs increased with the decrease of porosity caused by sintering of coating at high temperature. Nath et al. evaluated the resistance to plastic deformation of TBC at room temperature and with thermal exposure [15]. Baiamonte et al. indicated that nanostructured coating had larger hardness and modulus than the conventional coating by dynamic indentation tests at room and high temperatures [18]. The effect of microstructure on hardness and modulus at room temperature have also been investigated [19–22]. Jang et al. discovered that the hardness and Young's moduli at the side regions of TBCs decreased with increasing the porosity [19], which is consistent with reference [20]. Lima et al. studied the effect of roughness on the microhardness and elastic modulus of nanostructured zirconia coatings, their results showed that the microhardness and elastic modulus increased with decreasing roughness [21]. Zeng et al. found that the nanocrystalline zirconia coating had a larger microhardness than that of conventional coating, and they used the Hall–Petch model assuming that a dislocation network density slipped through the grain boundaries to explain the phenomenon [22].

In order to study the mechanical properties (hardness and modulus) of TBCs, several tests have been presented and widely used, such as beam bending tests [18,23–25] and indentation tests [5,14–23,26,27]. Thompson et al. indicated that the modulus given by indentation test was much higher than the value based on the beam bending technique [23], which is consistent with Baiamonte reported [18], because indentation test gave an indication of local modulus, and beam bending test gave a global value which contained the effects of micro-crack and porosity. All modulus values measured by beam bending methods were relatively small for TBCs [24,25].

The measured mechanical properties based on beam bending tests are average values on macroscopic scales. However, the measured mechanical properties based on the nanoindentation tests are local values on microscopic scales. The above two types of mechanical properties obtained by the above two scale tests are extremely different. Many studies have shown that an apparent indentation size effect (ISE) of hardness for different kinds of materials was observed, such as brittle materials [26,27], metal materials [28,29], and biomaterials [30]. In order to describe the ISE, a series of different theoretical methods were presented. Li et al. proposed an energy balance analysis based on elastic/plastic indentation model to analyze the apparent ISE of hardness [26]. Zotov et al. used a new empirical equation based on the concept of elastic recovery to explain the phenomenon of ISE [27]. Wei et al. presented the strain gradient plasticity theory based on the thermodynamics frame to describe the ISE and cross-scale fracture [28,29], and they predicted the micrometer-sized parameter of strain gradient plasticity theory by the comparison of theoretical predictions with experimental measurements. Song et al. [30] used the trans-scale mechanics theory presented by Wu et al. [31] to model the nanoindentation size effect for limnetic shell, and they considered both the strain gradient effect and surface/interface effects.

In this study, the microstructures of micron-/nano-grain coatings were observed by using scanning electron microscopy (SEM) and atomic force microscopy (AFM); the porosity and grain size of micron-/nano-grain coatings were calculated by quantitative image analysis using Image-Pro Plus software. The mechanical properties (hardness and modulus) of micron-/nano-grain coatings were measured by using nanoindentation tests. The effect of microstructure on mechanical properties was analyzed by comparing the microstructure of micron-grain coating with that of nano-grain coating. The trans-scale mechanics theory presented by Wu et al. [31] was adopted to characterize the nanoindentation size effect for the micron-/nano-grain coatings.

2. Experimental Procedure

2.1. Materials and Plasma Spray

Ni-based superalloy plates (GH3128) with a dimension of 250 mm × 50 mm × 1.2 mm were used as substrates. A commercial CoNiCrAlY powder (AMDRY 995, Sulzer Metco, Pfäffikon, Switzerland) with a grain size of 5–37 µm was used for depositing the bond coat. Two kinds of spray-dried and sintered ZrO_2-8 wt % Y_2O_3 (8YSZ) powders were used as the feedstock for spraying on the bond coat to form the top coat with 300 µm and 500 µm in thickness. The particle size of the first feedstock ranged from 60 to 100 µm, the small particles inside the feedstock were submicron particles with a grain size of 320–650 nm. In the present research, the micron-grain coating was produced from the first feedstock. The particle size of the second feedstock varied from 30 to 50 µm, the small particles inside the feedstock were nanoparticles with a grain size of 10–50 nm. The nano-grain coating was fabricated from the second feedstock. The spraying process was carried out by atmospheric plasma spraying (APS) system (A-2000, Sulzer-Metco F4 gun, Pfäffikon, Switzerland). The coating samples were fabricated by the Shanghai Institute of Ceramics, Chinese Academy of Sciences, Shanghai, China. The parameters for plasma spraying were given in Table 1.

Table 1. Plasma spray parameters.

Parameters	NiCrCoAlY Bond Coat	8YSZ Coating
Primary Ar (slpm)	57	30
Second H_2 (slpm)	8	12
Spray distance (mm)	120	120
Gun current (A)	620	620
Power (kW)	39	42
Carrier gas (slpm)	3.5	3.5
Powder feed rate (g/min)	40	40

2.2. Microstructure Observation Tests

The samples for microstructure observation on surface were circle plates with polished coating surface, diameters of 14 mm and top coat thicknesses of 300 µm. The samples for microstructure observation on cross section were cuboids with polished (fractured) cross section, lengths of 20 mm, widths of 3 mm, and top coat thicknesses of 500 µm (300 µm). The surface and cross-sectional microstructure for micron-/nano-grain coatings were observed by using SEM (JSM-7500F, JEOL, Tokyo, Japan), and the micrographs of unmelted nanoparticles were observed by using AFM (MultiMode 8, Bruker, Billerica, MA, USA).

2.3. Nanoindentation Tests

The samples for nanoindentation experiments were circle plates with polished coating surface and diameters of 14 mm. Thicknesses of top coat, bond coat, and substrate were 300 µm, 15 µm, and 1.2 mm, respectively. Hardness and elastic modulus were measured on the coating surface of micron-/nano-grain coatings based on the nanoindentation tests. The tests were carried out using a Nano Indenter G200 (Agilent, Santa Clara, CA, USA) with a diamond Berkovich tip (tip curvature radius is about 20 nm). The force and displacement resolutions of G200 are about 1 nN and 0.0002 nm, respectively. The tests were performed under displacement control with a maximum indent depth of 1 µm. Hardness (H) and elastic modulus (E) were obtained based on the Oliver–Pharr method [32], the corresponding expression was

$$H = \frac{P}{A}, \quad \frac{1}{E_r} = \frac{1-v^2}{E} + \frac{1-v_i^2}{E_i}, \tag{1}$$

where P was the indent load, A was the contact area, E (E_i) and v (v_i) were the elastic modulus and Poisson's ratio for the sample (indenter), respectively. The material parameters of the diamond indenter were taken as $E_i = 1141$ GPa and $v_i = 0.07$. The reduced elastic modulus (E_r) was obtained using a relation including contact stiffness and contact area.

3. Experimental Results and Discussion

3.1. Microstructure Characteristics

Figure 1 shows the surface images of as-sprayed micron-/nano-grain coatings. Microscopically, the coating was non-uniform and non-dense, and had a large quantity of pores and micro-cracks. As shown in Figure 1, the nano-grain coating had a denser microstructure and fewer irregular pores and micro-cracks than the micron-grain coating. The surface microstructure characteristics are related to the formation mechanism of coating. In the process of forming the coating, the ceramic powders were first melted or partially melted under the plasma flame and rushed toward the surface of bond coat at high speed. The powders deformed and rapidly condensed and shrank to adhere on the bond coat in flat state. Because of insufficient deformation of some ceramic particles, pores generated among the particles, resulting in non-uniform and non-dense structure. The higher porosity of micron-grain coating may be attributed to the larger particle size of feedstock. The occurrence of micro-cracks on the coating surface was mainly due to the large tensile stress generated when the ceramic droplets condensed and shrank.

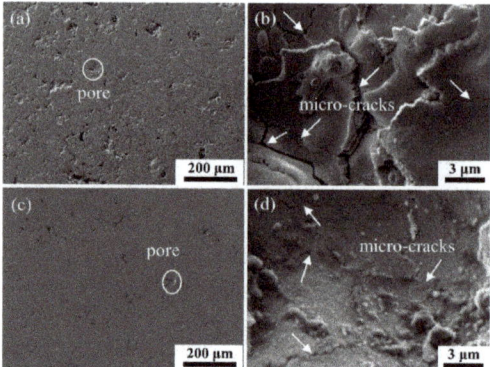

Figure 1. Microstructure of coatings: (**a**) polished surface of micron-grain coating; (**b**) a detailed view of the micro-cracks of micron-grain coating in circular region of (**a**); (**c**) polished surface of nano-grain coating; (**d**) a detailed view of the micro-cracks of nano-grain coating in circular region of (**c**). The magnifications of images (**a**–**d**) are 100, 5000, 100, 5000×, respectively.

Figure 2 shows the polished cross-section of representative specimens for micron-/nano-grain coatings. From the figure, it can be seen that the TBC system was clearly a multilayer structure, with many pores distributed randomly in the top coat. In this article, Image-Pro Plus software (Media Cybernetics, Silver Springs, MD) was utilized to calculate porosity by quantitative image analysis. The final statistical result of porosity was the average value based on 10 cross-sectional images for an individual sample [8,9]. The statistical results showed that the porosities of micron-/nano-grain coatings were approximately 6.5% ± 0.2% and 5.1% ± 0.3%, respectively.

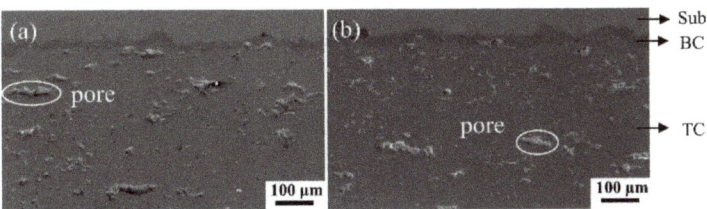

Figure 2. Images of polished cross-section of specimens: (**a**) for micron-grain coating; (**b**) for nano-grain coating. Sub, substrate; BC, bond coat; TC, top coat. The magnification of images (**a**) and (**b**) is 350×.

Figure 3 shows the images of fractured cross-section of micron-/nano-grain coatings. As seen from Figure 3a, the micron-grain coating consisted of layered structures (splats) with many inter-splat cracks, the micrometer-sized columnar crystals within the splats (see Figure 3b) were observed, and long cracks among the columnar crystals were visible. Rapid nucleation occurred at the cooler surface of the flattened droplet at large undercooling and the crystals grew rapidly opposite to the heat flow, leading to the formation of a columnar grain structure [8]. The columnar microstructure showed a strong anisotropy due to different heat flow directions, as reported by [19]. Figure 3c,d show the images of fractured cross-section of nano-grain coating. From the figures, columnar grain structures were also observed, inter-splat cracks and long cracks were also visible (see Figure 3c). The average inter-splat spacing of the nano-grain coating was smaller than that of the micron-grain coating. Fewer and smaller inter-splat cracks and long cracks existed in nano-grain coating, that is to say the nano-grain coating had a finer columnar grain structure. However, the nano-grain coating had a significantly different feature compared with the micron-grain coating, because some unmelted particles were loosely distributed in the recrystallization zone (splats) as shown in Figure 4, which is consistent with literatures [5,7–9]. These unmelted particles reserved the initial nanostructure of the agglomerated powder, as later confirmed by AFM (see Figure 5e). The complex structure of nano-grain coating was related to the formation mechanism of the coating. Individual submicron-sized and nanosized powders cannot be carried in a moving gas stream and deposited on a substrate because of their low mass and flowability. To overcome the above problem, the slurries prepared from the submicron-sized and nanosized powders was then spray dried to form micrometer-sized agglomerated powders. The two agglomerated powders were used as the feedstock for depositing the top coat. All sprayed feedstock were accelerated by the plasma flame, then rushed towards the surface of the bond coat at high speed and deposited. The melted particles re-nucleated and grew in the liquid state, forming the columnar grain structures (see Figure 3d); the unmelted particles reserved the initial nanostructure of the agglomerated powders (see Figure 4). The columnar grain structure acted as a binder and maintained the integrity of the coating.

During the plasma spraying, the particle temperature and velocity determined the melting state of particles. The plasma spray parameters affected the temperature and velocity of in-flight particles, such as powder feed rate, spray distance, spray power [9]. In the present research, we used the same spray parameters to deposit the two kinds of top coats, only considered the influence of particle size of feedstock powders on the microstructure of as-sprayed coatings. The effect of particle size of feedstock powders on the pore, micro-crack, and columnar grain structure was illustrated in Figures 1–3, and on grain size of as-sprayed coatings was demonstrated in Figure 5, as discussed in the following text.

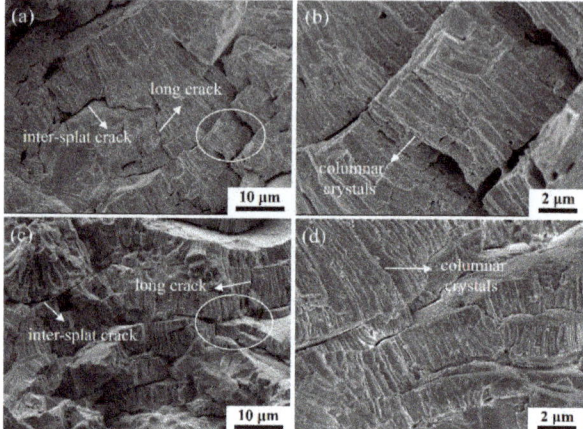

Figure 3. Microstructure of coatings: (**a**) fractured cross-section of micron-grain coating; (**b**) a detailed view of the columnar crystals of micron-grain coating in circular region of (**a**); (**c**) fractured cross-section of nano-grain coating; (**d**) a detailed view of the columnar crystals of nano-grain coating in circular region of (**c**). The magnifications of images (**a**–**d**) are 2000, 8000, 2000, 8000×, respectively.

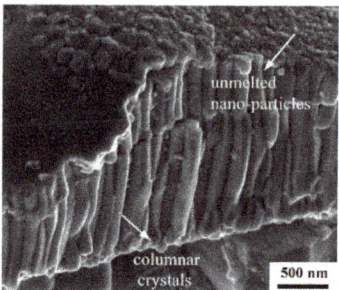

Figure 4. SEM image of columnar crystals and unmelted nano-particles in the nano-grain coating. The magnification of the image is 30,000×.

Figure 5 shows the representative images of grains in the micron-/nano-coatings, and distributions of grain size. The grain sizes were calculated by quantitative image analysis using Image-Pro Plus software. The final statistical result of grain sizes was the average value based on five images for an individual sample. The SEM image illustrated in Figure 5a indicates that micron-grain coating was composed of grains with different size (size ranged from 393 to 2507 nm), some micro-pores and micro-cracks were found at the intersection of multiple grains. The statistical result shown in Figure 5b demonstrates that the grain sizes were distributed mainly in 500–1500 nm, and the average value of grain size of micron-grain coating was approximately 1123 ± 310 nm. Figure 5c gives the SEM image of nano-grain coating surface after thermal etching, the image reveals that the coating was composed of small grains (size ranged in 75–96 nm) and large grains (size varied from 101 to 984 nm). Figure 5d reveals that the grain sizes were distributed mainly in 75–400 nm, and the average value of grain size of nano-grain coating in the recrystallization zone was approximately 242 ± 87 nm. Figure 5e shows the AFM image of nanograins in the nano-grain coating. The sizes of unmelted nanoparticles ranged in 10–50 nm, revealing these unmelted nanoparticles maintained the initial nanostructure of agglomerated powder. Figure 5f indicates that the grain sizes were distributed mainly in 10–30 nm, and the average value of grain size of nanoparticles was approximately 20 ± 6 nm. It was clear from Figure 5a–d that the grain size of nano-grain coating in the recrystallization zone was smaller than that of the micron-grain coating. The reduction of particles size of feedstock powders contributed to the reduction of grain size of the coating, leading to the improvement of microstructure of the coating.

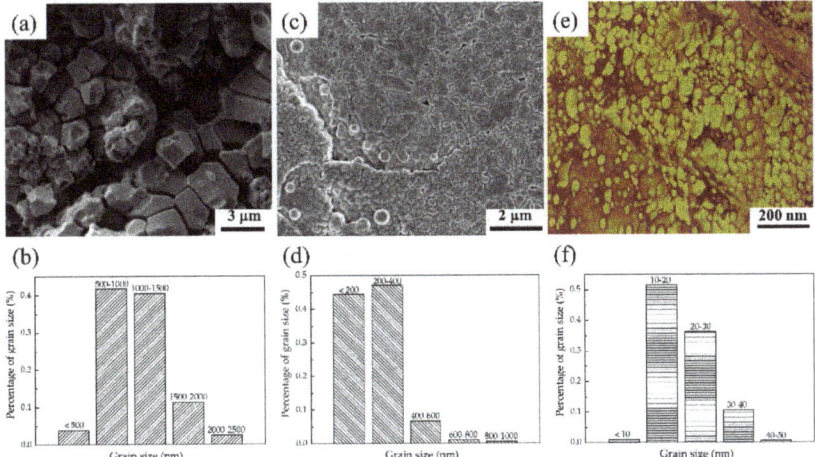

Figure 5. (**a**) SEM image of grains in the micron-grain coating; (**b**) statistical results on grain size of the micron-grain coating; (**c**) SEM image of grains in the recrystallization zone of nano-grain coating; (**d**) statistical results on grain size of nano-grain coating in the recrystallization zone; (**e**) AFM image of nanograins in the nano-grain coating; (**f**) statistical results on grain size of unmelted nanoparticles in the nano-grain coating. Numbers on the y-axis indicate the size of grains. The magnifications of images (**a**) and (**c**) are 5000 and 10,000×, respectively.

3.2. Hardness and Modulus Measurements

In the present research, the nanoindentation mechanical properties (hardness and elastic modulus) of micron-/nano-grain coating were measured and compared. It is worth noting that in the nanoindentation test, the influence of indenter tip curvature on the hardness measured is inevitable and a challenge question in that area. However, it is meaningful to distinguish mechanical properties between nano-grain coating and micron-grain coating by using the technique in averaging meaning. Figure 6 demonstrates the hardness-depth curves of the two kinds of TBCs corresponding to 10 indent points on the surface of top coat. Curves shown in Figure 6a,b correspond to the cases of micron- and nano-grain coating, respectively. When the indent depth was very shallow, the hardness-depth curves had a dramatic variation resulted from many factors, such as the unsteady contact between the indenter and the specimen surface at the beginning of the test, surface morphology of the indenter and specimens, etc. However, as the indent depth increased, the hardness-depth curves tended to keep stable rapidly. From Figure 6, generally the measured hardness-depth curves varied with indent depth non-monotonically because of radius of curvature existed in the tip of the indenter. When the indent depth was larger than about 100 nm, the influence of indenter on variation of the curve can be ignored, and the hardness increased with the decrease of indent depth, this phenomenon was called "nanoindentation size effect". The measured hardness values of nano-grain coating were larger than those of micron-grain coating at the same indent depth by comparison of curves in Figure 6a with those in Figure 6b. In order to compare the mechanical properties of the two coatings more clearly, we further derived the average value of hardness and modulus based on the curves and displayed the results in Figure 7. Figure 7 shows that both the measured average values of hardness and modulus of nano-grain coating were larger than those of micron-grain coating.

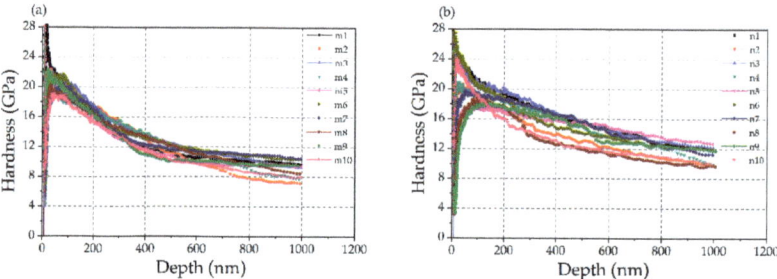

Figure 6. Hardness–depth curves of the two kinds of TBCs corresponding to 10 indent points on the surface of top coat, m and n denote micron-grain coating and nano-grain coating, respectively, (**a**) for micron-grain coating, (**b**) for nano-grain coating.

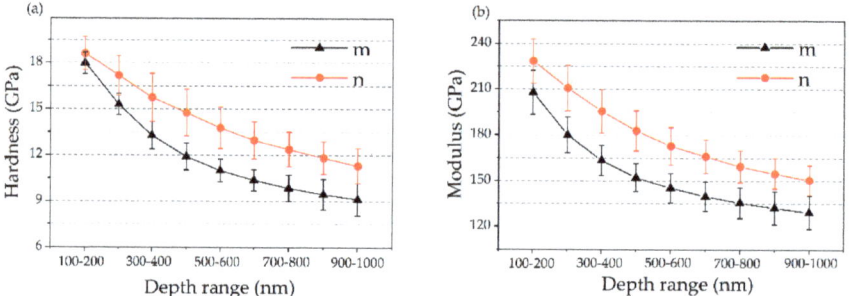

Figure 7. Average hardness and modulus of two kinds of TBCs in different depth ranges based on 10 indent points, m and n denote micron-grain coating and nano-grain coating, respectively, (**a**) for hardness and (**b**) for modulus.

As we all know, the mechanical properties of materials are closely related to their microstructures. In the present research, the feedstock powders used for depositing nano-grain (OSe) coating was smaller than those for depositing micron-grain coating, which contributed to the improvement of microstructure (such as reduction of contents of pores and micro-cracks and reduction of grain size) of the nano-grain coating, as Liang reported [6]. Furthermore, microstructure observation shows that the nano-grain coating was denser and had finer columnar grain structure and fewer pores and micro-cracks than the micron-grain coating (see Figures 1–3), so hardness and modulus of nano-grain coating were larger than those of the micron-grain coating, as reported by references [5,18]. In addition, the nanostructures and reduction of grain size of nano-grain coating (see Figure 5) contributed to the increase of hardness of the coating according to the Hall–Petch equation.

4. Modeling of Nanoindentation Size Effect for Micron-/Nano-Grain Coating

4.1. Theoretical Model Based on the Trans-Scale Mechanics Theory

Considering that the coatings has a certain inelastic deformation when the nanoindentation depth is at the nanoscale, and assuming that the coating is continuous and homogeneous, without considering the effects of pores and micro-cracks, we use the trans-scale mechanics theory [31] to describe the nanoindentation size effect, including the strain gradient and surface/interface effects, and the strain gradient effect is described by strain gradient plasticity theory.

For micron-grain coating, only the strain gradient effect is considered assuming that the surface/interface effects can be neglected in this case. For nano-grain coating, both strain gradient and surface/interface effects are considered.

The trans-scale mechanics theory only considering strain gradient effect was studied by Wei et al. [28], the expression of solution can be written as

$$\frac{H_{SGP}}{H_0} = g_0(\frac{h}{l}, \frac{E}{H_0}, \nu, N, \beta) \tag{2}$$

where H_0 is the reference hardness value corresponding to deep indentation hardness, l is material length scale, h is indent depth, N is strain hardening exponent of material, β describes the angle of the circular conic indenter through area equivalency with true pyramid indenter.

The trans-scale mechanics theory considering strain gradient and surface/interface effects was investigated by Song et al. [30], the nanoindentation hardness H with respect to indent depth h for the case of circular conic indenter and TBCs can be expressed as

$$\frac{H}{H_0} = f(\frac{h}{l}, \frac{E}{H_0}, \nu, N, \beta, \frac{\gamma}{H_0 h}, \frac{\Gamma}{H_0 d_0}) \tag{3}$$

where γ is the surface energy density, Γ is the interface energy density, d_0 is the representative size of the nanoparticle. The solution for Equation (3) can be obtained by quantity level analysis. Since the values of the dimensionless quantities both $\gamma/H_0 h$ and $\Gamma/H_0 d_0$ are much smaller than unity within an effective indent depth range, the dimensionless hardness can be approximately written as follows through small parameter expansion for the case of circular conic indenter and equal-size cubic grain [30]

$$\frac{H}{H_0} \approx g_0(\frac{h}{l}, \frac{E}{H_0}, \nu, N, \beta) + \frac{2}{\cos\beta}\frac{\gamma_s}{H_0 h} + \frac{6\sqrt{3}\Gamma}{H_0 d_0 \tan\beta}(1 - \frac{d_0 \tan\beta}{2\sqrt{3}h}) \tag{4}$$

where the first term of right-hand side is the solution only considering the strain gradient effect (see Equation (2)), and the other terms are the contributions resulted from the surface and interface effects.

4.2. Interpretation to Experimental Hardness Size Effect Based on the Theoretical Model

Hardness-depth curves shown in Figure 6 can be also simulated by using the trans-scale mechanics theory through considering a curvature radius existing at indenter tip. H_0 for deep indent is taken as 2.8 GPa for the present case. For the circular conic indenter, β is taken as 30° [28]. For the nano-grain coating used in experiments, the nanoparticle size d_0 can be measured based on the AFM figure and is about 20 nm. Coating can be considered as a medium-level hardening material when the nanoindentation depth is at the nanoscale, so we take $N = 0.3$.

Figure 8 shows the theoretical simulations to experimental results for two kinds of TBCs. In Figure 8a, two experimental curves denoted by symbols corresponding to the upper bound and lower bound curves in Figure 6a for micron-grain coating, respectively, theoretical results denoted by black solid lines are based on Equation (2). In Figure 8b, two experimental curves denoted by symbols corresponding to the upper bound and lower bound curves in Figure 6b for nano-grain coating, respectively, theoretical results denoted by black solid lines are based on Equation (4). For the micron-grain coating, we only consider the strain gradient effect, one can obtain the value of material length scale by comparison of theoretical results with experimental ones (l = 4.81–7.51 µm), and the values are within reasonable ranges [33]. For the nano-grain coating, substituting the material and geometric parameters into Equation (4), and letting the values of surface energy density and interface energy density the same in this paper, one can obtain the values of material length scale and surface/interface energy density by comparison of theoretical results with experimental ones (l = 1.73–7.26 µm, $\gamma = \Gamma$ = 2.88–7.12 J/m^2), and the values are within reasonable ranges [33]. Therefore, the trans-scale theory can effectively model the nanoindentation size effect for the two kinds of TBCs.

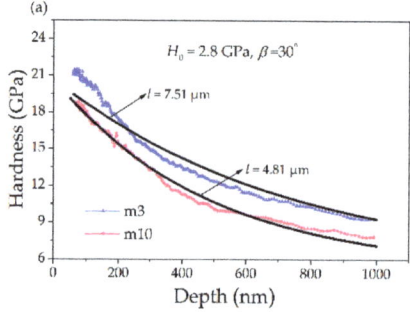 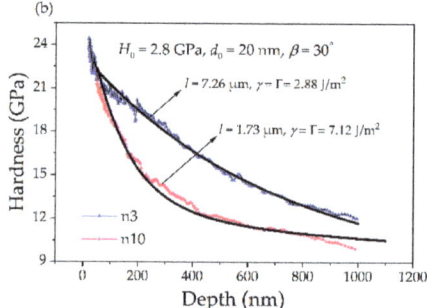

Figure 8. Theoretical simulations to experimental results for two kinds of TBCs. (**a**) For micron-grain coating, two experimental curves denoted by symbols corresponding to the upper bound and lower bound curves in Figure 6a, respectively, theoretical results denoted by black solid lines are based on Equation (2). (**b**) For nano-grain coating, two experimental curves denoted by symbols corresponding to the upper bound and lower bound curves in Figure 6b, respectively, theoretical results denoted by black solid lines are based on Equation (4).

5. Conclusions

For two kinds of TBCs, experimental microstructure observation, hardness, and modulus measurement by using the nanoindentation tests were performed, and nanoindentation size effect was described by using the trans-scale mechanics model. The main conclusions can be summarized as follows:

- Scanning electron microscope images reveal that the micron-grain coating had a columnar grain structure with an average grain size of about 1123 nm; the nano-grain coating also had a columnar grain structure formed by recrystallization of the melted particles with an average grain size of about 242 nm, while it possessed the initial nanostructure of the agglomerated powders reserved by the unmelted particles with an average grain size of about 20 nm.
- The hardness and modulus of two kinds of TBCs were measured by using nanoindentation tests. The measurement relation between hardness and indent depth showed a strong size effect. The measured results indicated that the nano-grain coating had larger hardness and modulus than the micron-grain coating, the improved properties of nano-grain coating were associated with retained nanostructure and reduction of porosity, micro-crack, and grain size of the coating.
- The nanoindentation size effect of hardness for the two kinds of TBCs was effectively described by utilizing the trans-scale mechanics theory, the values of material parameters were obtained by comparison of theoretical predictions with experimental measurements and they were within reasonable range.

Author Contributions: Data curation, Y.W. (Yingbiao Wang) and Y.L.; Methodology, J.S. and H.L. (Hao Long); Writing—original draft, H.L. (Haiyan Liu); Writing—review and editing, Y.W. (Yueguang Wei) and L.L. All authors have read and agreed to the published version of the manuscript.

Funding: This research was funded by the National Natural Science Foundation of China, Nos.: 11890681, 11672301, 11511202, 11672296, 91860102, 11432014.

Conflicts of Interest: The authors declare no conflict of interest.

References

1. Strangman, T.E. Thermal barrier coatings for turbine airfoils. *Thin Solid Films* **1985**, *127*, 93–106. [CrossRef]
2. Evans, A.; Mumm, D.; Hutchinson, J.; Meier, G.; Pettit, F. Mechanisms controlling the durability of thermal barrier coatings. *Prog. Mater. Sci.* **2001**, *46*, 505–553. [CrossRef]

3. Padture, N.P.; Gell, M.; Jordan, E.H. Thermal barrier coatings for gas-turbine engine applications. *Science* **2002**, *296*, 280–284. [CrossRef]
4. Koch, C.C. *Nanostructured Materials—Processing, Properties, and Applications*, 2nd ed.; Noyes Publications, William Andrew Publishing: Norwich, NY, USA, 2002.
5. Ghasemi, R.; Razavi, R.S.; Mozafarinia, R.; Jamali, H. Comparison of microstructure and mechanical properties of plasma-sprayed nanostructured and conventional yttria stabilized zirconia thermal barrier coatings. *Ceram. Int.* **2013**, *39*, 8805–8813. [CrossRef]
6. Liang, B.; Ding, C. Thermal shock resistances of nanostructured and conventional zirconia coatings deposited by atmospheric plasma spraying. *Surf. Coat. Technol.* **2005**, *197*, 185–192. [CrossRef]
7. Wang, W.; Sha, C.; Sun, D.; Gu, X. Microstructural feature, thermal shock resistance and isothermal oxidation resistance of nanostructured zirconia coating. *Mater. Sci. Eng. A* **2006**, *424*, 1–5. [CrossRef]
8. Wang, Y.; Bai, Y.; Liu, K.; Wang, J.W.; Kang, Y.X.; Li, J.R.; Chen, H.Y.; Li, B.Q. Microstructural evolution of plasma sprayed submicron-/nano-zirconia-based thermal barrier coatings. *Appl. Surf. Sci.* **2016**, *363*, 101–112. [CrossRef]
9. Bai, Y.; Zhao, L.; Qu, Y.; Fu, Q.; Wang, Y.; Liu, K.; Tang, J.; Li, B.Q.; Han, Z. Particle in-flight behavior and its influence on the microstructure and properties of supersonic-atmospheric-plasma-sprayed nanostructured thermal barrier coatings. *J. Alloys Compd.* **2015**, *644*, 873–882. [CrossRef]
10. Liu, K.; Tang, J.; Bai, Y.; Yang, Q.; Wang, Y.; Kang, Y.; Zhao, L.; Zhang, P.; Han, Z. Particle in-flight behavior and its influence on the microstructure and mechanical property of plasma sprayed $La_2Ce_2O_7$ thermal barrier coatings. *Mater. Sci. Eng. A* **2015**, *625*, 177–185. [CrossRef]
11. Yin, Z.; Tao, S.; Zhou, X.; Ding, C. Particle in-flight behavior and its influence on the microstructure and mechanical properties of plasma-sprayed Al_2O_3 coatings. *J. Eur. Ceram. Soc.* **2008**, *28*, 1143–1148. [CrossRef]
12. Zhao, Y.; Li, D.; Zhong, X.; Zhao, H.; Wang, L.; Shao, F.; Liu, H.; Tao, S. Thermal shock behaviors of YSZ thick thermal barrier coatings fabricated by suspension and atmospheric plasma spraying. *Surf. Coat. Technol.* **2014**, *249*, 48–55. [CrossRef]
13. Fan, Z.; Wang, K.; Dong, X.; Duan, W.; Mei, X.; Wang, W.; Cui, J.; Lv, J. Influence of columnar grain microstructure on thermal shock resistance of laser re-melted ZrO_2-7wt.% Y_2O_3 coatings and their failure mechanism. *Surf. Coat. Technol.* **2015**, *277*, 188–196. [CrossRef]
14. Wu, J.; Guo, H.; Gao, Y.-Z.; Gong, S.-K. Microstructure and thermo-physical properties of yttria stabilized zirconia coatings with CMAS deposits. *J. Eur. Ceram. Soc.* **2011**, *31*, 1881–1888. [CrossRef]
15. Nath, S.; Manna, I.; Majumdar, J.D. Nanomechanical behavior of yttria stabilized zirconia (YSZ) based thermal barrier coating. *Ceram. Int.* **2015**, *41*, 5247–5256. [CrossRef]
16. Li, C.; Wang, T.; Liu, X.; Zheng, Z.; Li, Q. Evolution of mechanical properties of thermal barrier coatings subjected to thermal exposure by instrumented indentation testing. *Ceram. Int.* **2016**, *42*, 10242–10250. [CrossRef]
17. Kim, C.H.; Heuer, A.H. A high-temperature displacement-sensitive indenter for studying mechanical properties of thermal barrier coatings. *J. Mater. Res.* **2004**, *19*, 351–356. [CrossRef]
18. Baiamonte, L.; Marra, F.; Pulci, G.; Tirillò, J.; Sarasini, F.; Bartuli, C.; Valente, T. High temperature mechanical characterization of plasma-sprayed zirconia–yttria from conventional and nanostructured powders. *Surf. Coat. Technol.* **2015**, *277*, 289–298. [CrossRef]
19. Jang, B.-K.; Matsubara, H. Influence of porosity on hardness and Young's modulus of nanoporous EB-PVD TBCs by nanoindentation. *Mater. Lett.* **2005**, *59*, 3462–3466. [CrossRef]
20. Lamuta, C.; Di Girolamo, G.; Pagnotta, L. Microstructural, mechanical and tribological properties of nanostructured YSZ coatings produced with different APS process parameters. *Ceram. Int.* **2015**, *41*, 8904–8914. [CrossRef]
21. Lima, R.; Kucuk, A.; Berndt, C. Evaluation of microhardness and elastic modulus of thermally sprayed nanostructured zirconia coatings. *Surf. Coat. Technol.* **2001**, *135*, 166–172. [CrossRef]
22. Zeng, Y.; Lee, S.; Gao, L.; Ding, C. Atmospheric plasma sprayed coatings of nanostructured zirconia. *J. Eur. Ceram. Soc.* **2002**, *22*, 347–351. [CrossRef]
23. Thompson, J.; Clyne, T. The effect of heat treatment on the stiffness of zirconia top coats in plasma-sprayed TBCs. *Acta Mater.* **2001**, *49*, 1565–1575. [CrossRef]
24. Malzbender, J.; Steinbrech, R. Mechanical properties of coated materials and multi-layered composites determined using bending methods. *Surf. Coat. Technol.* **2004**, *176*, 165–172. [CrossRef]

25. Hayase, T.; Waki, H. Measurement of Young's modulus and Poisson's ratio of thermal barrier coating based on bending of three-layered plate. *J. Therm. Spray Technol.* **2018**, *27*, 983–998. [CrossRef]
26. Li, M.H.; Hu, W.Y.; Sun, X.F.; Guan, H.R.; Hu, Z.Q. Apparent indentation size effect of EB-PVD thermal barrier coatings. *Rare Metal. Mat. Eng.* **2006**, *35*, 186–188. [CrossRef]
27. Zotov, N.; Bartsch, M.; Eggeler, G. Thermal barrier coating systems—Analysis of nanoindentation curves. *Surf. Coat. Technol.* **2009**, *203*, 2064–2072. [CrossRef]
28. Wei, Y.; Wang, X.; Wu, X.; Bai, Y. Theoretical and experimental researches of size effect in micro-indentation test. *Sci. China Ser. A Math.* **2001**, *44*, 74–82. [CrossRef]
29. Wei, Y.; Hutchinson, J.W. Hardness trends in micron scale indentation. *J. Mech. Phys. Solids* **2003**, *51*, 2037–2056. [CrossRef]
30. Song, J.; Fan, C.; Ma, H.; Wei, Y. Hierarchical structure observation and nanoindentation size effect characterization for a limnetic shell. *Acta Mech. Sin.* **2015**, *31*, 364–372. [CrossRef]
31. Wu, B.; Liang, L.; Ma, H.; Wei, Y. A trans-scale model for size effects and intergranular fracture in nanocrystalline and ultra-fine polycrystalline metals. *Comput. Mater. Sci.* **2012**, *57*, 2–7. [CrossRef]
32. Oliver, W.; Pharr, G. An improved technique for determining hardness and elastic modulus using load and displacement sensing indentation experiments. *J. Mater. Res.* **1992**, *7*, 1564–1583. [CrossRef]
33. Fleck, N.; Muller, G.; Ashby, M.; Hutchinson, J. Strain gradient plasticity: Theory and experiment. *Acta Metall. Mater.* **1994**, *42*, 475–487. [CrossRef]

© 2020 by the authors. Licensee MDPI, Basel, Switzerland. This article is an open access article distributed under the terms and conditions of the Creative Commons Attribution (CC BY) license (http://creativecommons.org/licenses/by/4.0/).

Article

Microstructure and Sliding Wear Resistance of Plasma Sprayed Al_2O_3-Cr_2O_3-TiO_2 Ternary Coatings from Blends of Single Oxides

Maximilian Grimm [1,*], Susan Conze [2], Lutz-Michael Berger [2], Gerd Paczkowski [1], Thomas Lindner [1] and Thomas Lampke [1]

[1] Materials and Surface Engineering Group, Institute of Materials Science and Engineering, Chemnitz University of Technology, D-09107 Chemnitz, Germany; gerd.paczkowski@mb.tu-chemnitz.de (G.P.); lindt@hrz.tu-chemnitz.de (T.L.); thomas.lampke@mb.tu-chemnitz.de (T.L.)

[2] Fraunhofer Institute for Ceramic Technologies and Systems, Fraunhofer IKTS, D-01277 Dresden, Germany; susan.conze@ikts.fraunhofer.de (S.C.); lutz-michael.berger@ikts.fraunhofer.de (L.-M.B.)

* Correspondence: maximilian.grimm@mb.tu-chemnitz.de; Tel.: +49-371-531-36581

Received: 2 December 2019; Accepted: 30 December 2019; Published: 3 January 2020

Abstract: Al_2O_3, Cr_2O_3, and TiO_2 are most commonly used oxide materials for thermal spray coating solutions. Each oxide shows unique properties comprising behavior in the spray process, hardness, corrosion, and wear resistance. In order to exploit the different advantages, binary compositions are often used, while ternary compositions are not studied yet. Atmospheric plasma spraying (APS) of ternary compositions in the Al_2O_3-Cr_2O_3-TiO_2 system was studied using blends of plain powders with different ratios and identical spray parameters. Coatings from the plain oxides were studied for comparison. For these powder blends, different deposition rates were observed. The microstructure, roughness, porosity, hardness, and wear resistance were investigated. The formation of the splats from particles of each oxide occurs separately, without interaction between the particles. The exception are the chromium oxide splats, which contained some amounts of titanium. The predominant oxide present in each blend has a decisive influence on the properties of the coatings. While TiO_x causes a low coating porosity, the wear resistance can be increased by adding Cr_2O_3.

Keywords: atmospheric plasma spraying; Al_2O_3; Cr_2O_3; TiO_2; microstructure; sliding wear; phase transformation; reactivity

1. Introduction

Thermally sprayed oxide ceramic coatings have an outstanding importance in many technological areas. Coatings sprayed from the single oxides and some commercially available binary compositions of the Al_2O_3-Cr_2O_3-TiO_2 system have multi-functional properties and are widely used as wear and as (sealed) corrosion resistant coatings. Depending on the composition, they are electrically insulating or conductive [1,2].

Plasma spraying and, in particular, conventional atmospheric plasma spraying (APS), is the most widely used thermal spray process for oxide coating manufacturing. Water stabilized plasma spraying (WSP) is a special high-energy process for coating manufacturing and is applied for special purposes, such as manufacturing of ceramic tubes. Coatings with excellent properties can be deposited by high velocity oxy-fuel spraying (HVOF), but often lower powder feed rates and deposition efficiencies are considered as drawbacks. More detailed descriptions of the spray processes used for oxide coating manufacturing are given elsewhere [2–4].

Most commonly feedstock powders with a typical particle size in the range 15–45 µm are used. The main feedstock powder manufacturing method for the single oxides is fusing and crushing.

For avoiding the formation of metallic chromium for Cr_2O_3 sintering and crushing is also common. For the commercially available binary compositions, the variety of the manufacturing methods are significantly broader. This includes mechanical blends of separately fused and crushed single oxide powders, jointly fused and crushed powders, but also agglomerated and sintered powders. Depending on the oxide particle or grain size, there are large differences in the homogeneity of mixing of the metallic elements in these methods. The homogeneity is lowest in case of the powder blends from single oxide powders and highest in the case of agglomerated and sintered powders, where finely dispersed oxide powders are mixed in the manufacturing process. Suspensions as feedstock become increasingly important, but are currently limited to single oxides [5]. Suspensions with two components are under investigation [6].

Each of the single oxides shows a specific material behavior during spraying, which is detrimental to the processing and/or coating properties [1,2].

For alumina, the detrimental phase transformation from α-Al_2O_3 (corundum) existing in the feedstocks to metastable phases, predominantly γ-Al_2O_3, in the coatings is well known [7–10]. The reason is the high cooling rate and nucleation of undercooled melt. Some content of the remaining α-Al_2O_3 in the coating is usually explained by the occurrence of non-molten particles [7]. An increase of the α-Al_2O_3 content in the coatings is described as an important measure to improve the coating properties, such as wear, electrical, and corrosion resistance. Except the addition of other oxides (e.g., Cr_2O_3 [8–10]), there are several technological measures for this, as the selection of special spray process conditions [11], use of suspensions [1,5] or heat post-treatments [12] as well as plasma-electrolytic oxidation of arc-sprayed and flame-sprayed aluminum coatings [13]. However, each of these technological measures has certain limitations and the stabilization of the α-Al_2O_3 by a tailored powder would be favored.

Chromia has a low deposition efficiency due to oxidation from Cr_2O_3 to volatile CrO_3, which immediately reconverts to Cr_2O_3 when cooling down [1,2,14]. Formation of Cr(VI)-oxyhydrates in wet atmospheres is another detrimental reaction [14,15]. Although the appearance of hexavalent chromium is below the maximum allowable concentrations under normal process conditions, increased safety regulations are permanently under discussion. By adding Al_2O_3 [16] or TiO_2 [1,2] to Cr_2O_3 and forming respective solid solutions, both the formation of hexavalent chromium is suppressed and the deposition efficiency is increased.

Since titania TiO_2 readily loses oxygen in a reducing environment such as during the fusion step of feedstock manufacturing, fused and crushed feedstock powders are non-stoichiometric and preferably designated as TiO_x. The oxygen content can also change during the spray process. The oxygen defects are often disordered in coatings under the strong nonequilibrium conditions of the thermal spray process, but also often forms ordered-structures for certain O/Ti ratios (Magnéli-phases) in feedstock powders, manufactured with lower cooling rates. Due to a eutectic with an oxygen content corresponding to x of about 1.78 in the Ti-O phase diagram (in the two-phase region of the Magnéli-phases Ti_4O_7 and Ti_5O_9). The temperature of appearance of a melt is decreased from 1857 °C for TiO_2 down to 1679 °C at the eutectic point [2,17].

It was found that the addition of a second oxide can improve the coating properties [1,2]. There are some indications in the literature that the coating properties can be further improved by ternary compositions and by the addition of the third oxide of the system. Examples of such potential improvements can be the stabilization of target phases, an acceleration of the formation of a compound and improved sintering properties. This can relate to the feedstock powder manufacturing step and/or the spray process.

This addition can be made in different ways, e.g., during the feedstock manufacturing process (joint processing in fused and crushed as well as agglomerated and sintered powder manufacturing) or by blending single oxide powders. The latter case is the simplest one, and blended powders have a high importance in the current industrial practice of thermal spraying. This is valid for commercially available blends in the Al_2O_3-TiO_2 system, such as Al_2O_3-40%TiO_2 and Al_2O_3-13%TiO_2 [18,19], but also for blending at the production site [20]. In general, an interaction between components of the blends will occur only during the spray process and is, in general, expected to be low due to the large particle size. However, an intensive interaction during the spray process was found in the case of an Al_2O_3-40%TiO_2 blend [21]. Another example are blends of alumina-rich compositions of the Al_2O_3-Cr_2O_3 system, which can lead to a stabilization of the α-Al_2O_3 during WSP [10]. Blends of the binary Cr_2O_3-TiO_2 system were investigated as well [17,20]. However, ternary compositions of the Al_2O_3-Cr_2O_3-TiO_2 system were not studied neither as powder blends, pre-alloyed powders, nor suspensions.

In this work, the interaction of single oxide ternary powder blends of the Al_2O_3-Cr_2O_3-TiO_2 system during APS and their sliding wear properties are studied. Coatings from the single oxides are investigated for comparison.

2. Materials and Methods

Commercial fused and crushed Al_2O_3, Cr_2O_3, and TiO_x powder grades, compiled in Table 1, were used in this work. The particle size distribution was determined by laser diffraction analysis in a Cilas 930 device (Cilas, Orléans, France). To prepare the powder blends, dried powders were mixed using a tumble mixer, according to the compositions given in Table 2.

Table 1. Commercial single oxide feedstock powder grades, their particle size given by the supplier, and their granulometric data (own measurement). All powders were produced by fusing and crushing.

Material	Supplier	Particle Size	Granulometric Data		
			d_{10}	d_{50}	d_{90}
Al_2O_3	Saint Gobain Coating Solutions, Avignon, France	−45 + 15 µm	21	34	56
Cr_2O_3	GTV, Luckenbach, Germany	−45 + 15 µm	19	34	54
TiO_x	Ceram, Albbruck-Birndorf, Germany	−45 +20 µm	22	39	61

Table 2. Compositions of the powder blends.

Designation of the Blend/Coating	Components (at %)			Components (wt %)		
	Al_2O_3	Cr_2O_3	TiO_2	Al_2O_3	Cr_2O_3	TiO_2
ACT	50	25	25	47	35	18
CAT	25	50	25	21	63	16
TAC	25	25	50	25	36	39

Feedstock powders were investigated with a scanning electron microscope (SEM) LEO 1455VP (Zeiss, Jena, Germany) with an acceleration voltage of 25 kV. By using a backscattered electron detector (BSD), the material contrast is visualized by different grey levels. In addition, phase composition was studied by X-ray diffraction (XRD) using a Bragg-Brentano geometry operating with Cu K_α radiation with a D8 Advance diffractometer (Bruker AXS, Billerica, MA, USA) in a range of 2θ = 15°–120° with a step size of 0.02° and 3 s/step.

Low carbon steel (S235) samples with a diameter of 40 mm were used as substrates, which were grit blasted with alumina (EK-F 24) (3 bar, 20 mm distance, 70° angle) and cleaned in an ultrasonic ethanol bath before applying the coating. The coatings were produced by atmospheric plasma spraying using an F6 torch (GTV, Luckenbach, Germany) and the spraying parameters, according to Table 3. In order to ensure good comparability, all coatings in this work were sprayed with this parameter set. The parameter set was chosen in such a way that a shift of the composition of the powder blends during processing was avoided. Interruptions of the coating process ensured that the substrate did not heat up to more than 200 °C.

The cross sections of the coatings were prepared by the standard metallographic procedure. The analysis of the microstructure was conducted using an optical microscope GX51 (Olympus, Shinjuku, Japan) equipped with a SC50 camera (Olympus, Shinjuku, Japan) as well as by SEM with the same device used for powder analysis. The coating thickness was ascertained at 10 evenly distributed points. To determine the porosity, five images of the coating were evaluated by the image analysis method with the software ImageJ. Furthermore, the hardness of the coatings was measured on the cross sections using a Wilson Tukon 1102 device (Buehler, Uzwil, Switzerland). For this purpose, 10 Vickers indentations with a test load of 2.94 N were examined. The XRD patterns of the coatings were recorded with 3003 TT diffractometer from GE Inspection Technologies in a range of $2\theta = 15°–80°$ with a step size of $0.03°$ and 4 s/step. The high-resolution microstructure and the local chemical composition of individual splats in the coatings from blends were determined using a FESEM Ultra (Zeiss, Jena, Germany) equipped with an EDS detector X-Max80 and using a voltage of 13 kV. At least five measurements for the three typical individual splat compositions in each coating were performed. The calibration for the quantitative EDS analyses were performed using a cobalt standard.

Table 3. The parameters of the APS process with the F6-torch.

Argon	Hydrogen	Current	Spraying Distance	Traverse Speed	No. of Passes	Powder Feed Rate
41 L/min	11 L/min	600 A	110 mm	0.4 m/s	10	30 g/min

The processing of the coatings by grinding and polishing was investigated. All coated samples were simultaneously ground on grinding wheels up to a number of 1200 for 3 min at 300 rpm and 25 N. Polishing was then carried out with diamond suspensions down to 1 µm at 200 rpm and a slightly reduced force in order to obtain comparable surfaces for the wear tests. The roughness of as-sprayed and ground coatings was recorded with a tactile profilometer (Jenoptik, Jena, Germany) with five measuring tracks. The characterisation of the sliding wear behavior of the coatings was carried out using a ball on disc test (Tetra, Ilmenau, Germany) using the parameters summarized in Table 4. For each coating, three wear tracks were generated. The evaluation of the wear tracks with regard to the wear depth and the wear volume was carried out using a 3D profilometer MikroCAS (LMI, Teltow, Germany).

Table 4. Ball-on-disk test parameters.

Force	Radius	Speed	Cycles	Wear Distance	Counter Body	
					Material	Diameter
10 N	5 mm	0.05 m/s	15916	500 m	Al_2O_3	6 mm

3. Results

3.1. Feedstock Powder Characterisation

The typical angular morphology of the fused and crushed powder particles is shown in the SEM micrograph in Figure 1. The granulometric parameters d_{10}, d_{50}, and d_{90}, which are also given in Table 1, confirm only minor differences of their particle sizes, which makes them suitable for blending.

Figure 1. SEM images of the feedstock powders. (a) Al_2O_3. (b) Cr_2O_3. (c) TiO_x.

According to the XRD measurements, as shown in Figure 2, the alumina and chromia feedstock powders consist only of α-Al$_2$O$_3$ (corundum) and Cr$_2$O$_3$ (eskolaite), respectively. In the SEM investigation of the chromia powder, small amounts of metallic chromium were detected, which do not appear in the XRD pattern. A large number of peaks in addition to those of rutile were detected for the TiO$_x$ powder. These additional peaks, which are not labelled in the pattern in Figure 2, belong to substoichiometric titanium oxide phases. The presence of the (110) peak of rutile at 2θ = 27.4° indicates that local concentrations of oxygen are closed to stoichiometry, while, in other regions, disordered oxygen-deficient structures exist. According to the gravimetric measurements in another study, x in TiO$_x$ powder is about 1.9 [19].

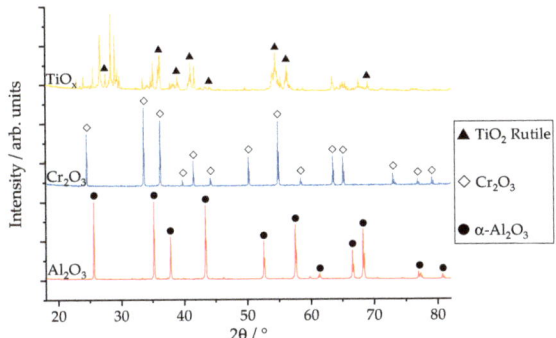

Figure 2. XRD pattern of the feedstock powders.

3.2. Coating Characterisation

It was shown that, with the selected parameter set, it was possible to produce well adhering and dense coatings from all single oxides and their ternary blends. The results of the roughness measurements for the as-sprayed coatings shown in Figure 3 reveal clear differences. In the case of single oxides, the TiO$_x$ coating has the highest as-sprayed Rz value with approximately 70 ± 2 µm. The Cr$_2$O$_3$ coatings show the lowest roughness with approximately 33 ± 2 µm. Coatings produced from the powder blends show roughness values between these limits. The as-sprayed Rz value of the ACT, CAT, and TAC coatings are 50 ± 4 µm, 49 ± 3 µm, and 55 ± 6 µm, respectively. By grinding and polishing the roughness of the samples, it can be reduced by more than 90%. As can be seen in Figure 3, the Rz values after polishing are in the range of 1.3 ± 0.2 µm for Cr$_2$O$_3$ and 3.2 ± 0.2 µm for Al$_2$O$_3$. The roughness of the TiO$_x$ coating can even be reduced by almost 98% by polishing and indicates particularly good machinability. The coatings from the blends show a similar behavior. The Cr$_2$O$_3$-rich CAT coating with 2.3 ± 0.5 µm have the lowest roughness. After polishing, all coatings have a comparable roughness, which makes them suitable for wear testing.

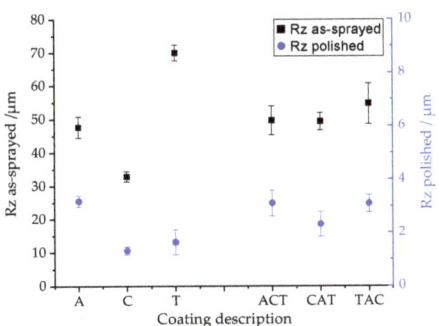

Figure 3. Roughness values of as-sprayed and polished coated samples.

The results of coating thickness and porosity after 10 passes are shown in Figure 4. The coating thickness gives an indication of the deposition rate of the single oxides and the blends. In the case of the individual oxides, the highest coating thickness (469 ± 16 µm) was measured for the TiO$_x$ coating, while the chromia coating has the lowest coating thickness. A similar observation was made by others [17]. For the coatings from the blended powders, the titanium oxide-rich TAC coating with 430 ± 18 µm shows significantly higher thickness than the ACT and CAT coatings, where the coating thickness is 342 ± 14 µm and 369 ± 16 µm, respectively.

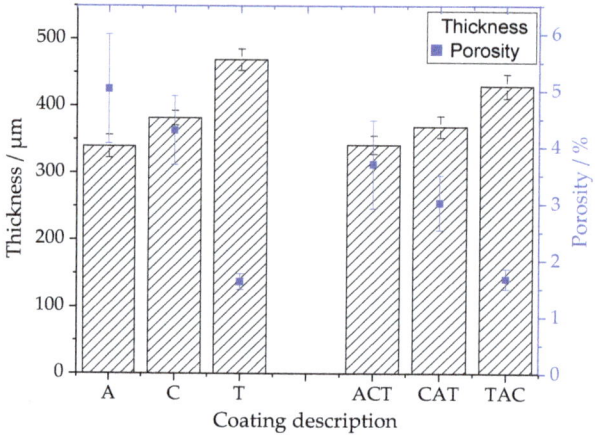

Figure 4. Thickness and porosity of plain oxide coatings (A, C, T) and coatings from blends (ACT, CAT, and TAC).

When studying the porosity, it is noticeable that different levels exist between the individual oxides as a result of the identical spray parameters. While the porosity of the Al$_2$O$_3$ coating is about 5.0% ± 1.0%, the TiO$_x$ coating with a porosity of 1.6% ± 0.1% is significantly denser. As illustrated in Figure 4, the alumina-rich ACT and chromium oxide-rich CAT coatings have similar porosities of 3.7% ± 0.8% and 3.0% ± 0.5%, respectively. Only the TAC coating with a high content of TiO$_x$ shows a lower porosity of 1.7% ± 0.2%.

Low magnification SEM images of the cross sections of the coatings from the blends are shown in Figure 5a,c,e. The characteristic microstructure of thermal spray coatings, characterized by unmolten particles, pores, microcracks, and a lamellar structure, is observed. Splats from the individual oxide particles are clearly distinguishable. The high magnification images in Figure 5b,d,f show that the coatings of each blend consist of individual splats of a comparable grayscale denoted as I, II, and III in each of the images.

The results of the measurements of local chemical composition of these individual splats by EDS are compiled in Figure 6. The dark areas (SEM area I) in all coatings have a composition of approximately 60 at % oxygen and 40 at % aluminum, while the elements titanium and chromium were not detected. The bright areas (SEM area II) assigned to Cr$_2$O$_3$ contain a significant proportion of titanium in the range between 0.9 at %. and 2.2 at %. It is also noticeable that these Cr$_2$O$_3$ lamellae contain more oxygen than expected (slightly above 60 at %). Variations in the grayscales in these lamellas were found to be caused by variations of the oxygen content. The areas III with an intermediate grayscale relate to the titanium oxide splats.

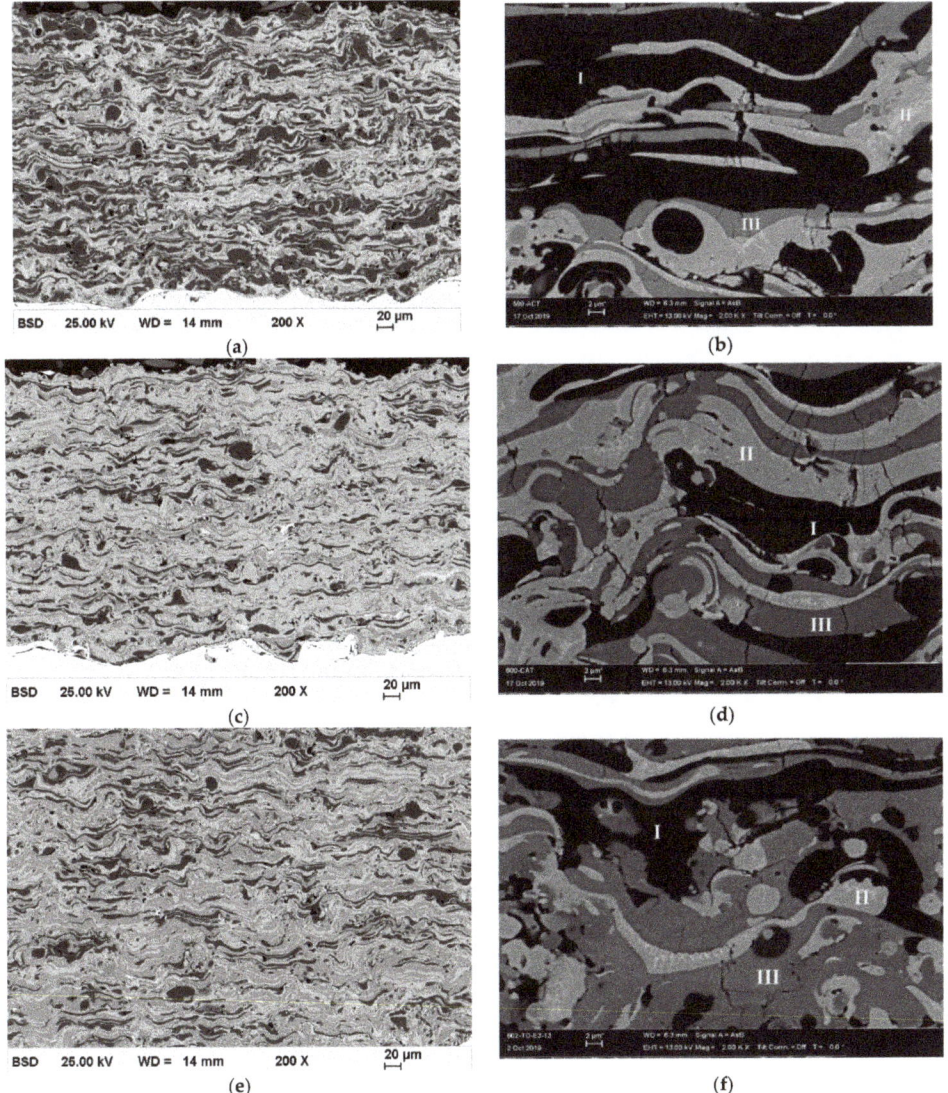

Figure 5. SEM images (BSD detector) of APS Al_2O_3-Cr_2O_3-TiO_2 composite coatings: top row: ACT-coating: (**a**) overview image, (**b**) detailed image with marked areas for the EDS measurement. middle row: CAT-coating: (**c**) overview image, (**d**) detailed image with marked areas for the EDS measurement. bottom row: TAC-coating: (**e**) overview image, (**f**) detailed image with marked areas for the EDS measurement. The EDS measurements of the marked areas are shown in Figure 6.

The results of the hardness measurements are presented in Figure 7 and show that all single oxide coatings have high hardness values above 1000 HV0.3, while the Cr_2O_3 coating has the highest hardness of 1250 ± 79 HV0.3. The coatings from powder blends do not reach the hardness of the plain chromia and alumina coatings. Only the chromium oxide-rich CAT coating has a hardness of 1074 ± 65 HV0.3, which is higher than the hardness of the TiO_x coating. The lowest hardness was found for the ACT coating.

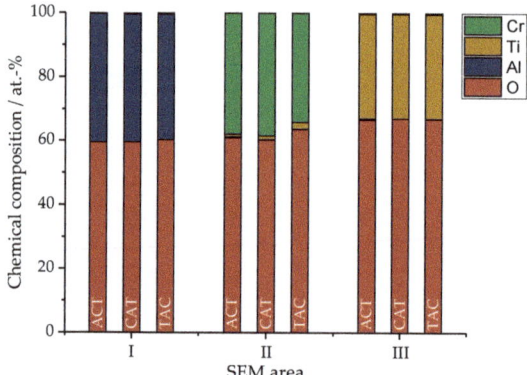

Figure 6. EDS analysis of composite coatings: chemical composition of the material areas marked in Figure 5b,d,f.

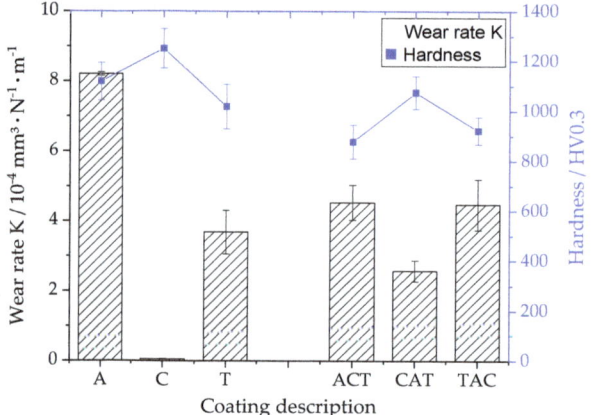

Figure 7. Hardness and sliding wear rates of plain oxide coatings (A, C, T) and composite coatings (ACT, CAT, TAC).

The results of the sliding wear tests are also shown in Figure 7. The Al_2O_3 coating shows the highest wear rate (8.2 ± 0.1 × 10^{-4} mm³·N⁻¹·m⁻¹), which is more than twice than that of the TiO_x coating. The wear resistance of the chromium oxide coating is so high that no meaningful wear rate was measured. The coatings from the blends also show a high wear resistance. The ACT and TAC coatings have a very similar wear rate. The chromium oxide-rich CAT coating proves to be more wear resistant.

The presentation of the XRD patterns, displayed in Figure 8, is limited to the range $2\theta = 18°–82°$. For the powder blends, the presence of α-Al_2O_3 (corundum), Cr_2O_3 (eskolaite), and different titanium oxide phases, as expected from the pattern of the individual oxides (see Figure 2) was found. The XRD patterns of the coatings from the blends, shown in Figure 8, reveal the presence of α-Al_2O_3, γ-Al_2O_3, Cr_2O_3 (eskolaite), and different titanium oxide phases. A shift of the peak positions relative to the standards did not occur. Due to the inhomogeneous distribution of oxygen, the intensity of the suboxide peaks is low. A quantitative determination of the phase is not possible. A change in the intensity of some peaks was observed. For the TAC coating, the TiO_x peaks of $2\theta = 26°–30°$ lose significantly in intensity in the coating compared to the powders. The intensity of the TiO_2 rutile peaks at $2\theta = 27.4°$ (110) and $2\theta = 54.2°$ (211) increases markedly. For all coatings, especially for the CAT and TAC coatings, a decrease of peak intensity of α-Al_2O_3 from powder to coating was detected.

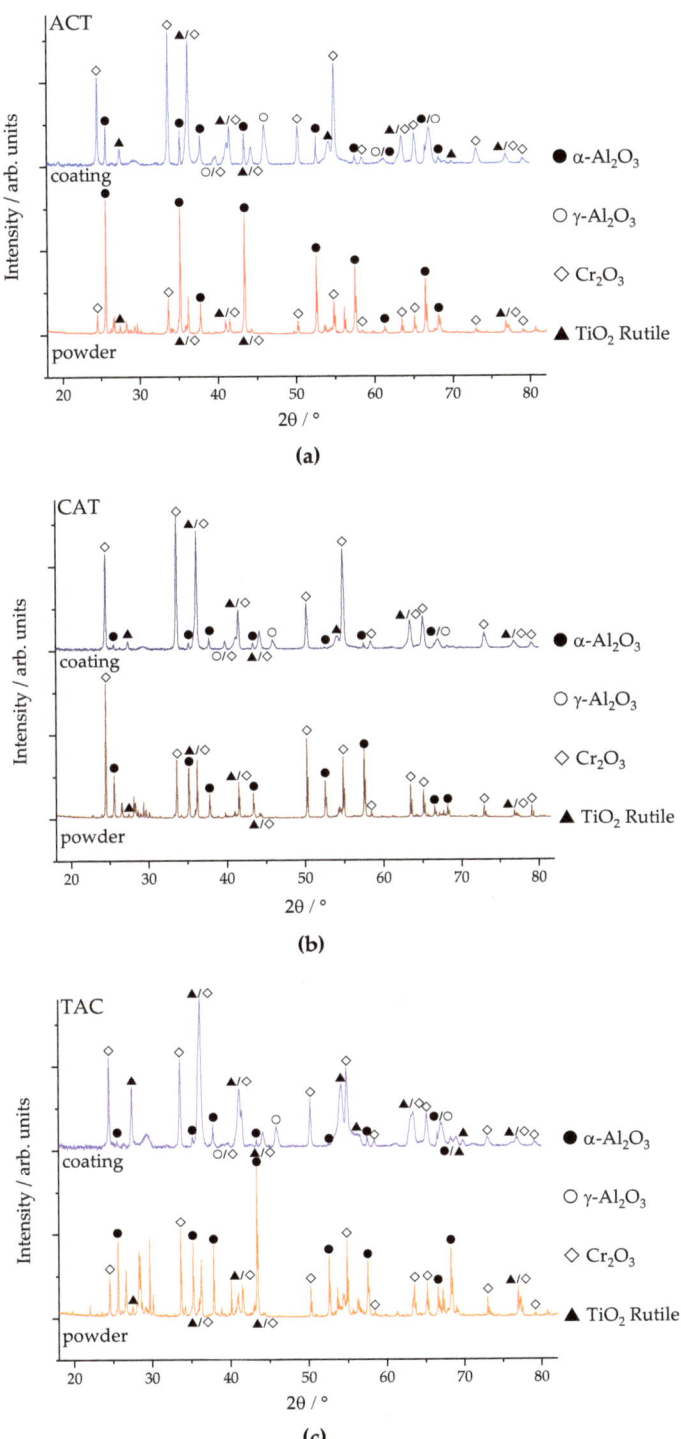

Figure 8. Diffraction patterns of the powder blends and corresponding coatings: (**a**) ACT, (**b**) CAT, and (**c**) TAC.

4. Discussion

Spraying of blends containing large powder particles has the disadvantage that only one component can be sprayed with a truly optimized parameter set. For ternary blends, this is naturally even more difficult as for binary ones. By this reason, for this study, an identical parameter set was used, which represents a compromise for the processability of all oxides. In case of the blends of the Al_2O_3-Cr_2O_3-TiO_2 system, two high melting components (Al_2O_3 and Cr_2O_3) are combined with one lower melting oxide TiO_x. Thus, the liquid phase occurs at lower temperatures compared to TiO_2 and the addition of TiO_x to Cr_2O_3 and Al_2O_3 decreases the coating porosity. For Al_2O_3-rich and Cr_2O_3-rich powder blends (ACT and CAT), a part of the large particle fraction is not melted to the extent that it can contribute to coating build up due to the higher melting temperature. As indicated by the coating thickness after 10 passes (see Figure 4), the highest deposition efficiency is found for the TAC coating.

The XRD investigations show that the typical phase transformation from α-Al_2O_3 to γ-Al_2O_3 as described in Section 1 occurs for all powder blends. The additions of Cr_2O_3 and TiO_2 for the given spray conditions do not have an effect on this phase transformation, which aligns with the literature data [2]. In addition, reduced peak intensities related to sub-stoichiometric phases and an increase of the rutile peaks intensity at $2\theta = 27.4°$ (110) and $2\theta = 54.2°$ (211) can be an indication for oxygen gain during the processing and formation of stoichiometric TiO_2 or near stoichiometric TiO_x as a result of the thermal spray process.

Recently, a surprising high dissolution of titanium atoms in γ-Al_2O_3 in APS coatings obtained from a commercial Al_2O_3-40% TiO_2 powder blend was observed [19]. Surprisingly, this was not observed in this study for any of the ternary blends.

In the literature, the formation of a Cr_2O_3-rich solid solution $(Cr,Ti)_2O_3$ is described for binary suspensions [14], while, for blends of large particles, contradictory results are reported [17,20]. In this study, the EDS measurements have shown the presence of small amounts of titanium in the chromium oxide lamellae. The titanium content in the chromia splats of the ACT and TAC coatings was below 2 at % but is slightly higher in the TiO_x-rich TAC coating. The higher titania content increases the number of interfaces between titanium and chromium oxide splats. Therefore, more titanium diffuses into the chromium oxide lamellae.

Since the hardness of the Cr_2O_3 coating was the highest, it is not surprising that the hardness is increased with a rising content of Cr_2O_3. The Cr_2O_3-rich CAT coating shows the highest hardness for the ternary blends. As expected, the Al_2O_3-rich ACT coating and the plain Al_2O_3 coating show higher hardness values than TiO_2-rich coatings. However, the titanium oxide-rich TAC coating has a similar hardness compared to the ACT coating. The lower porosity of the TAC coating is assumed to be responsible for this effect.

The sliding wear rate of Al_2O_3 coatings can be significantly reduced by adding TiO_x and Cr_2O_3. Both coatings have a higher wear resistance than the plain Al_2O_3 coating. Possible reasons could be the improved toughness of the coating expected by adding TiO_x [21] and/or addition of Cr_2O_3, which has, by far, the highest wear resistance. Except for the Al_2O_3 coating, coating hardness and sliding wear rates show some correlation especially for coatings from the blends. Whereas ACT and TAC coatings have similar hardness values and wear rates, the higher content of Cr_2O_3 in the CAT coating leads to significantly higher hardness values and a reduction in the wear rate. It should also be taken into account that the sliding wear rates of the plain oxide could be influenced by the identical spray parameter set, which is applied to the deposition of all coatings in this study.

All investigations reveal that the individual properties of the plain oxide strongly influence the properties of the coating from the blends. Therefore, for example, increasing the TiO_x content leads to denser coatings and higher deposition rates as well as higher as-sprayed roughness. The increase of the Cr_2O_3 content positively influences the sliding wear resistance.

5. Summary and Conclusions

It was shown that firmly adhering coatings can be produced by atmospheric plasma spraying from powder blends containing Al_2O_3, Cr_2O_3, and TiO_2 using an identical spray parameter set. Investigations with XRD have shown that the phase transformation from α-Al_2O_3 to γ-Al_2O_3 also occurs in the powder blends and is not influenced by the addition of Cr_2O_3 or TiO_x. Furthermore, for titania, a gain of oxygen content was found. EDS measurements have shown the existence of small amounts of titanium in the chromium oxide lamellaes. The investigations have shown that the respective dominant single oxide has a significant influence on the coating properties. Whereas a high TiO_x content leads to higher deposition rates, higher as-sprayed roughness, and low porosities, the hardness and wear resistance of the coatings can be improved by increasing the Cr_2O_3 content. By using powder blends, the sliding wear rate can be improved when compared to plain Al_2O_3 coatings. Thus, the use of powder blends presents a promising approach to adapt or extend the property profile of plain oxide coatings. The reactivity between the materials involved needs to be further investigated in order to exploit further improvement potentials.

Author Contributions: M.G., S.C., and G.P. conceived, designed, and performed the experiments. M.G., L.-M.B. and T.L. (Thomas Lindner) analyzed the data and wrote the paper. T.L. (Thomas Lampke) directed the research and contributed to the discussion and interpretation of the results. All authors have read and agreed to the published version of the manuscript.

Funding: This project was funded under contracts 100310631/100310633 via Sächsische Aufbaubank by the European Structural Fonds EFRE and by the Free State of Saxony. The German Research Foundation/DFG-392676956 and the Technische Universität Chemnitz in the funding program Open Access Publishing funded the publication costs of this paper.

Acknowledgments: The authors would like to thank Marc Pügner for the XRD measurements and his support in interpreting them, Frank Trommer for supporting the coating process and Kerstin Sempf for EDS analyses.

Conflicts of Interest: The authors declare no conflict of interest.

References

1. Berger, L.-M.; Toma, F.-L.; Scheitz, S.; Trache, R.; Börner, T. Thermisch gespritzte Schichten im System Al_2O_3-Cr_2O_3-TiO_2—ein Update. *Mater. Werkst.* **2014**, *45*, 465–475. [CrossRef]
2. Berger, L.-M. Tribology of Thermally Sprayed Coatings in the Al_2O_3-Cr_2O_3-TiO_2 System. In *Thermal Sprayed Coatings and Their Tribological Performances*; Roy, M., Davim, J.P., Eds.; Engineering Science Reference: Hershey, PA, USA, 2015; pp. 227–267. [CrossRef]
3. Fauchais, P.L.; Heberlein, J.V.R.; Boulos, M.I. *Thermal Spray Fundamentals: From Powder to Part*; Springer: New York, NY, USA; Heidelberg, Germany; Dordrecht, The Netherlands; London, UK, 2014. [CrossRef]
4. Pawlowski, L. *The Science and Engineering of Thermal Spray Coatings*, 2nd ed.; Wiley: Hoboken, NJ, USA, 2008.
5. Toma, F.-L.; Potthoff, A.; Berger, L.-M.; Leyens, C. Demands, Potentials, and Economic Aspects of Thermal Spraying with Suspensions: A Critical Review. *J. Therm. Spray Technol.* **2015**, *24*, 1143–1152. [CrossRef]
6. Potthoff, A.; Kratzsch, R.; Barbosa, M.; Kulissa, N.; Kunze, O.; Toma, F.-L. Development and Application of Binary Suspensions in the Ternary System Cr_2O_3-TiO_2-Al_2O_3 for S-HVOF Spraying. *J. Therm. Spray Technol.* **2018**, *27*, 710–717. [CrossRef]
7. McPherson, R. On the formation of thermally sprayed alumina coatings. *J. Mater. Sci.* **1980**, *15*, 3141–3149. [CrossRef]
8. Chráska, P.; Dubsky, J.; Neufuss, K.; Pisacka, J. Alumina-base plasma-sprayed materials part I: Phase stability of alumina and alumina-chromia. *J. Therm. Spray Technol.* **1997**, *6*, 320–326. [CrossRef]
9. Dubsky, J.; Chraska, P.; Kolman, B.; Stahr, C.C.; Berger, L.-M. Phase formation control in plasma sprayed alumina-chromia coatings. *Ceram. Silik.* **2011**, *55*, 294–300.
10. Stahr, C.C.; Saaro, S.; Berger, L.-M.; Dubský, J.; Neufuss, K.; Herrmann, M. Dependence of the stabilization of α-Alumina on the spray process. *J. Therm. Spray Technol.* **2007**, *16*, 822–830. [CrossRef]
11. Heintze, G.N.; Uematsu, S. Preparation and structures of plasma-sprayed γ- and α-Al_2O_3 coatings. *Surf. Coat. Technol.* **1992**, *50*, 213–222. [CrossRef]

12. Marple, B.R.; Voyer, J.; Béchard, P. Sol infiltration and heat treatment of alumina-chromia plasma-sprayed coatings. *J. Eur. Ceram. Soc.* **2001**, *21*, 861–868. [CrossRef]
13. Lampke, T.; Meyer, D.; Alisch, G.; Nickel, D.; Scharf, I.; Wagner, L.; Raab, U. Alumina coatings obtained by thermal spraying and plasma anodising—A comparison. *Surf. Coat. Technol.* **2011**, *206*, 2012–2016. [CrossRef]
14. Toma, F.-L.; Potthoff, A.; Barbosa, M. Microstructural Characteristics and Performances of Cr_2O_3 and Cr_2O_3-15%TiO_2 S-HVOF Coatings Obtained from Water-Based Suspensions. *J. Therm. Spray Technol.* **2018**, *27*, 344–357. [CrossRef]
15. Yu, S.H.; Wallar, H. Chromia Spray Powders and a Process for Making the Same. U.S. Patent 6,774,076, 10 August 2004.
16. Richter, A.; Berger, L.-M.; Conze, S.; Sohn, Y.J.; Vaßen, R. Emergence and impact of Al_2TiO_5 in Al_2O_3-TiO_2 APS coatings. In Proceedings of the IOP Conference Series: Materials Science and Engineering 21th Chemnitz Seminar on Materials Engineering, Chemnitz, Germany, 6–7 March 2019; Volume 480, p. 012007. [CrossRef]
17. Sert, Y.; Toplan, N. Tribological behavior of a plasma-sprayed Al_2O_3-TiO_2-Cr_2O_3 coating. *Mater. Tehnol.* **2013**, *47*, 181–184.
18. Vernhes, L.; Bekins, C.; Lourdel, N.; Poirier, D.; Lima, R.S.; Li, D.; Klemberg-Sapieha, J.E. Nanostructured and Conventional Cr_2O_3, TiO_2, and TiO_2-Cr_2O_3 Thermal-Sprayed Coatings for Metal-Seated Ball Valve Applications in Hydrometallurgy. *J. Therm. Spray Technol.* **2016**, *25*, 1068–1078. [CrossRef]
19. Richter, A.; Berger, L.-M.; Sohn, Y.J.; Conze, S.; Sempf, K.; Vaßen, R. Impact of Al_2O_3-40 wt.% TiO_2 feedstock powder characteristics on the sprayability, microstructure and mechanical properties of plasma sprayed coatings. *J. Eur. Ceram. Soc.* **2019**, *39*, 5391–5402. [CrossRef]
20. Ctibor, P.; Píš, I.; Kotlan, J.; Pala, Z.; Khalakhan, I.; Štengl, V.; Homola, P. Microstructure and properties of plasma-sprayed mixture of Cr_2O_3 and TiO_2. *J. Therm. Spray Technol.* **2013**, *22*, 1163–1169. [CrossRef]
21. Fervel, V.; Normand, B.; Coddet, C. Tribological behavior of plasma sprayed Al_2O_3-based cermet coatings. *Wear* **1999**, *230*, 70–77. [CrossRef]

© 2020 by the authors. Licensee MDPI, Basel, Switzerland. This article is an open access article distributed under the terms and conditions of the Creative Commons Attribution (CC BY) license (http://creativecommons.org/licenses/by/4.0/).

Article

Low Frictional MoS$_2$/WS$_2$/FineLPN Hybrid Layers on Nodular Iron

Pior Kula [1,2], Robert Pietrasik [1,2,*], Sylwester Pawęta [1,2] and Adam Rzepkowski [1,2]

1. Hart-Tech Ltd., 45 Niciarniana Street, 92-320 Lodz, Poland; p.kula@hart-tech.pl (P.K.); s.paweta@hart-tech.pl (S.P.); adam.rzepkowski@p.lodz.pl (A.R.)
2. Institute of Materials Science and Engineering, Lodz University of Technology, 1/15 Stefanowski Street, 90-924 Lodz, Poland
* Correspondence: robert.pietrasik@p.lodz.pl

Received: 1 March 2020; Accepted: 19 March 2020; Published: 21 March 2020

Abstract: The paper presents the new concept of low frictional hybrid composite coatings on nodular cast iron. The structure of it is multilayer and consists of MoS$_2$ and/or WS$_2$ nanoinclusions embedded in the iron nitrides' zone and relatively deep hard diffusion zone. It offers a low friction coefficient as well as high wear resistance of coated parts. The details of technology as well as the mechanism of layer's growth have been presented and discussed. The presented technology may be an interesting alternative for chromium-based galvanic coatings of piston rings made of nodular iron using Cr^{6+}.

Keywords: nitriding; low friction; piston ring

1. Introduction

Nodular cast iron is the most popular material used for piston rings. The extremely high and intricate thermal and mechanical load of them [1–4] require the application of advanced solutions, both for a bulk material [5] and surface layer [6,7], as well as for a lubrication regime [1,2,8,9]. In order to improve cooperation between the piston ring and the cylinder sleeve, various coatings are applied. Such coatings include chromium and/or chromium–molybdenum galvanic coatings [10] as well as flame [11] or plasma sprayed [12–16], laser cladded, and PVD/CVD ones [17–19]. The PVD coatings of machining tools [20,21], which are still developed, are generally not applicable for piston ring improvement, due to an adhesive nature of the interface between a coating and the original material. The most common protection of piston rings against scuffing and wear are chromium-based galvanic coatings. However, they are also the most dangerous ones for the employees as well as the natural environment, since Cr^{6+} is used [22].

The nitriding process is also widely used for surface treatment of rings made mainly of steel [17,23]. It improves the friction coefficient of the surface layer as well as its resistance against hydrogen wear both at dry and lubricated regime [24–28]. More intricate is the issue of cast iron nitriding, due to the presence of graphite precipitation in the microstructure [29,30] Recently, the new non-equilibrium, low-pressure nitriding process (FineLPN) has been developed. It may be used for creation of fully controlled phase structure of nitrided case both on steel and cast iron due to dedicated neural network computer support [31].

Layered materials, like MoS$_2$, WS$_2$, etc., are well known as efficient solid lubricants [32,33]. They may be used in frictional contacts also at extremely high contact pressures as well as at elevated temperatures. The critical issue with the application of them for the improvement of piston rings frictional performance is the necessity to incorporate of them into a hard and strong matrix of a surface layer. Similar trials of manufacturing of multiphase, gradient surface layers have been presented in numerous papers [10,14,17,18].

The new concept of low frictional hybrid composite coatings [34] on nodular cast iron has been presented in the paper. The structure of it is multilayer and it consists of MoS_2 and/or WS_2 nanoinclusions embedded in iron nitrides zone and relatively deep hard diffusion zone. It offers low friction coefficient, as well as high wear resistance and fatigue strength of coated parts. The details of the multi-stage new technology, as well as mechanism of the layer's growth, have been presented and discussed.

2. Materials and Methods

The substrate material specimens for research were pieces of industrially manufactured piston rings (Φ117.5 mm × 2.68 mm × 4.60 mm) made of S14 grade nodular cast iron. The standardised chemical composition of used cast iron is shown in Table 1.

Table 1. Standardised Chemical Composition of S14 Nodular Cast Iron [% by Weight].

C	Si	Mn	P	S	Cr
3.6–4.0	2.1–3.3	0.2–0.5	0.3	Max. 0.05	Max. 0.2

According to the industrial technology, the rings have been hardened to obtain the matrix microstructure of tempered martensite free of carbide precipitations. The ca. 2 μm of micro-particles diameter of tungsten disulphide and molybdenum disulphide have been used as low frictional reinforcing phases. The creation of a low friction surface layer was multi-stage and the process has been conducted as follows. First, the slurry 1 g of MoS_2 or MoS_2 + WS_2 suspended in 10 mL of C_3H_7OH + 1 mL of $C_5H_{11}OH$ has been prepared. Then, it has been used for the uniform spray coating of piston rings specimens. After the natural drying, the coated specimens have been annealed at 300 °C for 1 h in an inert N_2 atmosphere to sinter the cladded particles and to adhere them to the metallic substrate. Next, such preliminary prepared green compacts were thermo-chemically treated by FineLPN low-pressure nitriding [31] or alternatively, additionally treated by gas sulphonitriding in an active atmosphere containing ammonia gas and sulphur vapors [35]. Two options A and B of the multi-stage surface engineering process have been conducted and compared. The technological schedule and parameters of them are presented in the Table 2.

Table 2. Technological Details of Compared Processes Options.

Option A	Option B
Low frictional particles—MoS_2 + WS_2	Low frictional particles—MoS_2
Sintering—300 °C, 1 h, N_2	Sintering—300 °C, 1 h, N_2
Two-stage thermochemical treatment —FineLPN – 8 h, 540 °C, 40–60 mbar —Gas Sulphonitriding – 4 h, 540 °C, sulphur evaporation at 180 °C	One-stage thermochemical treatment —Fine – 12 h, 540 °C, 40–60 mbar

The cross-section microstructures of specimens were observed having used the optical microscope Nikon MA200 (Nikon Instech Co., Ltd., Tokyo, Japan). The microstructure and chemical composition of surface layers were investigated also by using the scanning electron microscope (SEM) JEOL JSM-6610 LV (JEOL Ltd., Tokyo, Japan) equipped with the energy dispersion spectroscope (EDS) X-MAX 80 Oxford Instruments (Oxford Instruments Group, Abingdon, UK).

Dry friction tribological tests at oscillating movement have been conducted using ball on disc tribometer SRV Optimol Instruments Prüftechnik (Optimol Instruments Prüftechnik GmbH, Munich, Germany). The parameters of tests were as follows: 20 N regular load, 1 mm stroke, 20 Hz frequency, 1800 s total time of test, at temperature of 25 °C. Dry friction coefficient has been registered during the test, and the maximum depth of frictional tracks has been measured after it alike. Three specimens were investigated for each option. The average values of frictional coefficient have been determined from all plots' courses (total duration 5400 s). The average values of the maximum depth of frictional tracks as well as the standard deviation of results have been calculated too.

3. Results and Discussion

The external appearance of piston rings specimens after spray coating and preliminary annealing are shown in Figure 1a. The metallic surface has been homogenously coated by a tight coating of low frictional particles which adhere themselves strong enough, and also to the substrate. An exemplary final microstructure cross-section of the obtained hybrid low frictional layer is presented in Figure 1b. It consists of the following subzones: an externally grown compound zone containing low-friction particles of MoS_2 and optionally WS_2 embedded in iron nitrides ε – 3 µm, white iron nitrides zone (−17.8 µm), partially containing FeS fine inclusions (−5.8 µm), and relatively deep (−154 µm) dark diffusive zone in the original microstructure. Such a multizone gradient structure of the surface layer should be beneficial from the point of view of frictional and antiwear properties of piston rings made of nodular cast iron.

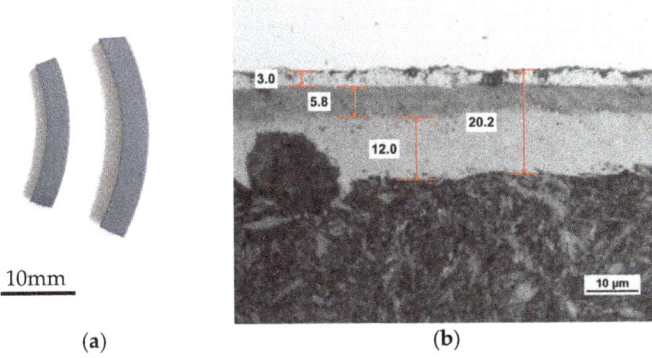

Figure 1. (a) External appearance of samples after spray coating and preliminary annealing. Sintering parameters: temperature: 300 °C, time: 2 h ramp 5 °C/min withstand for 1 h in N atmosphere; (b) Exemplary cross section of final microstructure of hybrid layer – option A.

SEM cross sections and EDS pictures are presented in Figures 2 and 3 for layers options A and B, respectively.

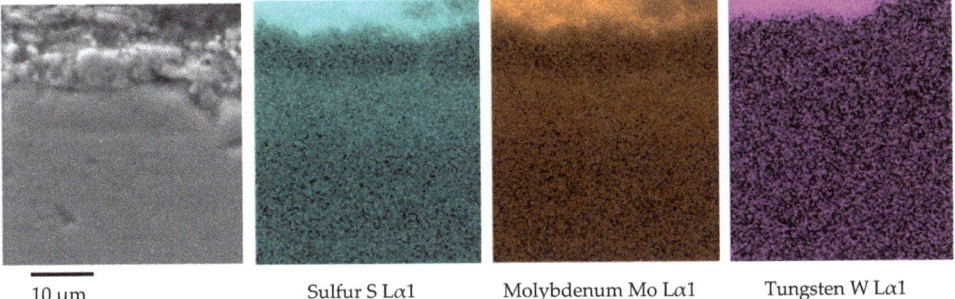

Figure 2. SEM Cross Sections and corresponding EDS maps of S, Mo and W distribution for layer option A.

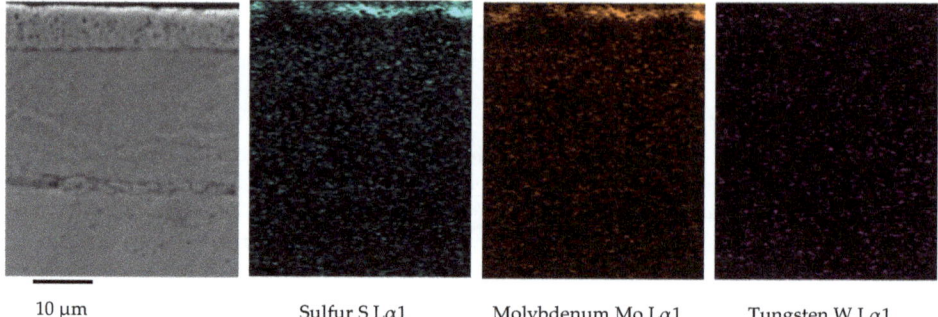

Figure 3. SEM Cross Sections and corresponding EDS of S, Mo and W distribution maps for layer option B.

The analysis of SEM pictures and EDS maps confirmed the incorporation phenomenon of low frictional particles MoS_2 and WS_2 into the structure of outer zone of hybrid layers.

The probable incorporation mechanism of them is the growth of ε iron nitrides during FineLPN and sulphonitriding processes outside of the original metallic surface due to reciprocal diffusion of iron ions through cationic defects in nitrides structure [35]. After the B process, the outer compound zone that contains sulphur and molybdenum is relatively shallow and tight (Figure 2). The application of additional sulphonitriding process after FineLPN treatment (option A) causes an important thickening as well as structural loosening of composite's outer zone that contains numerous and very fine inclusions of low frictional particles MoS_2 and WS_2, which are embedded in ε-nitrides porous matrix. That porous composite outer zone based on ε – iron nitrides matrix should be beneficial from a tribological point of view for both dry [25] and liquid friction conditions [26,36]. Additionally, a relatively deep (ca. 150 µm) and relatively hard (up to 700 HV) diffusive zone should decrease the development of frictional contact area and protect a nodular iron against hydrogen wear [24].

The low frictional and antiwear properties for optimized structure of hybrid layer (option A) have been confirmed by dry friction and oscillation tribological tests. The results of them are presented in Figure 4.

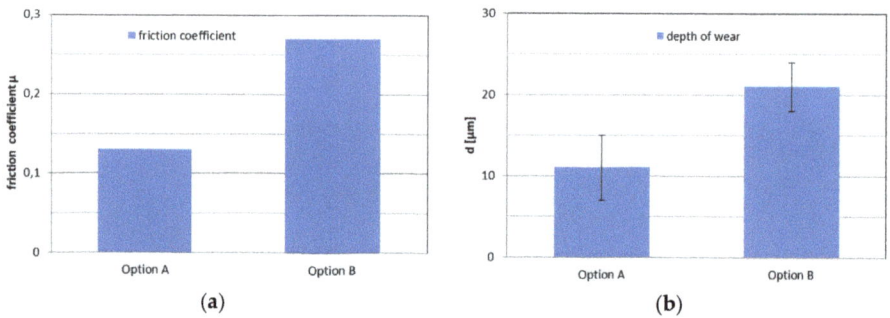

Figure 4. (a) Comparison of friction coefficients obtained during tribological tests; (b) Comparison of the maximum depth of frictional tracks.

The specimens treated in the hybrid process according to the schedule and parameters of option A have shown an extremely low value of dry friction coefficient, 0.13, which is two-fold than the non-optimum layers, e.g., those treated to the schedule and parameters of option B (Figure 4). Also, the linear wear results after tribological tests have been twice as less for specimens treated according to the parameters A. That low frictional effect is the result of layered microstructure of numerous and relatively deep distributed MoS_2 and WS_2 inclusions. They have been incorporated relatively deep into hard ε – iron nitrides matrix. Therefore, they are durable structural sources of solid lubricant for long frictional action in piston rings – engine cylinders systems.

The obtained tribological test results allow to conclude that optimized hybrid layers contain both MoS_2 and WS_2 low frictional inclusions, and are well aerated by double thermo-chemical treatment (FineLPN + sulphonitriding). Thus, such new technology may represent an interesting alternative to replace and exclude the chromium-based galvanic coatings using Cr^{6+} from the manufacturing of piston rings made of nodular iron.

4. Conclusions

- The new hybrid, multistage technology—that consists of following stages: slurry coating, drying, sintering, FineLPN low pressure nitriding, and sulphonitriding thermochemical treatments—may be an interesting alternative for chromium-based galvanic coatings of piston rings made of nodular iron using Cr^{6+}.
- The optimum tribological properties of hybrid layers have been obtained for option of two low frictional particles MoS_2 and WS_2 and additional sulphonitriding heat treatment for important thickening, as well as the structural loosening of outer composite zone.
- The extremely low dry friction coefficient (0.13) and low linear wear have been revealed for the optimum hybrid layer, which were ca. twice as less in comparison to the benchmark technological solutions.
- The optimised low frictional effect is the result of a layered microstructure composed of numerous MoS_2 and WS_2 inclusions, which are and relatively deeply distributed in a hard ε – iron nitrides matrix.

Author Contributions: Conceptualization, P.K. and R.P.; writing—original draft preparation, P.K.; design and implementation of processes, tribology, R.P.; SEM and EDS research, A.R.; review and editing, visualization, literature, S.P. All authors have read and agreed to the published version of the manuscript.

Funding: The work has been done under Measure 1.2—Sectoral Research & Development programs of "Program Operacyjny Inteligentny Rozwój" 2014–2020 (Smart Growth Operational Program 2014–2020) co-funded by European Regional Development Fund. The project: "Gradient low-friction coats produced by means of a hybrid FineLPN process, nanostructured with MoS_2 and rGO particles for use in aircraft sealing." Contract Number: POIR.01.02.00-00-0011/15 (NIWAG).

Conflicts of Interest: The authors declare no conflict of interest.

References

1. Williams, J. Boundary lubrication and friction. In *Engineering Tribology*; Cambridge University Press: Cambridge, UK, 2005; pp. 348–380. [CrossRef]
2. Neville, A.; Morina, T.; Haque, M.; Voong, M. Compatibility between tribological surfaces and lubricant additives—How friction and wear reduction can be controlled by surface/lubes synergies. *Tribol. Int.* **2007**, *40*, 1680–1695. [CrossRef]
3. Ostapski, W. Analysis of thermo-mechanical response in an aircraft piston engine by analytical, FEM, and test-stand investigations. *J. Therm. Stress.* **2011**, *34*, 285–312. [CrossRef]
4. Zabala, B.; Igartua, A.; Fernández, X.; Priestner, C.; Ofner, H.; Knaus, O.; Nevshupa, R. Friction and wear of a piston ring/cylinder liner at the top dead centre: Experimental study and modeling. *Tribol. Int.* **2017**, *106*, 23–33. [CrossRef]
5. Yamagata, H. *The Science and Technology of Materials in Automotive Engines*; Woodhead Publishing Limited: Cambridge, UK, 2005.
6. Chida, C.; Okano, M. Piston Ring. U.S. Patent US10385971B2, 20 August 2019.
7. Esser, P.K. Coated Piston Ring Having A Protective Layer. U.S. Patent Application US2019/0100836A1, 4 April 2019.

8. Sherrington, I.; Smith, E.H. Experimental methods for measuring the oil-film thickness between the piston-rings and cylinder-wall of internal combustion engines. *Tribol. Int.* **1985**, *18*, 315–320. [CrossRef]
9. Ali, M.K.A.; Hou, X.; Mai, L.; Cai, Q.; Turkson, R.F.; Chen, B. Improving the tribological characteristics of piston ring assembly in automotive engines using Al_2O_3 and TiO_2 nanomaterials as nano-lubricant additives. *Tribol. Int.* **2016**, *103*, 540–554. [CrossRef]
10. Bindumadhavan, P.N.; Makesh, S.; Gowrishnkar, N.; Wah, H.K.; Prabhakar, O. Aluminizing and subsequent nitriding of plain carbon low alloy steels for piston ring applications. *Surf. Coat. Tech.* **2000**, *127*, 251–258. [CrossRef]
11. Babu, M.V.; Kumar, R.K.; Prabhakar, O.; Shankar, N.G. Simultaneous optimization of flame spraying process parameters for high quality molybdenum coatings using Taguchi methods. *Surf. Coat. Tech.* **1996**, *79*, 276–288. [CrossRef]
12. Li, Y.; Dong, S.; He, P.; Yan, S.; Li, E.; Liu, X.; Xu, B. Microstructure characteristics and mechanical properties of new-type FeNiCr laser cladding alloy coating on nodular cast iron. *J. Mater. Process. Tech.* **2019**, *269*, 163–171. [CrossRef]
13. Zhou, Y.X.; Zhang, J.; Xing, Z.G.; Wang, H.D.; Lv, Z.L. Microstructure and properties of NiCrBSi coating by plasma cladding on gray cast iron. *Surf. Coat. Tech.* **2019**, *361*, 270–279. [CrossRef]
14. Arps, J.H.; Page, R.A.; Dearnaley, G. Reduction of wear in critical engine components using ion-beam-assisted deposition and ion implantation. *Surf. Coat. Tech.* **1996**, *84*, 579–583. [CrossRef]
15. Dahm, K.L.; Dearnley, P.A. Novel plasma-based coatings for piston rings. *Tribol. Ser.* **2002**, *40*, 243–246.
16. Karamış, M.B.; Yıldızlı, K.; Çakırer, H. An evaluation of surface properties and frictional forces generated from Al–Mo–Ni coating on piston ring. *Appl. Surf. Sci.* **2004**, *230*, 191–200. [CrossRef]
17. Kula, P.; Dybowski, K.; Lipa, S.; Batory, D.; Sawicki, J.; Wołowiec, E.; Klimek, L. Hybrid surface layers, made by nitriding with DLC coating, for application in machine parts regeneration. *Arch. Mater. Sci. Eng.* **2013**, *60*, 32–37.
18. Lin, J.; Wei, R.; Bitsis, D.C. Development and evaluation of low friction TiSiCN nanocomposite coatings for piston ring applications. *Surf. Coat. Tech.* **2016**, *298*, 121–130. [CrossRef]
19. Friedrich, C.; Berg, G.; Broszeit, E.; Rick, F.; Holland, J. PVD Cr_xN coatings for tribological application on piston rings. *Surf. Coat. Tech.* **1997**, *97*, 661–668. [CrossRef]
20. Fernández-Abia, A.I.; Barreiro, J.; Fernández-Larrinoa, J.; López de Lacalle, L.N.; Fernández-Valdivielso, A.; Pereira, O.M. Behaviour of PVD coatings in the turning of austenitic stainless steels. *Proc. Eng.* **2013**, *63*, 133–141. [CrossRef]
21. Rodriguez-Barrero, S.; Fernández-Larrinoa, J.; Azkona, I.; López de Lacalle, L.N.; Polvorosa, R. Enhanced performance of nanostructured coatings for drilling by droplet elimination. *Mater. Manuf. Process.* **2016**, *31*, 593–602. [CrossRef]
22. Prado, F.E.; Hilal, M.; Chocobar-Ponce, S.; Pagano, E.; Rosa, M.; Prado, C. Chapter 6—Chromium and the plant: A dangerous affair? In *Plant Metal Interaction*; Ahmad, P., Ed.; Elsevier: Amsterdam, The Netherlands, 2016; pp. 149–177.
23. Dayanç, A.; Karaca, B.; Kumruoğlu, L.C. Plasma nitriding process of cast camshaft to improve wear resistance. *Acta Phys. Pol. A* **2019**, *135*, 793–799. [CrossRef]
24. Kula, P. The comparison of resistance to "hydrogen wear" of hardened surface layers. *Wear* **1994**, *178*, 117–121. [CrossRef]
25. Kula, P. The "self-lubrication" by hydrogen during dry friction of hardened surface layers. *Wear* **1996**, *201*, 155–162. [CrossRef]
26. Kula, P.; Pietrasik, R.; Wendler, B.; Jakubowski, K. The effect of hydrogen in lubricated frictional couples. *Wear* **1997**, *212*, 199–205. [CrossRef]
27. Gawronski, Z.; Sawicki, J. Technological surface layer selection for small module pitches of gear wheels working under cyclic contact loads. *Mater. Sci. Forum* **2006**, *513*, 69–74. [CrossRef]
28. Sawicki, J.; Gorecki, M.; Kaczmarek, Ł.; Gawronski, Z.; Dybowski, K. Increasing the durability of pressure dies by modern surface treatment methods. *Chiang. Mai. J. Sci.* **2013**, *40*, 886–897.
29. Writzl, V.; Rovani, A.C.; Pintaude, G.; Lima, M.S.F.; Guesser, W.L.; Borges, P.C. Scratch resistances of compacted graphite iron with plasma nitriding, laser hardening, and duplex surface treatments. *Tribol. Int.* **2020**, *143*, 106081. [CrossRef]

30. Ampaw, E.K.; Arthur, E.K.; Badmos, A.Y.; Obayemi, J.D.; Adewoye, O.O.; Adetunji, A.R.; Olusunle, S.O.O.; Soboyejo, W.O. Sliding wear characteristics of pack cyanided ductile iron. *J. Mater. Eng. Perform.* **2019**, *28*, 7227–7240. [CrossRef]
31. Wołowiec-Korecka, E.; Kula, P.; Pawęta, S.; Pietrasik, R.; Sawicki, J.; Rzepkowski, A. Neural computing for a low-frictional coatings manufacturing of aircraft engines' piston rings. *Neur. Comput. Appl.* **2019**, *31*, 4891–4901. [CrossRef]
32. Watanabe, S.; Noshiro, J.; Miyake, S. Tribological characteristics of WS$_2$/MoS$_2$ solid lubricating multilayer films. *Surf. Coat. Tech.* **2004**, *183*, 347–351. [CrossRef]
33. Wong, K.C.; Lub, X.; Cotter, J.; Eadie, D.T.; Wong, P.C.; Mitchell, K.A.R. Surface and friction characterization of MoS$_2$ and WS$_2$ third body thin films under simulated wheel/rail rolling–sliding contact. *Wear* **2008**, *264*, 526–534. [CrossRef]
34. Kula, P.; Pietrasik, R.; Paweta, S.; Patent Office of the Republic of Poland. Low-friction layer from Nanocomposite Gradient Material and Method for Producing It. Poland Patent PL233113B1, 30 September 2019.
35. Kocemba, I.; Szynkowska, M.I.; Mackiewicz, E.; Goralski, J.; Rogowski, J.; Pietrasik, R.; Kula, P.; Kaczmarek, L.; Jóźwik, K. Adsorption of gas-phase elemental mercury by sulphonitrided steel sheet. Effect of hydrogen treatment. *J. Hazard Mater.* **2019**, *368*, 722–731. [CrossRef]
36. Liskiewicz, G.; Kula, P.; Neville, A.; Pietrasik, R.; Morina, A.; Liskiewicz, T. Hydrogen influence on material interaction with ZDDP and MoDTC lubricant additives. *Wear* **2013**, *297*, 966–971. [CrossRef]

© 2020 by the authors. Licensee MDPI, Basel, Switzerland. This article is an open access article distributed under the terms and conditions of the Creative Commons Attribution (CC BY) license (http://creativecommons.org/licenses/by/4.0/).

Article

Al$_2$O$_3$ + Graphene Low-Friction Composite Coatings Prepared By Sol–Gel Method

Bożena Pietrzyk [1,*], Sebastian Miszczak [1], Ye Sun [2] and Marcin Szymański [1]

[1] Institute of Materials Science and Engineering, Faculty of Mechanical Engineering, Lodz University of Technology, Stefanowskiego Str. 1/15, 90-924 Lodz, Poland; sebastian.miszczak@p.lodz.pl (S.M.); m.szymanski@mail.com (M.S.)
[2] Wanhua Research Institute (NERP), NO. 59. Chongqing Road, Yantai 264006, Shandong, China; sunye@whchem.com
* Correspondence: bozena.pietrzyk@p.lodz.pl

Received: 15 July 2020; Accepted: 2 September 2020; Published: 4 September 2020

Abstract: In this work, Al$_2$O$_3$ + graphene coatings were prepared using the sol–gel method. The aim of the study was preliminary determination of the influence of size and amount of graphene nanoplatelets on morphology, chemical structure, and basic tribological properties of Al$_2$O$_3$ + graphene composite coatings. Two types of reduced graphene oxide (rGO) nanoplatelets with different lateral size and thickness were used to prepare the coatings. To characterize them, scanning electron microscope (SEM), glow discharged optical emission spectrometer (GDOES), Fourier-transform infrared (FTIR), reflectance spectrometer, and ball-on-disk tribological tests were used. It was found that the presence of graphene in the Al$_2$O$_3$ + graphene coatings did not fundamentally change the chemical transformation of ceramic Al$_2$O$_3$ matrix. Morphology examinations of coatings containing larger graphene nanoplatelets revealed a tendency to their parallel arrangement in relation to the coated surface. The tribological properties of Al$_2$O$_3$ + graphene coatings turned out to be strongly dependent on the size of graphene nanoplatelets as well as on the heat treatment temperature. The friction coefficient as low as 0.11 and good durability were obtained for the Al$_2$O$_3$ + graphene coating with larger nanoplatelets and heat-treated at 500 °C. The results of conducted research indicate the potential use of Al$_2$O$_3$ + graphene composite coatings prepared by the sol–gel method as low-friction ceramic coatings.

Keywords: ceramic coating; alumina coating; sol–gel; composite coating; graphene oxide; graphene nanoplatelets (GNP); rGO

1. Introduction

Alumina (Al$_2$O$_3$) as a technical ceramic is one of the most commonly used oxide materials [1]. Thanks to its good mechanical strength and high hardness, chemical stability, high thermal and very low electrical conductivity, and cost-effectiveness, alumina has been utilized for a large variety of applications, such as electric and electronic devices, reinforcing components, catalytic supports, medical implants, and thermal and wear resistant components, among others [2–8].

Aluminum oxide can also be used in the form of coatings produced by various methods: oxidation [9,10], thermal spraying [4,11], chemical vapor deposition [12,13], sputtering [14,15], sol–gel deposition [16,17], and so on. These coatings can also have a variety of applications, but among the most popular are the following: protection against electrochemical and high temperature corrosion [18–22], as well as reducing wear [23–25]. The latter application is justified by high hardness and thermal/chemical stability of Al$_2$O$_3$ coatings, despite the fact that achieved coefficients of friction (COF) of pure alumina coatings are relatively high—between 0.65 and 0.9—regardless of manufacturing methods [11,26–30].

As lowering the COF is crucial for durability and wear resistance of coatings, efforts are being made to reduce it. The basic direction of modification is the implementation of additives in the coating, for example, particles of oxides [17,31], sulphides and fluorides [32–34], carbon nanotubes [35–37], graphite [38,39], or recently graphene [40–42]. Graphene, a two-dimensional allotropic form of carbon consisting of hexagonally arranged atoms in the form of a single flat layer [43], aroused great interest thanks to its excellent physico-chemical, electrical, and mechanical properties and resulting applications [44–47]. The tribological properties of graphene justify its use to reduce wear [48,49], provided that it has been effectively introduced into the volume of the coating in an appropriate amount. In case of alumina + graphene coatings, despite the efforts, the results of incorporating graphene into alumina vary. In most cases, the improvement of wear resistance can be seen, but with the use of commonly used production methods (mainly plasma spraying and complex CVD methods), a large spread of the friction coefficient is obtained, fluctuating between 0.3 and 0.7 [40–42]. It can be presumed that the reason for this may be insufficient control over graphene dispersion and/or the adverse effect of the physico-chemistry of the deposition processes itself. Considering the above, the sol–gel method may be an interesting alternative for the production of alumina + graphene composite coatings. The sol–gel method is a well-known and well-established technique of obtaining ceramic materials [50,51], which can also be used for the production of ceramic coatings [52]. The method is a simple, convenient, and low-cost chemical route for preparation of high quality coatings, including alumina [53]. The most important advantage of the sol–gel method is the presence of the liquid phase stage-sol, which can be deposited on substrates using various methods [54], and allows the introduction of specific additives into the coating, for example, particles of other phases, such as graphene.

In this study, Al_2O_3 + graphene coatings were prepared using a simple three-step sol-gel method: (1) preparation of Al_2O_3 + graphene sols, (2) dip-coating deposition of prepared sols onto the substrates, and (3) furnace heat treatment of obtained coatings at various temperatures. The aim of the study was preliminary determination of the influence of amount and type (size) of graphene nanoplatelets on surface morphology, chemical structure, and basic tribological properties of Al_2O_3 + graphene composite coatings.

2. Materials and Methods

2.1. Preparation of Sols

An alumina sol was prepared by dissolving aluminum isopropoxide (Sigma-Aldrich, >98%, Steinheim, Germany) in boiling water. During mixing, nitric acid solution (1 mol/L) was added to the precursor solution as a hydrolysis catalyst. The concentration of precursor in the sol was 0.7 mol/L.

Two kinds of reduced graphene oxide (rGO) in the form of graphene nanoplatelets (GNPs) with different size and thickness were used for the preparation of composite coatings:

(1) G1 graphene with average GNP thickness of 12 nm and average lateral size 4500 nm (AO-3, Graphene Laboratories Inc., New York, NY, USA);
(2) G2 graphene with average GNP thickness of 1–5 nm and average lateral size less than 2000 nm (0540DX, SkySpring Nanomaterials, Inc., Houston, TX, USA).

The suspensions of both G1 and G2 graphenes were prepared using Tween 20 (Sigma-Aldrich, ≥40%, Steinheim, Germany) as surfactant. For the preparation of each suspension, the surfactant (0.02 mL) was dissolved in 50 mL of water and then added to the proper mass of graphene powder. These solutions were sonicated using an ultrasonic bath for 15 min to prepare suspensions of both G1 and G2 graphene with 2 and 4 wt.% concentration. Subsequently, the same volumes of the prepared graphene suspensions and the alumina sol were mixed and sonicated in an ultrasonic bath for 15 min. As a result, Al_2O_3 + graphene sols were obtained—the final concentrations of each type of graphene were 1 wt.% and 2 wt.%. The final concentration of alumina precursor in Al_2O_3 + graphene sols was about 0.35 mol/L. The sedimentation processes of graphene flakes in alumina sol were observed: for suspensions of G1 in alumina sol, the layer of graphene was visible on the bottom of the vessel in a few minutes after mixing; the suspension of G2 in alumina sol was much more stable. For homogeneity of suspension, the 2 min of sonication of each Al_2O_3 + graphene sols was carried out directly before each deposition process.

For deposition of Al_2O_3 coatings without GNPs, the obtained alumina sol was diluted using distilled water to the concentration of precursor of 0.35 mol/L in order to maintain the same deposition conditions as for the rest of the coatings.

2.2. Deposition of Coatings

Al_2O_3 and Al_2O_3 + graphene coatings were deposited by the sol–gel dip-coating method. The discs of 316 stainless steel with diameter of 16 mm and height of 5 mm as well as monocrystalline silicon wafers (100) were used as substrates. The surface of stainless steel substrates was ground on abrasive paper with grade 800 up to 2000. Before the deposition, substrates (both stainless steel and silicon) were washed in ethanol for 15 min using ultrasonic cleaner and dried in compressed air.

The substrates were immersed in previously prepared alumina sols as well as in suspensions of graphene in alumina sol, and withdrawn at a controlled constant speed of 120 mm/min using dip-coater (TLO 0.1 MTI Corporation, Richmond, CA, USA). After deposition, coatings were dried at ambient temperature in air for 30 min and heat-treated in the furnace at the temperature of 300 °C or 500 °C for 15 min. The same technological process of coatings production was the basis for comparing their properties.

2.3. Characterization of Coatings

The evaluation of Al_2O_3 and Al_2O_3 + graphene coatings morphology was performed using scanning electron microscope (SEM) (S-3000M, Hitachi, Tokyo, Japan) for coatings deposited on silicon.

The elemental analysis of coatings was carried out using glow discharged optical emission spectrometer (GDOES) (LECO GDS850A, St. Joseph, MI, USA) for coatings deposited on silicon.

The thickness as well as the refractive index of coatings deposited on stainless steel substrates were determined with use of reflectance spectrometer (Thin Film Analyzer UV20, Filmetrics, San Diego, CA, USA), using the Cauchy model.

The chemical structure of the coatings was studied using a Fourier-transform infrared (FTIR) spectrometer (Nicolet iS50, Thermo Fisher Scientific, Waltham, MA, USA) in the range of 4000–400 cm^{-1}. The study was carried out in transmission mode, recording the absorbance of IR radiation passing through coatings deposited on silicon or through graphene powder.

2.4. Tribological Tests

Friction tests were performed for Al_2O_3 and Al_2O_3 + graphene coatings deposited on the 316 stainless steel using a ball-on-disk tribometer (CSM THT, Needham, MA, USA) in rotary mode. The tribological tests were performed at room temperature, linear speed was 0.1 m/s, and radius was 5 mm. The applied normal load was 1N and sliding distance was fixed at 100 m. The alumina (α-Al_2O_3) testing ball measuring 6.35 mm in diameter was used as static element. The friction coefficients were automatically measured and recorded in real time by computer system of the tribometer.

3. Results and Discussion

3.1. Morphology and Thickness of Coatings

Figure 1 presents the morphology of Al_2O_3 coating heat treated at 500 °C and Al_2O_3 + graphene (both G1 and G2) coatings after drying at ambient temperature as well as after heat treatment at 500 °C. SEM observations revealed that the surface of coatings was free of cracks, delamination, or other discontinuities. The surface of Al_2O_3 coating was smooth (Figure 1a), while Al_2O_3 + graphene coatings contained uniformly distributed particles whose shape and size depended on the type of graphene used. In the Al_2O_3 + G1 graphene coating (Figure 1b,c), the shape of particles was rather flat and their size was mostly between 2 and 5 μm, corresponding to the size of G1 graphene nanoplatelets. Al_2O_3 + G2 graphene coating (Figure 1d,e) contained spherical particles that looked like small agglomerates of G2 GNPs with different sizes, but mostly smaller than 2 μm. The particle size was roughly correlated with the size of each type of graphene.

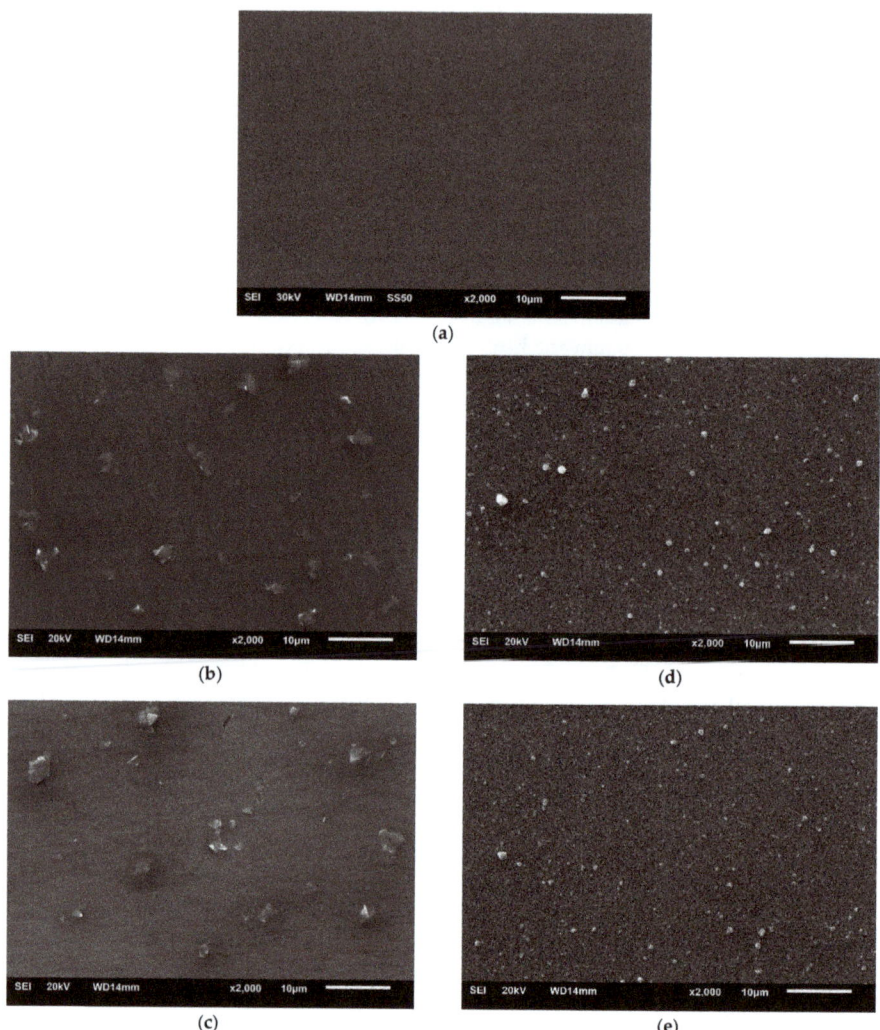

Figure 1. Scanning electron microscope (SEM) images of Al_2O_3 and Al_2O_3 + graphene coatings with 1% of graphene in sol suspension: (**a**) Al_2O_3 coating heat treated at 500 °C; (**b**) Al_2O_3 + G1 coating dried at ambient temperature; (**c**) Al_2O_3 + G1 coating heat treated at 500 °C; (**d**) Al_2O_3 + G2 coating dried at ambient temperature; (**e**) Al_2O_3 + G2 coating heat treated at 500 °C.

It should be pointed out that, whether coatings were heat treated or not, particles of both G1 and G2 graphene could be observed up to the temperature of 500 °C and did not change their shape and average size.

Figure 2 shows SEM images of Al_2O_3 + G1 and Al_2O_3 + G2 graphene coatings taken at an inclination of approximately 75 degrees. The analysis of the Al_2O_3 + G1 coating image (Figure 2a) indicates that the G1 graphene particles are in the form of thin nanoplatelets stacks located under the surface of the coating within the oxide matrix as well as protruding slightly above the surface (scheme in Figure 2c). In the image obtained at higher magnification (Figure 2b), the flat nature of the G1 nanoplatelet stack and its arrangement parallel to the surface of the coating and substrate is clearly visible.

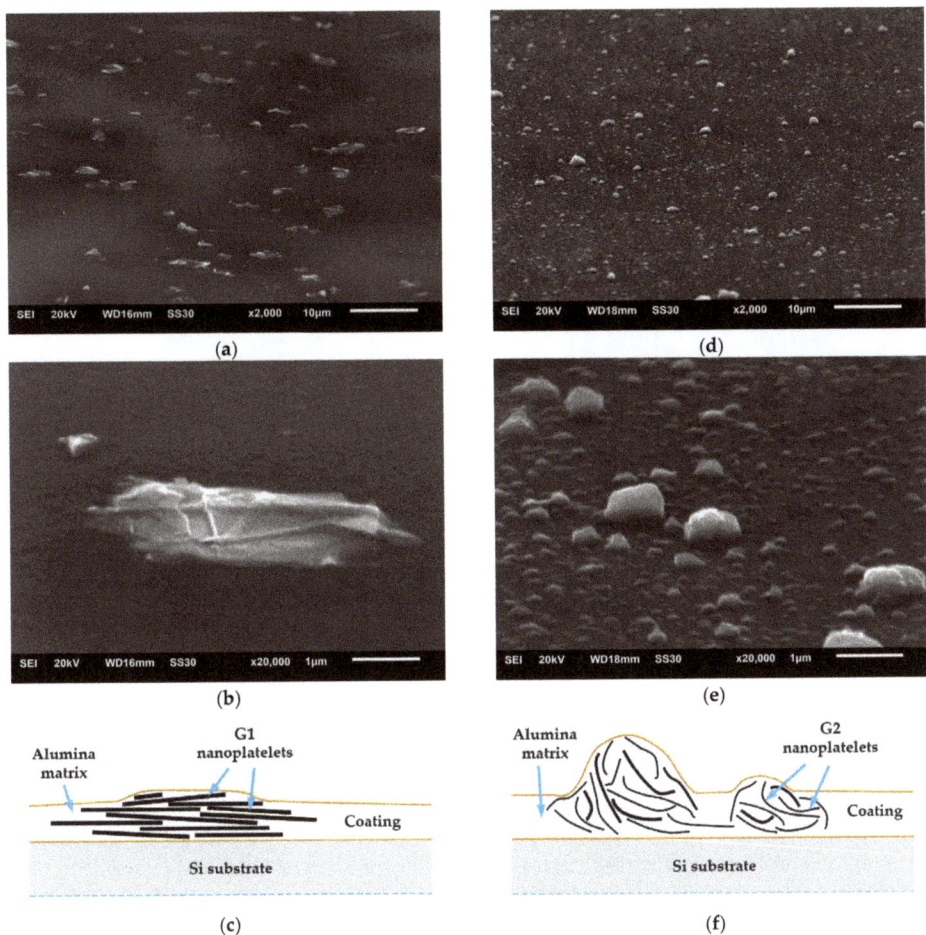

Figure 2. SEM images of Al$_2$O$_3$ + graphene coatings with 1% of graphene in sol suspension: (a,b) Al$_2$O$_3$ + G1 coatings; (d,e) Al$_2$O$_3$ + G2 coatings, and proposed schemes of their internal structures (c,f).

The images of the Al$_2$O$_3$ + G2 coating (Figure 2d,e) show a large number of bulges protruding above the coating surface. Taking into account their varied size and height (up to 1 μm), they can be identified as agglomerates of G2 graphene nanoplatelets of different lateral size and spatial orientation with respect to the substrate plane. These agglomerates consist of nanoplatelets (scheme in Figure 2f) surrounded by an oxide matrix, and their size significantly exceeds the thickness of the coating. The analysis of the shape of graphene agglomerates shows that G1 graphene (larger lateral size and thickness) forms agglomerates with GNPs oriented parallel to the surface, while G2 graphene (smaller lateral size and thinner) forms agglomerates with a globular shape and more random arrangement.

Figure 3 shows depth profiles of GDOES elemental analysis carried out for Al_2O_3 and Al_2O_3 + graphene coatings deposited on silicon from a suspensions with 2% of G1 and 2% of G2 graphene and heat treated at 300 °C. The results indicate that both the actual carbon content and the thickness of the coatings depend on the type of graphene. The Al_2O_3 coating had a carbon content of approximately 6%, which comes mainly from remnants of organometallic precursor. The carbon contents in Al_2O_3 + graphene coatings were higher: approximately 11% in the Al_2O_3 + G1 coating and approximately 20% in the Al_2O_3 + G2 coating. Despite the same original concentrations of G1 and G2 graphene in the suspensions, the carbon content of Al_2O_3 + G2 coating was much higher than that of Al_2O_3 + G1. This is a consequence of the observed sedimentation of G1 graphene particles in the suspension, so that its effective amount (by weight) in the coating is lower compared with G2 graphene.

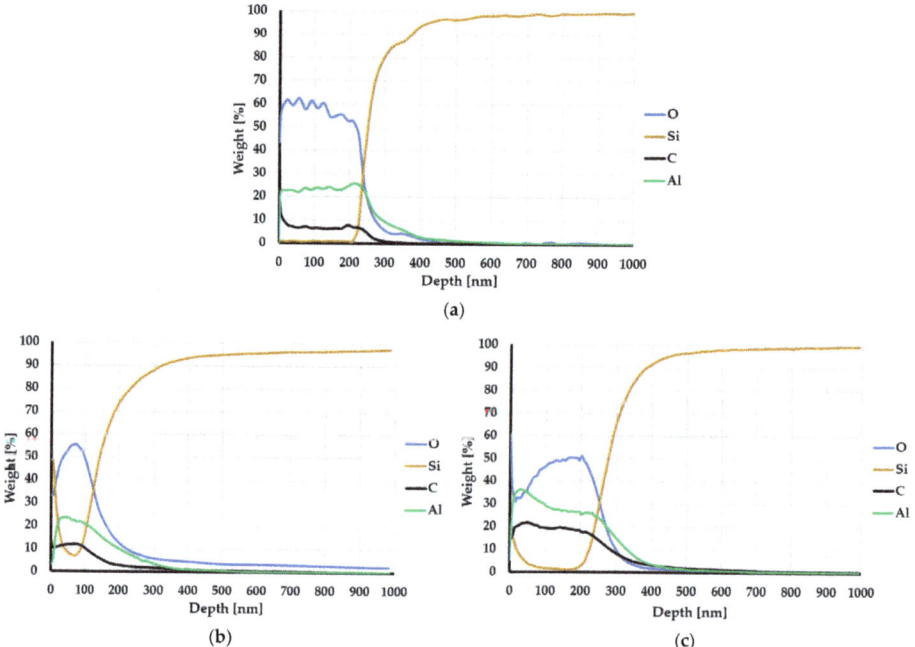

Figure 3. Elemental analysis of Al_2O_3 (**a**), Al_2O_3 + G1 (**b**), and Al_2O_3 + G2 (**c**) coatings with 2% of graphene in sol suspension, heat treated at a temperature of 300 °C.

On the basis of the profiles obtained, it is also possible to estimate the thickness of the coatings: approximately 200 nm for the Al_2O_3 coating, approximately 110 nm for the Al_2O_3 + G1 coating, and approximately 200 nm for the Al_2O_3 + G2 coating.

Similar dependence of thickness on the type of graphene was observed for coatings deposited on a steel substrate. The results of the thickness measurements of these coatings are shown in Table 1.

Table 1. The thickness of coatings deposited on steel substrates.

Type of Coating	Temperature of Heat Treatment (°C)	Concentration of Graphene in Suspension (%)	Thickness (nm)	n (Refractive Index)
Al_2O_3 + G1	300	1	134 ± 2	1.49 ± 0.005
		2	130 ± 1	1.50 ± 0.01
	500	1	125 ± 3	1.46 ± 0.03
		2	123 ± 2	1.43 ± 0.01
Al_2O_3 + G2	300	1	184 ± 1	1.55 ± 0.01
		2	171 ± 1	1.56 ± 0.01
	500	1	152 ± 1	1.50 ± 0.01
		2	147 ± 1	1.55 ± 0.005
Al_2O_3	300	-	163 ± 8	1.56 ± 0.06
	500	-	152 ± 1	1.58 ± 0.005

It was found that the thickness of Al_2O_3 coatings was similar to the thickness of coatings with G2 graphene. In contrast, the thickness of coatings with G1 graphene was lower than the corresponding coatings with G2 graphene or without any graphene. This may be owing to the shape and size of G1 graphene nanoplatelets, which, owing to the tendency to orientate parallel to the surface of the substrate, affect the transition between capillary and draining mechanisms characteristic for dip-coating deposition [55] and change the thickness of the wet layer formed during pulling out from the sol.

Analyzing the data in Table 1, it can also be seen that the thickness of the coatings decreases as the heat treatment temperature increases. This is a consequence of sintering of the Al_2O_3 matrix. The change of thickness is typical for sol–gel coatings below the temperature of obtaining a stable, dense crystal structure [52].

3.2. Chemical Structure

In order to determine the chemical structure, FTIR spectra of graphene powder as well as Al_2O_3 and Al_2O_3 + graphene coatings deposited on silicon substrate from a suspensions with 2% of G1 or 2% of G2 graphene, heat treated at different temperatures, were recorded.

FTIR spectra of G1 and G2 graphene, demonstrated in Figure 4a, were found to exhibit several absorption bands of carbon and oxygen-containing groups characteristic for graphene oxide: the absorption band around 1000 cm^{-1} for C–O stretching vibration bonds, band around 1200 cm^{-1} for C–OH stretching, and band around 1600 cm^{-1} for carbon–carbon bonds conjugated with carbon–oxygen bonds [56,57]. The broad peak at 3400 cm^{-1} was characteristic for stretching vibrations of -OH groups. The small peak at 2930 cm^{-1} in the spectra was attributed to the C–H stretching vibration. The FTIR spectra of G1 and G2 graphene were similar, but the most characteristic bands for graphene oxide, visible at 1200 cm^{-1} and 1600 cm^{-1}, exhibited higher intensity in G2 spectrum.

In the spectra of the Al_2O_3 coatings shown in Figure 4b, three main ranges of absorption bands can be distinguished: 2900–3700 cm^{-1}, 1300–1700 cm^{-1}, and 400–1100 cm^{-1}.

In the range of 2900–3700 cm^{-1}, the absorption bands for hydroxyl groups deriving from water absorbed in the coating were visible. The intensity of these bands decreased as the heat treatment temperature increased, which can indicate the removal process of absorbed water.

In the range of 1200–1700 cm^{-1}, bands of chemically bonded –OH groups (1200–1500 cm^{-1}) as well as C–O and C–H bonds (1300–1700 cm^{-1}) originating from the aluminum isopropoxide precursor were visible. The intensity of bands in this region also decreased with the increasing temperature, but some of them were visible even in the spectrum of coating heat treated at 500 °C.

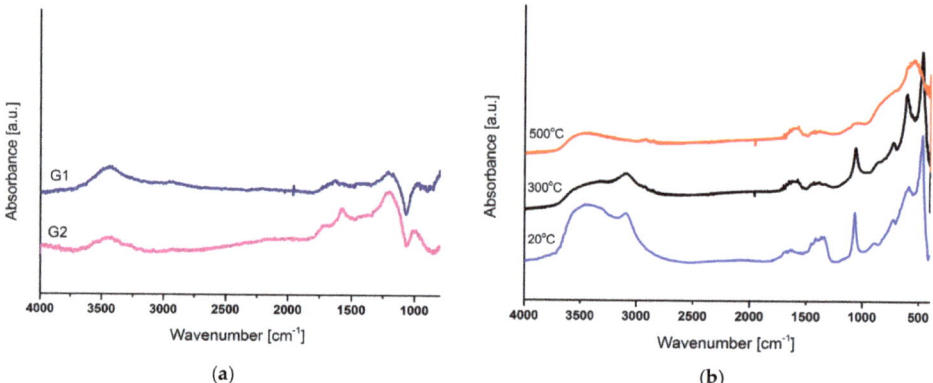

Figure 4. Fourier-transform infrared (FTIR) spectra of (**a**) G1 and G2 graphene powders and (**b**) Al_2O_3 coatings heat treated at different temperatures.

The range of 400–1100 cm^{-1} was characteristic for bands derived from Al–O bonds. The typical boehmite (AlOOH) band system (482, 620, 740, 1070, and 3090 cm^{-1}) [16] was visible in the spectrum obtained at room temperature as well as heat treated at 300 °C. In the spectrum of coating heat treated at 500 °C, one can observe that the discussed band became wider and showed maxima at 560 cm^{-1} and 800 cm^{-1}, which indicated the transformation of the matrix structure from AlOOH to Al_2O_3 [16].

In the spectra of Al_2O_3 + graphene coatings (Figure 5), the characteristic bands of graphene were hardly visible because they overlapped with the bands present in the spectrum of the Al_2O_3 coating in the range 1000–1700 cm^{-1} (Figure 4b).

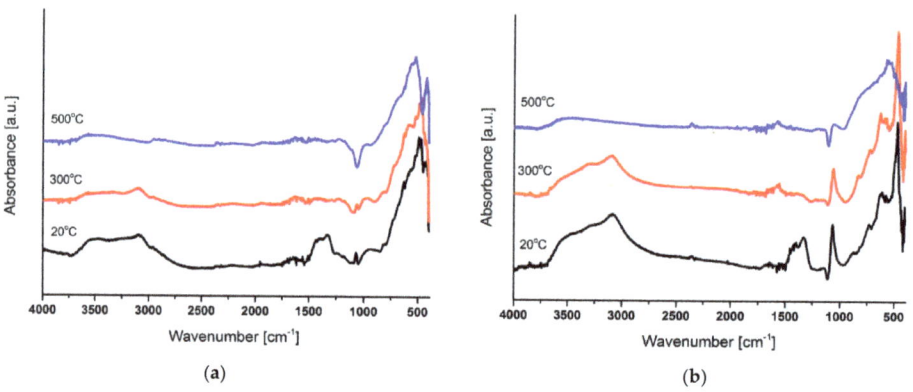

Figure 5. FTIR spectra of Al_2O_3 + graphene coatings with (**a**) G1 graphene and (**b**) G2 graphene, heat treated at different temperatures.

The similarity of the Al_2O_3 + G2 coating spectra (Figure 5b) to that of Al_2O_3 coating in the range of 400–1100 cm^{-1} indicated no effect of the addition of graphene G2 on the changes taking place in the Al_2O_3 matrix during its heat treatment. The conversion of AlOOH into Al_2O_3 was analogous to that in the alumina coating. In the spectra of the Al_2O_3 + G1 coatings treated at 20 °C and 300 °C (Figure 5a), bands in the range of 400–1100 cm^{-1} were less intensive and broadened. This may indicate a lower ordering of Al–O and Al–OH bonds structure caused by the presence of G1 graphene. However, in the spectrum of the coating treated at 500 ° C, further broadening of the bands and their shift towards higher wavenumbers, typical for the structure of the Al_2O_3 coating, were observed.

Similarly as for the Al_2O_3 coating, for both (G1 and G2) types of the analyzed Al_2O_3 + graphene coatings, the intensity of absorption band of hydroxyl groups in the range of 2800–3700 cm^{-1} decreased as the heat treatment temperature increased. It can be noticed that the intensity of this band for coatings with G2 graphene was higher than for coatings with G1 graphene. This was especially visible in the spectra of Al_2O_3 + G2 coatings obtained at lower temperatures: 20 °C and 300 °C. This may be owing to the adsorption of –OH groups on GNPs, whose amount in Al_2O_3 + G2 coatings was greater than in Al_2O_3 + G1 coatings.

3.3. Tribological Properties

To investigate the effects of graphene on the friction and wear of Al_2O_3 + graphene coatings, ball-on-disc tribological tests were performed for Al_2O_3 and Al_2O_3 + graphene coatings deposited on 316 steel substrates.

The results of the friction tests of uncoated 316 steel substrate and Al_2O_3 coatings heat treated at different temperatures are shown in Figure 6. In a test carried out on uncovered steel substrate, the COF quickly reached 0.88 and remained at this level (steady state) until the end of the test. A very similar shape of COF changes was obtained for the Al_2O_3 coating heat treated at 300 °C. The increase of COF over a short sliding distance to values typical for uncovered substrate is the result of rapid destruction of the coating. In the case of coating heat treated at 500 °C, the increase of COF was slightly slower and the destruction process was prolonged, which is probably the result of a better densification of the coating.

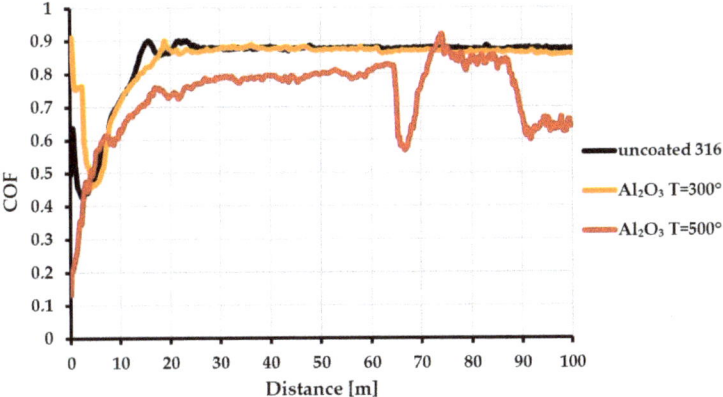

Figure 6. The ball-on-disc tests of uncoated 316 steel and Al_2O_3 coatings heat treated at 300 °C and 500 °C. COF, coefficient of friction.

Figure 7 shows the friction curves of Al_2O_3 + G1 coatings heated at 300 °C and 500 °C. Analyzing changes in the friction coefficient as a function of the friction path, significant differences can be observed. Compared with alumina coatings, no rapid initial increase/decrease of COF can be observed, regardless of the amount of graphene addition or heat treatment temperature. This may indicate a rapid development of a tribofilm containing graphene originated from the upper layer of the coating. The coating containing 1% of G1 graphene and heat treated at 300 °C lost its anti-friction properties over a sliding distance of over 12 m (~500 laps) and was destroyed, which is indicated by a rapid increase in COF to a value close to that of an uncovered steel substrate. Similar behavior was observed for the coating heat treated at 500 °C, except that the loss of anti-friction properties occurred over sliding distance of about 41 m (~1700 laps). In both cases, despite initially very low COF, the G1 graphene content appears to be too low to achieve and maintain a satisfactory self-lubricating effect. A slightly higher durability of the coating heat-treated at 500 °C compared with that of 300 °C may be a consequence of better densification of alumina matrix, which is typical for thermal processing of sol–gel derived coatings [52,58].

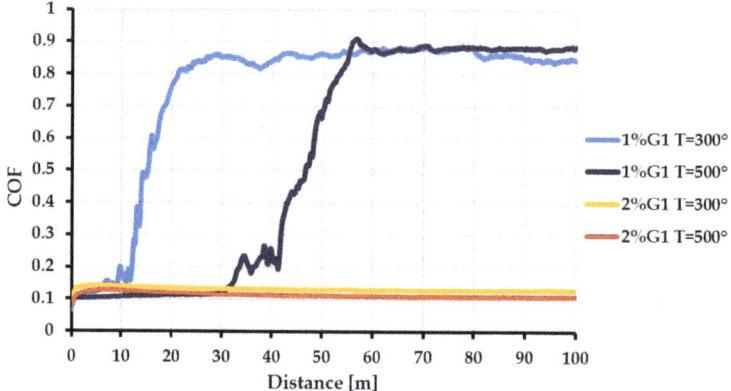

Figure 7. The ball-on-disc test of Al$_2$O$_3$ + G1 graphene coatings heat treated at 300 °C and 500 °C.

Different behavior was observed for coatings containing 2% of G1 graphene. Regardless of the heat treatment temperatures, these coatings showed a practically constant and very stable course of friction curves. The average COF of steady state for coatings heat-treated at 300 °C and 500 °C were 0.13 and 0.11, respectively. This behavior indicates the rapid development and maintenance of a tribofilm containing graphene nanoplatelets, the content of which was sufficient to provide a very low COF until the end of the test. Slightly lower average COF of coating heat-treated at 500 °C may again indicate the beneficial effect of better densification of Al$_2$O$_3$ matrix.

Comparison of the friction curves of coatings containing 1% and 2% of G1 graphene (Figure 7) showed major differences. Coatings with 1% of G1 graphene quickly lost their anti-friction properties (rapid increase in COF) and were destroyed. Despite initially very low COF, the 1% of G1 graphene content appears to be too low to achieve and maintain a satisfactory self-lubricating effect. Meanwhile, coatings containing 2% of G1 graphene showed a low and stable COF to the very end of the friction test distance. This indicates that, regardless of other parameters, the amount of graphene is crucial for proper friction properties of Al$_2$O$_3$ + G1 coatings.

Figure 8 shows friction curves of Al$_2$O$_3$ + G2 coatings heat treated at 300 °C and 500 °C. In the case of a coating containing 1% of G2 graphene and heat treated at 300 °C, the sliding distance before wearing out was 47 m (~1900 laps), while the coating treated at 500 °C showed almost immediate growth of COF towards values above 0.8, characteristic for uncoated steel substrate. Similarly, the coating containing 2% of G2 graphene and heat treated at 300 °C had better durability than the coating treated at 500 °C. The coating with 2% of G2 graphene heat treated at 300 °C revealed low, stable COF of around 0.17 throughout the test, while the coating with 2% of G2 graphene heat treated at 500 °C showed a progressive increase of COF above 0.8 over a sliding distance of 70 m. The observed behavior is the opposite of that for coatings with the addition of graphene G1. Assuming a similar effect of higher heat treatment temperature on the densification of Al$_2$O$_3$ matrix, the difference in the behavior of the Al$_2$O$_3$ + G1 and Al$_2$O$_3$ + G2 coatings may be owing to the shape and spatial orientation of graphene nanoplatelets in the coating matrix. For Al$_2$O$_3$ + G1 graphene coatings, SEM images (Figure 2a,b) indicate parallel arrangement of the GNPs relative to the surface of the substrate and the coating. The shape and spatial arrangement of GNPs seem to be extremely conducive to the formation of graphene tribofilm. In work by Cheng et al. [59], it is proposed that "under high normal stress and localized heat during friction test, graphene sheets from nanoplatelets are pulled under shear and become welded with overlapping areas to form an ultrathin graphene tribofilm". This type of confinement is indicated as crucial in the frictional behavior of graphene, provided the thickness of the GNPs is maintained [60]. Graphene sheets attached to sliding surfaces facilitate their movement because graphene shears easily owing to the weak van der Waals forces at the sliding contact interface [61–63]. This phenomenon can be particularly effective when the GNPs are parallel to the plane of sliding surfaces—in combination with the multilayer structure of GNPs, it would allow to effectively use the slip mechanism proposed in the literature [64–66].

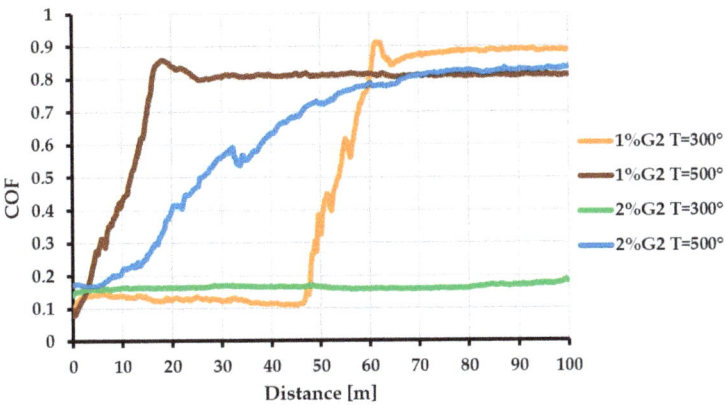

Figure 8. The ball-on-disc test of Al_2O_3 + G2 graphene coatings heat treated at 300 °C and 500 °C.

In the case of G2 graphene, there is a visible tendency to coagulate into small agglomerates (Figure 2d,e) without a clear spatial orientation. Owing to the smaller size, the spatial orientation of graphene flakes inside agglomerates is more random, which can deteriorate the utilization of slip mechanisms between them. As a result, despite the seemingly better dispersion of G2 graphene, its lubrication capabilities are worse than that of G1 graphene.

Another factor that may affect tribological efficiency of G2 graphene in Al_2O_3 + G2 coatings is its size and susceptibility to oxidation during heat treatment. The sol–gel method, based on the aqueous hydrolysis of organometallic precursors, is characterized by the presence of a large amount of hydroxyl groups derived from water absorbed in a coating. During heat treatment, a significant amount of them is still present in the coating. The atmosphere of air and presence of hydroxyl groups create favorable conditions for thermal oxidation of previously reduced graphene (rGO). High temperature oxidation of rGO can occur at a temperature of about 400 °C [67], and the increase in temperature can lead to a loss of lubricating properties of multilayer graphene [68]. Because the oxidation process occurs mainly at the edges of graphene flakes and in their defects, it will be more intense in the case of flakes with a smaller diameter—because of the worse circumference to the flake surface ratio. The smaller size of the G2 graphene flakes may make them more susceptible to this phenomenon, which may hinder or even prevent the formation of tribofilm, and—as a consequence—deteriorate tribological properties of the Al_2O_3 + G2 coating compared with the Al_2O_3 + G1 one.

4. Summary and Conclusions

Composite Al_2O_3 coatings containing 1% and 2% of GNPs were successfully prepared by the sol–gel method. The coatings were of good quality—free of cracks and discontinuities. The morphology of coatings containing G1 graphene revealed a tendency to parallel arrangement of GNP in relation to the coated surface, while in the coatings with smaller size G2 graphene, no spatial organization was observed. GDOES chemical analysis showed lower carbon content and lower thickness of Al_2O_3 + G1 graphene coatings compared with the Al_2O_3 and Al_2O_3 + G2. This is the result of a different dispersion of G1 graphene-flat arrangement of thin nanoplatelets during dip-coating deposition of the coatings.

The graphene nanoplatelets in Al_2O_3 + graphene coatings do not fundamentally change the nature of the chemical transformation of ceramic Al_2O_3 matrix. Regardless of the presence or absence of graphene, these changes rely on elimination of precursor functional groups and OH groups from the structure of the coatings during increasing of the temperature [16]. However, this process appears to vary depending on the size of the graphene nanoplatelets.

Tribological tests conducted on coatings showed different behavior depending on the type and quantity of GNPs. For both (G1 and G2) types of graphene, a content of 1% was insufficient to obtain and/or maintain a self-lubricating effect. Coatings with 2% of graphene had better properties. In the case of Al_2O_3 + G1 coatings, regardless of heat treatment temperatures, the coatings showed low COF (0.11 ÷ 0.13) and were not destroyed until the end of the tests. Al_2O_3 + G2 coatings behaved differently—only the coating heat treated at a lower temperature (300 °C) showed a satisfactory COF (0.17) and durability, while the coating heat treated at 500 °C showed a progressive increase of COF up to 0.8 and above, indicating its destruction. This is most likely associated with the random spatial orientation of G2 GNPs compared with G1, hindering the use of slip mechanisms between its layers, as well as the adverse effect of higher heat treatment temperature, which contributes to faster degradation of G2 graphene lubricating properties.

The results of this study show that, using the sol–gel method, it is possible to produce ceramic composite coatings in which graphene can act as a solid lubricant significantly reducing friction. The observed dependencies between manufacturing parameters (type, amount and dispersion of graphene nanoplatelets, heat treatment temperature) create opportunities to control the properties of graphene containing alumina composite coatings, which will be the subject of further research.

Author Contributions: B.P.: concept, methodology, investigations, data and results analysis, discussion, writing and editing; S.M.: concept, investigations, results analysis, discussion, writing and editing, Y.S.: methodology, investigations, results analysis, discussion, writing; M.S.: investigation, data analysis. All authors have read and agreed to the published version of the manuscript.

Funding: This research received no external funding.

Acknowledgments: The authors acknowledge Anna Sobczyk-Guzenda for recording of FTIR spectra, Krzysztof Jakubowski and Paulina Kowalczyk for GDOES tests, and Marek Klich for tribological investigations.

Conflicts of Interest: The authors declare no conflict of interest.

References

1. Heimann, R.B. Oxide Ceramics: Structure, Technology, and Applications. In *Classic and Advanced Ceramics: From Fundamentals to Applications*; Wiley-VCH Verlag GmbH & Co. KGaA: Weindheim, Germany, 2010; pp. 175–252. ISBN 978-3-527-63017-2.
2. Said, S.; Mikhail, S.; Riad, M. Recent progress in preparations and applications of meso-porous alumina. *Mater. Sci. Energy Technol.* **2019**, *2*, 288–297. [CrossRef]
3. Kim, J.-H.; Yoo, S.-J.; Kwak, D.-H.; Jung, H.-J.; Kim, T.-Y.; Park, K.-H.; Lee, J.-W. Characterization and application of electrospun alumina nanofibers. *Nanoscale Res. Lett.* **2014**, *9*, 44. [CrossRef] [PubMed]
4. Shakhova, I.; Mironov, E.; Azarmi, F.; Safonov, A. Thermo-electrical properties of the alumina coatings deposited by different thermal spraying technologies. *Ceram. Int.* **2017**, *43*, 15392–15401. [CrossRef]
5. Ruan, M.; Wang, J.W.; Liu, Q.L.; Ma, F.M.; Yu, Z.L.; Feng, W.; Chen, Y. Superhydrophobic and anti-icing properties of sol–gel prepared alumina coatings. *Russ. J. Non-Ferr. Met.* **2016**, *57*, 638–645. [CrossRef]
6. Luo, R.; Li, P.; Wei, H.; Chen, H.; Yang, K. Structure and electrical insulation characteristics of plasma-sprayed alumina coatings under pressure. *Ceram. Int.* **2018**, *44*, 6033–6036. [CrossRef]
7. Denes, E.; Barrière, G.; Poli, E.; Lévêque, G. Alumina biocompatibility. *J. Long-Term Eff. Med Implant.* **2018**, *28*, 9–13. [CrossRef]
8. Sequeira, S.; Fernandes, M.H.; Neves, N.; Almeida, M.M. Development and characterization of zirconia–alumina composites for orthopedic implants. *Ceram. Int.* **2017**, *43*, 693–703. [CrossRef]
9. Niedźwiedź, M.; Skoneczny, W.; Bara, M. The Influence of Anodic Alumina Coating Nanostructure Produced on EN AW-5251 Alloy on Type of Tribological Wear Process. *Coatings* **2020**, *10*, 105. [CrossRef]

10. Huang, H.; Qiu, J.; Wei, X.; Sakai, E.; Jiang, G.; Wu, H.; Komiyama, T. Ultra-fast fabrication of porous alumina film with excellent wear and corrosion resistance via hard anodizing in etidronic acid. *Surf. Coat. Technol.* **2020**, *393*, 125767. [CrossRef]
11. Deng, W.; Li, S.; Hou, G.; Liu, X.; Zhao, X.; An, Y.; Zhou, H.; Chen, J. Comparative study on wear behavior of plasma sprayed Al_2O_3 coatings sliding against different counterparts. *Ceram. Int.* **2017**, *43*, 6976–6986. [CrossRef]
12. Hsain, Z.; Zeng, G.; Strandwitz, N.C.; Krick, B.A. Wear behavior of annealed atomic layer deposited alumina. *Wear* **2017**, *372–373*, 139–144. [CrossRef]
13. Lazar, A.-M.; Yespica, W.P.; Marcelin, S.; Pébère, N.; Samélor, D.; Tendero, C.; Vahlas, C. Corrosion protection of 304L stainless steel by chemical vapor deposited alumina coatings. *Corros. Sci.* **2014**, *81*, 125–131. [CrossRef]
14. Lin, J. High rate reactive sputtering of Al_2O_3 coatings by HiPIMS. *Surf. Coat. Technol.* **2019**, *357*, 402–411. [CrossRef]
15. Yushkov, Y.G.; Oks, E.M.; Tyunkov, A.V.; Zolotukhin, D.B. Alumina coating deposition by electron-beam evaporation of ceramic using a forevacuum plasma-cathode electron source. *Ceram. Int.* **2019**, *45*, 9782–9787. [CrossRef]
16. Pietrzyk, B.; Miszczak, S.; Szymanowski, H.; Kucharski, D. Plasma enhanced aerosol–gel deposition of Al_2O_3 coatings. *J. Eur. Ceram. Soc.* **2013**, *33*, 2341–2346. [CrossRef]
17. Tlili, B.; Barkaoui, A.; Walock, M. Tribology and wear resistance of the stainless steel. The sol–gel coating impact on the friction and damage. *Tribol. Int.* **2016**, *102*, 348–354. [CrossRef]
18. Yu, Y.; Zuo, Y.; Zhang, Z.; Wu, L.; Ning, C.; Zuo, C. Al_2O_3 Coatings on Zinc for Anti-Corrosion in Alkaline Solution by Electrospinning. *Coatings* **2019**, *9*, 692. [CrossRef]
19. Li, Y.; Chen, M.; Li, W.; Wang, Q.; Wang, Y.; You, C. Preparation, characteristics and corrosion properties of α-Al_2O_3 coatings on 10B21 carbon steel by micro-arc oxidation. *Surf. Coat. Technol.* **2019**, *358*, 637–645. [CrossRef]
20. Lu, D.; Huang, Y.; Jiang, Q.; Zheng, M.; Duan, J.; Hou, B. An approach to fabricating protective coatings on a magnesium alloy utilising alumina. *Surf. Coat. Technol.* **2019**, *367*, 336–340. [CrossRef]
21. Nofz, M.; Dörfel, I.; Sojref, R.; Wollschläger, N.; Mosquera-Feijoo, M.; Schulz, W.; Kranzmann, A. Thin Sol–Gel Alumina Coating as Protection of a 9% Cr Steel Against Flue Gas Corrosion at 650 °C. *Oxid. Met.* **2018**, *89*, 453–470. [CrossRef]
22. Balcaen, Y.; Radutoiu, N.; Alexis, J.; Beguin, J.-D.; Lacroix, L.; Samélor, D.; Vahlas, C. Mechanical and barrier properties of MOCVD processed alumina coatings on TI6Al4V titanium alloy. *Surf. Coat. Technol.* **2011**, *206*, 1684–1690. [CrossRef]
23. Hübert, T.; Schwarz, J.; Oertel, B. Sol-gel alumina coatings on stainless steel for wear protection. *J. Sol-Gel Sci. Technol.* **2006**, *38*, 179–184. [CrossRef]
24. Lampke, T.; Meyer, D.; Alisch, G.; Wielage, B.; Pokhmurska, H.; Klapkiv, M.; Student, M. Corrosion and wear behavior of alumina coatings obtained by various methods. *Mater. Sci.* **2011**, *46*, 591–598. [CrossRef]
25. Lorenzo-Martin, C.; Ajayi, O.O.; Hartman, K.; Bhattacharya, S.; Yacout, A. Effect of Al_2O_3 coating on fretting wear performance of Zr alloy. *Wear* **2019**, *426–427*, 219–227. [CrossRef]
26. Singh, V.P.; Sil, A.; Jayaganthan, R. A study on sliding and erosive wear behaviour of atmospheric plasma sprayed conventional and nanostructured alumina coatings. *Mater. Des.* **2011**, *32*, 584–591. [CrossRef]
27. Marsal, A.; Ansart, F.; Turq, V.; Bonino, J.P.; Sobrino, J.M.; Chen, Y.M.; Garcia, J. Mechanical properties and tribological behavior of a silica or/and alumina coating prepared by sol-gel route on stainless steel. *Surf. Coat. Technol.* **2013**, *237*, 234–240. [CrossRef]
28. Sarafoglou, C.I.; Pantelis, D.I.; Beauvais, S.; Jeandin, M. Study of Al_2O_3 coatings on AISI 316 stainless steel obtained by controlled atmosphere plasma spraying (CAPS). *Surf. Coat. Technol.* **2007**, *202*, 155–161. [CrossRef]
29. Riedl, A.; Schalk, N.; Czettl, C.; Sartory, B.; Mitterer, C. Tribological properties of Al_2O_3 hard coatings modified by mechanical blasting and polishing post-treatment. *Wear* **2012**, *289*, 9–16. [CrossRef]
30. Hübert, T.; Svoboda, S.; Oertel, B. Wear resistant alumina coatings produced by a sol–gel process. *Surf. Coat. Technol.* **2006**, *201*, 487–491. [CrossRef]
31. Piwoński, I.; Soliwoda, K. The effect of ceramic nanoparticles on tribological properties of alumina sol–gel thin coatings. *Ceram. Int.* **2010**, *36*, 47–54. [CrossRef]

32. Peng, Z.J.; Cheng, T.; Nie, X.Y. MoS$_2$/Al$_2$O$_3$ Composite Coatings on A$_{356}$ Alloy for Friction Reduction. *Adv. Mater. Res.* **2012**, *496*, 488–492. [CrossRef]
33. Kim, S. Effects of Solid Lubricants on Tribological Behaviour of APSed Al$_2$O$_3$-ZrO$_2$ Composite Coatings. *Taehan-Kŭmsok-Hakhoe-chi J. Korean Inst. Met. Mater.* **2014**, *52*, 263–270. [CrossRef]
34. Pietrzyk, B.; Miszczak, S.; Kaczmarek, Ł.; Klich, M. Low friction nanocomposite aluminum oxide/MoS2 coatings prepared by sol-gel method. *Ceram. Int.* **2018**, *44*, 8534–8539. [CrossRef]
35. Jambagi, S.C.; Bandyopadhyay, P.P. Improvement in Tribological Properties of Plasma-Sprayed Alumina Coating upon Carbon Nanotube Reinforcement. *J. Mater. Eng. Perform.* **2019**, *28*, 7347–7358. [CrossRef]
36. Hentour, K.; Marsal, A.; Turq, V.; Weibel, A.; Ansart, F.; Sobrino, J.-M.; Chen, Y.M.; Garcia, J.; Cardey, P.-F.; Laurent, C. Carbon nanotube/alumina and graphite/alumina composite coatings on stainless steel for tribological applications. *Mater. Today Commun.* **2016**, *8*, 118–126. [CrossRef]
37. Keshri, A.K.; Huang, J.; Singh, V.; Choi, W.; Seal, S.; Agarwal, A. Synthesis of aluminum oxide coating with carbon nanotube reinforcement produced by chemical vapor deposition for improved fracture and wear resistance. *Carbon* **2010**, *48*, 431–442. [CrossRef]
38. Hentour, K.; Turq, V.; Weibel, A.; Ansart, F.; Sobrino, J.-M.; Garcia, J.; Cardey, P.-F.; Laurent, C. Dispersion of graphite flakes into boehmite sols for the preparation of bi-layer-graphene/alumina coatings on stainless steel for tribological applications. *J. Eur. Ceram. Soc.* **2019**, *39*, 1304–1315. [CrossRef]
39. Marcinauskas, L.; Mathew, J.S.; Milieška, M.; Aikas, M.; Kalin, M. Effect of graphite concentration on the tribological performance of alumina coatings. *J. Alloy. Compd.* **2020**, *827*, 154135. [CrossRef]
40. Murray, J.W.; Rance, G.A.; Xu, F.; Hussain, T. Alumina-graphene nanocomposite coatings fabricated by suspension high velocity oxy-fuel thermal spraying for ultra-low-wear. *J. Eur. Ceram. Soc.* **2018**, *38*, 1819–1828. [CrossRef]
41. Yin, H.; Dai, Q.; Hao, X.; Huang, W.; Wang, X. Preparation and tribological properties of graphene oxide doped alumina composite coatings. *Surf. Coat. Technol.* **2018**, *352*, 411–419. [CrossRef]
42. Da, B.; Rongli, X.; Yongxin, G.; Yaxuan, L.; Aradhyula, T.V.; Yongwu, Z.; Yongguang, W. Tribological behavior of graphene reinforced chemically bonded ceramic coatings. *Ceram. Int.* **2020**, *46*, 4526–4531. [CrossRef]
43. Tiwari, S.K.; Sahoo, S.; Wang, N.; Huczko, A. Graphene research and their outputs: Status and prospect. *J. Sci. Adv. Mater. Devices* **2020**, *5*, 10–29. [CrossRef]
44. Yan, Y.; Gong, J.; Chen, J.; Zeng, Z.; Huang, W.; Pu, K.; Liu, J.; Chen, P. Recent Advances on Graphene Quantum Dots: From Chemistry and Physics to Applications. *Adv. Mater.* **2019**, *31*, 1808283. [CrossRef] [PubMed]
45. Lawal, A.T. Graphene-based nano composites and their applications. A review. *Biosens. Bioelectron.* **2019**, *141*, 111384. [CrossRef] [PubMed]
46. Nag, A.; Mitra, A.; Mukhopadhyay, S.C. Graphene and its sensor-based applications: A review. *Sens. Actuators A Phys.* **2018**, *270*, 177–194. [CrossRef]
47. Mohan, V.B.; Lau, K.; Hui, D.; Bhattacharyya, D. Graphene-based materials and their composites: A review on production, applications and product limitations. *Compos. Part. B Eng.* **2018**, *142*, 200–220. [CrossRef]
48. Penkov, O.; Kim, H.-J.; Kim, H.-J.; Kim, D.-E. Tribology of graphene: A review. *Int. J. Precis. Eng. Manuf.* **2014**, *15*, 577–585. [CrossRef]
49. Kasar, A.K.; Menezes, P.L. Synthesis and recent advances in tribological applications of graphene. *Int. J. Adv. Manuf. Technol.* **2018**, *97*, 3999–4019. [CrossRef]
50. Brinker, C.J.; Scherer, G.W. *Sol.-Gel Science: The Physics and Chemistry of Sol.-Gel Processing*; Academic Press Inc.: San Diego, CA, USA, 1990; ISBN 978-0-12-134970-7.
51. Levy, D.; Zayat, M. *The Sol.-Gel Handbook: Synthesis, Characterization and Applications*, 1st ed.; Wiley-VCH Verlag GmbH & Co. KGaA: Weindheim, Germany, 2015; ISBN 978-3-527-67084-0.
52. Brinker, C.J.; Ashley, C.S.; Cairncross, R.A.; Chen, K.S.; Hurd, A.J.; Reed, S.T.; Samuel, J.; Schunk, P.R.; Schwartz, R.W.; Scotto, C.S. Sol—gel derived ceramic films—fundamentals and applications. In *Metallurgical and Ceramic Protective Coatings*; Stern, K.H., Ed.; Springer: Dordrecht, The Netherlands, 1996; pp. 112–151. ISBN 978-94-009-1501-5.
53. Kobayashi, Y.; Ishizaka, T.; Kurokawa, Y. Preparation of alumina films by the sol-gel method. *J. Mater. Sci.* **2005**, *40*, 263–283. [CrossRef]
54. Kumar, A.; Singh, R.; Bahuguna, G. Thin Film Coating through Sol-Gel Technique. *Res. J. Chem. Sci.* **2016**, *6*, 65.

55. Schneller, T.; Waser, R.; Kosec, M.; Payne, D. *Chemical Solution Deposition of Functional Oxide Thin Films*; Springer: Wien, Austria, 2013; p. 796. ISBN 978-3-211-99311-8.
56. Manoratne, C.H.; Rosa, S.R.D.; Kottegoda, I.R.M. RD-HTA, UV Visible, FTIR and SEM Interpretation of Reduced Graphene Oxide Synthesized from High Purity Vein Graphite. *Mater. Sci. Res. India* **2017**, *14*, 19–30. [CrossRef]
57. Kaczmarek, Ł.; Warga, T.; Makowicz, M.; Kyzioł, K.; Bucholc, B.; Majchrzycki, Ł. The Influence of the Size and Oxidation Degree of Graphene Flakes on the Process of Creating 3D Structures during Its Cross-Linking. *Materials* **2020**, *13*, 681. [CrossRef] [PubMed]
58. Jing, C.; Zhao, X.; Zhang, Y. Sol–gel fabrication of compact, crack-free alumina film. *Mater. Res. Bull.* **2007**, *42*, 600–608. [CrossRef]
59. Cheng, Z.; Andy, N.; Arvind, A. Ultrathin graphene tribofilm formation during wear of Al_2O_3–graphene composites. *Nanomater. Energy* **2016**, *5*, 1–9.
60. Kumar, P.; Wani, M. Synthesis and tribological properties of graphene: A review. *J. Tribol.* **2017**, *13*, 36–71.
61. Berman, D.; Erdemir, A.; Sumant, A.V. Few layer graphene to reduce wear and friction on sliding steel surfaces. *Carbon* **2013**, *54*, 454–459. [CrossRef]
62. Pu, J.; Wan, S.; Zhao, W.; Mo, Y.; Zhang, X.; Wang, L.; Xue, Q. Preparation and Tribological Study of Functionalized Graphene–IL Nanocomposite Ultrathin Lubrication Films on Si Substrates. *J. Phys. Chem. C* **2011**, *115*, 13275–13284. [CrossRef]
63. Liang, H.; Bu, Y.; Zhang, J. Graphene Oxide Film as Solid Lubricant. *ACS Appl. Mater. Interfaces* **2013**, *5*, 6369–6375. [CrossRef]
64. Feng, X.; Kwon, S.; Park, J.Y.; Salmeron, M. Superlubric Sliding of Graphene Nanoflakes on Graphene. *ACS Nano* **2013**, *7*, 1718–1724. [CrossRef]
65. Liu, Y.; Grey, F.; Zheng, Q. The high-speed sliding friction of graphene and novel routes to persistent superlubricity. *Sci. Rep.* **2014**, *4*, 4875. [CrossRef]
66. Li, J.; Li, J.; Luo, J. Superlubricity of Graphite Sliding against Graphene Nanoflake under Ultrahigh Contact Pressure. *Adv. Sci.* **2018**, *5*, 1800810. [CrossRef] [PubMed]
67. Feng, Y.; Wang, B.; Li, X.; Ye, Y.; Ma, J.; Liu, C.; Zhou, X.; Xie, X. Enhancing thermal oxidation and fire resistance of reduced graphene oxide by phosphorus and nitrogen co-doping, Mechanism and kinetic analysis. *Carbon* **2019**, *146*, 650–659. [CrossRef]
68. Xu, Z.; Zhang, Q.; Jing, P.; Zhai, W. High-Temperature Tribological Performance of TiAl Matrix Composites Reinforced by Multilayer Graphene. *Tribol. Lett.* **2015**, *58*, 3. [CrossRef]

© 2020 by the authors. Licensee MDPI, Basel, Switzerland. This article is an open access article distributed under the terms and conditions of the Creative Commons Attribution (CC BY) license (http://creativecommons.org/licenses/by/4.0/).

Article

Titanium Dioxide Coatings Doubly-Doped with Ca and Ag Ions as Corrosion Resistant, Biocompatible, and Bioactive Materials for Medical Applications

Barbara Burnat [1,*], Patrycja Olejarz [2], Damian Batory [3], Michal Cichomski [4], Marta Kaminska [5] and Dorota Bociaga [6]

1. Department of Inorganic and Analytical Chemistry, Faculty of Chemistry, University of Lodz, 12 Tamka St., 91-403 Lodz, Poland
2. Department of Environmental Chemistry, Faculty of Chemistry, University of Lodz, 163 Pomorska St., 90-236 Lodz, Poland; patrycja.olejarz@o2.pl
3. Department of Vehicles and Fundamentals of Machine Design, Faculty of Mechanical Engineering, Lodz University of Technology, 1/15 Stefanowskiego St., 90-924 Lodz, Poland; damian.batory@p.lodz.pl
4. Department of Materials Technology and Chemistry, Faculty of Chemistry, University of Lodz, 163 Pomorska St., 90-236 Lodz, Poland; michal.cichomski@chemia.uni.lodz.pl
5. Division of Biophysics, Faculty of Mechanical Engineering, Lodz University of Technology, 1/15 Stefanowskiego St., 90-924 Lodz, Poland; marta.kaminska@p.lodz.pl
6. Division of Biomedical Engineering and Functional Materials, Faculty of Mechanical Engineering, Lodz University of Technology, 1/15 Stefanowskiego St., 90-924 Lodz, Poland; dorota.bociaga@p.lodz.pl
* Correspondence: barbara.burnat@chemia.uni.lodz.pl; Tel.: +48-426-355-800

Received: 27 December 2019; Accepted: 8 February 2020; Published: 13 February 2020

Abstract: The aim of this study was to develop a multifunctional biomedical coating that is highly corrosion resistant, biocompatible, and reveals the bioactive properties. For that purpose, titanium dioxide coatings doubly-doped with Ca and Ag ions were deposited by dip-coating onto M30NW biomedical steel. The influence of different ratios of Ca and Ag dopants on morphology, surface structure, corrosion resistance, bioactivity, wettability, and biological properties of TiO_2-based sol-gel coatings was studied and discussed. Comprehensive measurements were performed including atomic force microscopy (AFM), scanning electron microscopy (SEM), X-ray diffraction (XRD), X-ray reflectivity (XRR), corrosion tests, immersion test, contact angle, as well as biological evaluation. The obtained results confirmed that anatase-based coatings containing Ca and Ag ions, independently of their molar ratio in the coating, are anticorrosive, hydrophilic, and bioactive. The results of the biological evaluation indicated that investigated coatings are biocompatible and do not reduce the proliferation ability of the osteoblasts cells.

Keywords: sol-gel coating; corrosion resistance; cells viability; biocompatibility; hydrophilic coating

1. Introduction

Metallic biomaterials as materials implanted into a living system must fulfill stringent requirements, including good corrosion resistance, mechanical properties, and biocompatibility. Nowadays, studies on orthopedic biomaterials are focused mainly on enhancing their bioactivity and giving them new properties (e.g., antibacterial properties). It is well known that antimicrobial biomaterials prevent post-operative infections by reducing the ability of adhesion and permanent attachment of microorganisms and thus the development of biofilm, which is the main reason for the infections. Within the numerous methods used for the preparation of biomaterials with the desired properties, covering their surfaces with functionalized coatings appears to be particularly interesting with high capabilities. Titanium dioxide is one of most interesting materials, which can be applied as a coating of biomaterial.

It is a hard, corrosion resistant, and well biocompatible material with UV-induced hydrophilicity [1]. Literature studies show that several techniques for titanium dioxide coatings preparation, such as thermal and electrochemical oxidation (in case of titanium substrates) [2,3], magnetron sputtering [4,5], chemical vapor deposition [6–9], and sol-gel [2,7,10–13] have been investigated. But one of the most frequently used methods is sol-gel due to its high application potential [14]. The sol-gel TiO_2 coating was proven to be a versatile platform for surface functionalization of stainless steels and other biomedical metals. The desired properties of the biomaterial can be obtained by careful control of sol-gel reaction conditions or by the use of suitable additives. One of the possibilities is to improve biocompatibility, bioactivity (related to osseointegration), and antibacterial properties of biomaterials by means of ion doping procedure. In the literature, there are reports about enhanced bioactivity of TiO_2 coating by means of incorporation of Ca [15,16], Mg [17] and Sr [18,19] ions. Our previous papers [20,21] also confirmed improved bioactivity of Ca-doped TiO_2 sol-gel coatings. It was also proved that, in an analogous manner, the antibacterial properties may be achieved by doping of bactericidal molecules that in most cases have antibacterial properties per se, namely Ag ions/nanoparticles [22–24], Cu ions [25,26], and Zn ions [27]. Several research studies [28–30] have shown that F doping enhances the antibacterial properties of TiO_2 too. However, studies on the biocompatibility of the fluoride-modified surface have contradictory and inconclusive results. A very interesting idea is to obtain a multifunctional coating in a co-doping procedure [31–33]. In the present study, we applied the possibility of incorporating more than one modifier into the sol solution in order to achieve a multifunctional biomedical coating with bioactive and antibacterial behavior, and what is the most important exhibiting corrosion protection capability towards biomedical stainless steels. We intended to combine anticorrosion properties of titanium dioxide with the bioactive effect of Ca ions and the potential antibacterial effect of Ag ions widely reported in the literature. In the literature, there are reports on TiO_2 coatings co-doped with Ca and Ag ions [34,35], but those coatings were applied to titanium rather than steel substrates, and deposition methods other than sol-gel were used. However, since the method chosen for coating preparation or its modification determines the final properties as well as the applicability of the biomaterial, we have to remember that the improvement of a certain feature or function of a biomaterial as a result of surface modification may be accompanied by deterioration of other biomaterial features. Yetim et al. reported, for example, that, using plasma nitriding, it was possible to improve the wear resistance of AISI 316L steel, but the nitriding treatment did not bring the expected improvement in the corrosion resistance of the AISI 316L steel, it was even worse [36]. Junping et al. stated that the Ce-modified 316L steel exhibits the hormesis effect against *Staphylococcus aureus* (the higher the Ce content, the better the antibacterial efficacy), but it is difficult to simultaneously obtain good corrosion resistance, antibacterial performance, and processability [37]. Based on these literature examples, it is clear how important it is to control the impact of carried out modifications on corrosion resistance, especially in the case of biomaterials, as it determines their biocompatibility.

The aim of this study was to develop a multifunctional biomedical sol-gel coating that is highly corrosion resistant, biocompatible, and reveals the bioactive properties. For that purpose, titanium dioxide coatings doubly-doped with Ca and Ag ions were deposited by dip-coating onto M30NW biomedical steel, and subsequently annealed at 450 °C in air. Our previous studies have shown that under these thermal oxidation conditions an effective crystallization of titanium dioxide occurs, and formed anatase exhibits good corrosion protection ability in case of biomedical steels [21]. In the presented study, the influence of different ratios of Ca and Ag dopants on morphology, surface structure, corrosion resistance, bioactivity, wettability, and biological properties of the TiO_2-based coating was investigated.

2. Materials and Methods

2.1. Samples' Preparation

Commercially available M30NW biomedical alloy (AUBERT & DUVAL, Paris, France), with composition and properties specified by X4CrNiMnMo21-9-4 standard, was used as a substrate. The alloy plates (22 mm in diameter) were ground with SiC papers down to 1200 grits, polished with alumina slurry (0.3 μm), ultrasonically cleaned in distilled water, etched in acid mixture, passivated in boiling distilled water, rinsed with ethanol, and dried with argon according to the procedure described elsewhere [38].

In this study several sols were used for surface modification of M30NW alloy: TiO_2 sol without dopants, sol doped with calcium ions Ca^{2+}, sol doped with silver ions Ag^+, as well as sols containing both Ca^{2+} and Ag^+ ions in different molar ratios (Ca/Ag 3:1, 1:1, 1:3).

All sols were synthesized using the sol-gel method, in which titanium tetrabutoxide (TiBut, $Ti[O(CH_2)_3CH_3]_4$, 97%, Sigma-Aldrich) was used as a precursor for titania, ethanol (EtOH, C_2H_5OH, pure p.a., 96%, POCh) as a solvent, nitric acid (HNO_3, pure p.a., 65%, POCh) as a catalyst, calcium nitrate ($Ca(NO_3)_2$ $4H_2O$, pure, CHEMPUR) and silver nitrate ($AgNO_3$, pure p.a., POCh) solutions at a concentration of 2.056 mol/L as dopant sources. The composition of all sol solutions used in this study is presented in Table 1.

Table 1. Composition of sol solutions used for surface modification of M30NW alloy.

Coating	TiBut [mL]	EtOH [mL]	HNO_3 [mL]	H_2O [mL]	$Ca(NO_3)_2$ [mL]	$AgNO_3$ [mL]	Ca/Ag Ratio
TiO_2	7	20	0.64	0.5	–	–	–
Ca_TiO_2	7	20	0.64	–	0.500	–	–
75Ca25Ag_TiO_2	7	20	0.64	–	0.375	0.125	3:1
50Ca50Ag_TiO_2	7	20	0.64	–	0.250	0.250	1:1
25Ca75Ag_TiO_2	7	20	0.64	–	0.125	0.375	1:3
Ag_TiO_2	7	20	0.64	–	–	0.500	–

A titania-based coating (as a single layer) was applied onto the alloy surface with the dip-coating technique using a DCMono 75 dip-coater (NIMA Technology Ltd., Coventry, UK). The substrate was immersed in the sol for 30 s and withdrawn at a speed of 20 mm/min. Such modified alloy samples were dried in the oven at 100 °C for 2 h and then annealed at 450 °C for 1 h.

2.2. Surface Characterization

Each type of sample was characterized in terms of surface properties. A metallographic microscope MMT 800 BT (Mikrolab, Lublin, Poland) was used for preliminary assessment of the sol-gel coatings' quality, including detection of cracks and defects. An atomic force microscope (AFM) Dimension Icon (Bruker, Santa Barbara, CA, USA) was applied for investigation of surface topography and roughness of the prepared coatings within a scan size of 1 μm × 1 μm. The AFM measurements were made in tapping mode using standard silicon probes (TESPA, Bruker AFM Probes, Camarillo, CA, USA). The morphology of the samples was observed using a field emission scanning electron microscope (FE-SEM, FEI Nova NanoSEM 450 with EDS analyzer, Thermo Fisher Scientific, Hillsbro, OR, USA), operating with an accelerating voltage of 15 kV. The phase composition and thickness of prepared coatings were identified using an Empyrean X-ray diffractometer (XRD, PANalytical, Malvern, UK) working with Cu Kα radiation (λ = 0.15418 nm). The phase analysis was carried out using GIXRD (grazing incidence X-ray diffraction) mode with an incident beam angle of 0.3°, whereas the thickness was estimated with the use of X-ray reflectivity method (XRR). Further data processing was performed using HighScore Plus with ICDD PDF 4+ Database and X'Pert Reflectivity software with Fourier transform analysis, respectively.

2.3. Corrosion Tests

The anticorrosion ability of prepared coatings was evaluated by electrochemical measurements in phosphate buffered saline (PBS, NaCl 8.0 g/L, KH_2PO_4 0.2 g/L, $Na_2HPO_4 \cdot 12\,H_2O$ 2.9 g/L, KCl 0.2 g/L, pH 7.4) solution using a PGSTAT 30 potentiostat-galvanostat (EcoChemie Autolab, Utrecht, The Netherlands). All electrochemical experiments were performed at 37 °C, similar to human body temperature. Degassing of the electrolyte was achieved by argon bubbling through the solution. A conventional three-electrode cell was used with a platinum gauze as a counter electrode, a saturated calomel electrode (SCE, E = 0.236 V_{SHE}) as a reference, and sample with an exposed area of 0.64 cm^2 as a working electrode.

In order to establish the corrosion potential E_{cor}, each sample was kept in PBS solution (under open circuit conditions) for 2000 s. The linear polarization measurements were performed in a scanning range of ±20 mV versus E_{cor} potential, with a scan rate of 0.166 mV/s. Potentiodynamic polarization tests were conducted with a scan rate of 1 mV/s from the initial potential of −200 mV versus E_{cor} to the potential at which current density of 5 mA/cm^2 was reached, then, the potential sweep was reversed and the backward branch was registered up to the initial potential. The surface morphology of the samples after potentiodynamic polarization was analyzed using scanning electron microscopy in order to determine the type and scale of corrosion damage.

The results of linear polarization measurements and potentiodynamic polarization tests were analyzed using CorrView software (Scribner Associates Inc., Southern Pines, NC, USA) and several corrosion parameters were determined: Polarization resistance, R_p; corrosion rate, CR; pitting potential, E_{pit}; and repassivation potential, E_{rep}. Triplicate measurements were conducted to check the reproducibility of the results. Each data point presented here is given as mean ± standard deviation (SD). All the potentials reported here are with respect to a saturated calomel electrode.

2.4. Immersion Tests

To evaluate the bioactivity (as apatite formation ability) of the M30NW samples with prepared titania-based coatings, the immersion tests in a simulated body fluid (SBF) solution were carried out according to the procedure reported by Kokubo [39]. The samples were immersed in SBF (at 37 °C, similar to human body temperature) for 28 days, the solution was renewed every week. Then the SEM-EDS technique was employed to characterize the morphology and surface composition of samples. Based on SEM-EDS results, their ability to apatite formation was evaluated.

2.5. Wettability

Measurements of contact angle of M30NW alloy samples modified with TiO_2-based coatings were carried out using the DSA25 Drop Shape Analyzer goniometer- (Krüss GmbH, Hamburg, Germany). Each time, the measurement was performed for a minimum of three drops of water at an ambient temperature of approximately 20 ± 2 °C. The amount of deionized water drops applied by microsyringe was 5 µL. The values of the contact angle were determined using Advance software.

The surface free energy values were calculated based on two polar liquids-water and glycerine, and one apolar liquid-diiodomethane. Calculations were performed using van Oss Chaudhury–Good method.

2.6. Biological Evaluation (Cell Viability Assays)

Before biological evaluation, all samples were cleaned in ethanol and ultrapure water (0.055 µS/cm) for 10 min using ultrasonic cleaner. Then the steam sterilization was performed (121 °C, 31 min) using an autoclave. In order to conduct the biocompatibility assessment of the samples, the cell viability and proliferation assays were conducted. The human osteoblast cell line Saos-2 (ATCC, Manassas, VA, USA) was selected as a biological material for this purpose. Saos-2 cells were grown in McCoy's 5A medium (ATCC, Manassas, VA, USA) containing 15% fetal bovine serum (Biological Industries),

100 units/mL penicillin and 50 µg/mL streptomycin (Biological Industries). Cells were cultured in standard conditions (37 °C, humidified atmosphere of 5% CO_2 in air) and medium was replaced every 2–3 days (75% confluence). Cells were used between passages 5 and 8.

For the evaluation of proliferation and cytotoxicity marking method, a "live/dead" test (Viability/Cytotoxicity Kit, Molecular Probes) was applied. Cells were seeded at 6×10^4 cells/mL/well/sample in 2 mL of McCoy's 5A medium (ATCC, Manassas, VA, USA) and cultured for 48 h. After that time, a mixture of two fluorescent dyes was used. One of the fluorescent dyes within the live cells produces an intense uniform green fluorescence and the second one, when the membranes are damaged, penetrates the cells and binds to nucleic acids, thereby producing a bright red fluorescence in dead cells. The samples were examined in a fluorescence microscope Olympus GX 71 equipped with a digital camera (DP70).

The obtained results were analyzed by one-way ANOVA analysis with a significance level of $p < 0.05$. Statistical analysis was performed using OriginPro 9 software.

3. Results and Discussion

3.1. Surface Characterization

All prepared TiO_2-based coatings were blue in color, homogeneous, without any cracks on the surface, and they exhibited good adhesion to the substrate. Surface characterization carried out with scanning electron microscopy revealed fine crystalline structure of all TiO_2-based coatings (results not shown). In case of coatings doped with silver ions, SEM analysis revealed small white points on the surface, of which the amount was increasing with increasing concentration of silver. The SEM method does not allow conclusions to be drawn about the convexity or concavity of the surface elements, thus the topography of these white points was analyzed by atomic force microscopy. In addition, the AFM analysis made it possible to determine the roughness (by R_q parameter) of the synthesized coatings. Figure 1 presents the general view of the coated samples, and AFM images (scan sizes of 5 µm × 5 µm and 1 µm × 1 µm) for all types of coatings.

The AFM results (AFM 2D images and values of R_q) presented in Figure 1 are in good agreement with SEM results. For every sample, the coating is uniform and it reflects the topography of the substrate regardless of the coating composition. Based on AFM images, it can be observed that coatings are applied even inside the surface scratches. In addition, in case of coatings doped with calcium and silver ions, especially with the increasing amount of silver ions, holes appear on the surface of the coatings. These holes correspond to the white points observed on the SEM images. They are the result of the thermal decomposition of calcium nitrate and silver nitrate used in doping procedure. According to "CRC Handbook of Chemistry and Physics" edited by Lide [40], both nitrates undergo decomposition during heat treatment, but at different temperatures. Calcium nitrate tetrahydrate decomposes at a temperature of 132 °C, but this decomposition is not total, it only involves the removal of water molecules. The total thermal decomposition of alkaline earth metal nitrates leading to the formation of nitrogen dioxide undergoes at temperatures higher than 500 °C. Whereas, in the case of silver nitrate, such total decomposition undergoes at a temperature of 444 °C according to Equation (1):

$$2AgNO_3 \rightarrow 2Ag + 2NO_2 \uparrow + O_2 \uparrow. \tag{1}$$

The presence of these holes (pores) results in different coating roughness. Coatings with an increasing amount of silver ions are characterized by a greater surface development (higher R_q).

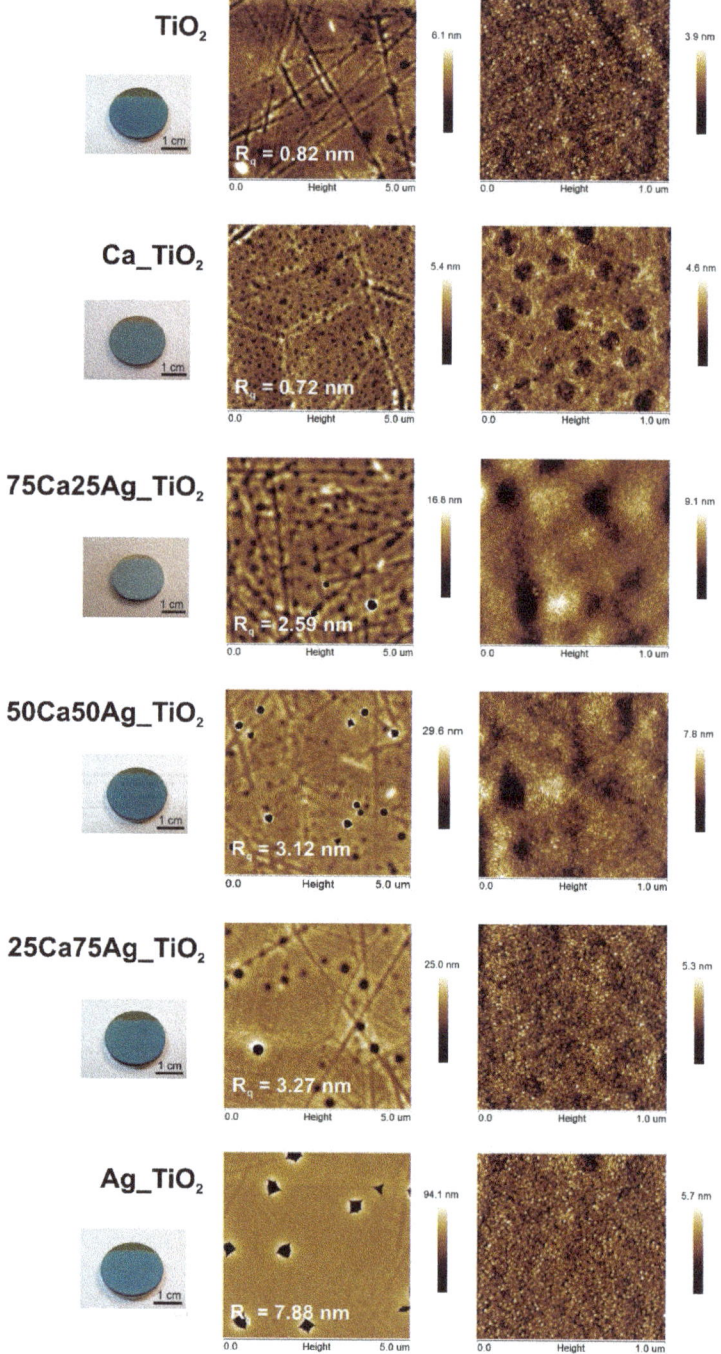

Figure 1. The general view of the coated samples, and atomic force microscopy (AFM) images (scan sizes of 5 μm × 5 μm and 1 μm × 1 μm) for all types of coatings.

The same coloration of the coatings implies similarity in the coating thickness. According to Velten et al. [2], the thickness of TiO_2 coatings that are blue in color should be in the range of 50–80 nm. The verification of this statement was performed via XRR analyses. Values of thickness of the investigated coatings were determined based on the Fourier transform analysis of the registered X-ray reflectivity curves. The obtained XRR results are presented in Table 2.

Table 2. The thickness of TiO_2-based coatings doped with Ca and Ag ions in different molar ratios.

Coating	Thickness/nm
TiO_2	77 ± 4
Ca_TiO_2	74 ± 4
$75Ca25Ag\_TiO_2$	80 ± 4
$50Ca50Ag\_TiO_2$	75 ± 4
$25Ca75Ag\_TiO_2$	77 ± 4
Ag_TiO_2	78 ± 4

The determined values are in good agreement with the literature-based predictions. Furthermore, the analysis of the results allowed to conclude that if the constancy of the sol composition (the ratio of individual reagents in doping procedure) is maintained, then the doping procedure does not significantly affect the thickness of the sol-gel coating.

The phase composition of the investigated coatings was determined by X-ray diffraction method. Figure 2 shows the XRD patterns obtained for TiO_2-based coatings. The results reveal that every single coating exhibits the anatase structure of TiO_2 (Ref. 00-064-0863). This is confirmed by peaks centered at 2theta, 25.37°, broad peak being a superposition of three peaks (centered at 2theta equal to 36.93°, 37.96°, 38.64°), 48.06°, 54.02° and 55.03° (marked with asterisks on the chart). The comparison of the intensity of the peaks for particular coatings shows positive influence of silver onto the crystallization process of titanium dioxide. The most intensive and well defined peaks were registered for TiO_2 coatings with the highest concentration of silver. In the case of calcium, no noticeable difference between TiO_2 and Ca_TiO_2 XRD spectra was observed, which means the incorporated Ca does not alter the crystallization process of anatase. Therefore, it can be stated that silver promotes the crystallization of titanium dioxide in the form of anatase. Such a finding corresponds to the report of García-Serrano et al. [41].

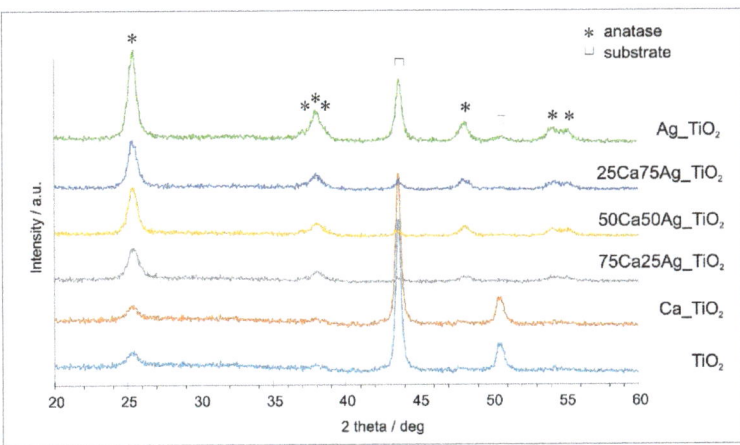

Figure 2. XRD patterns for TiO_2-based coatings doped with Ca and Ag ions in different molar ratios.

3.2. Corrosion Tests

Anticorrosion properties of TiO$_2$-based sol-gel coatings were determined via electrochemical methods based on polarization near the corrosion potential and polarization in wide anodic range. Such measurements allowed for the evaluation of the resistance of the samples against general and pitting corrosion in PBS solution.

The linear polarization measurements performed in a narrow scanning range (± 20 mV vs. E$_{cor}$), allowed for the calculation of the values of corrosion rate, CR, based on determined polarization resistance, R$_p$, values (according to the assumptions of standard ASTM G102-89 [42]). The mean values of E$_{cor}$, R$_p$, and CR with standard deviations for all investigated TiO$_2$-based coatings are given in Figure 3. In order to confirm the protective properties of TiO$_2$-based coatings, the results for uncoated M30NW alloy substrate are also included in Figure 3.

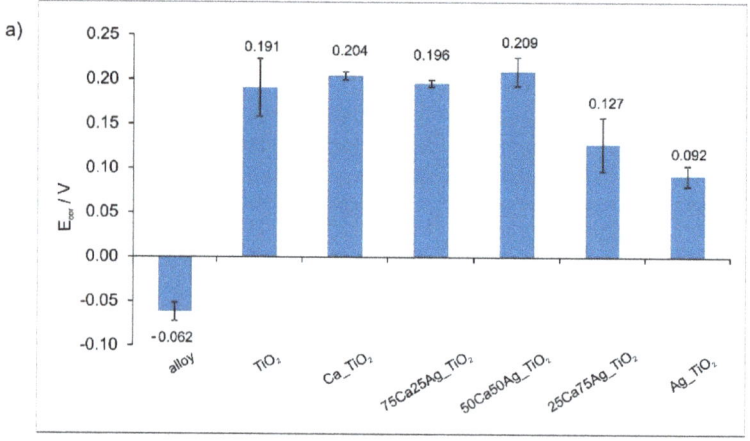

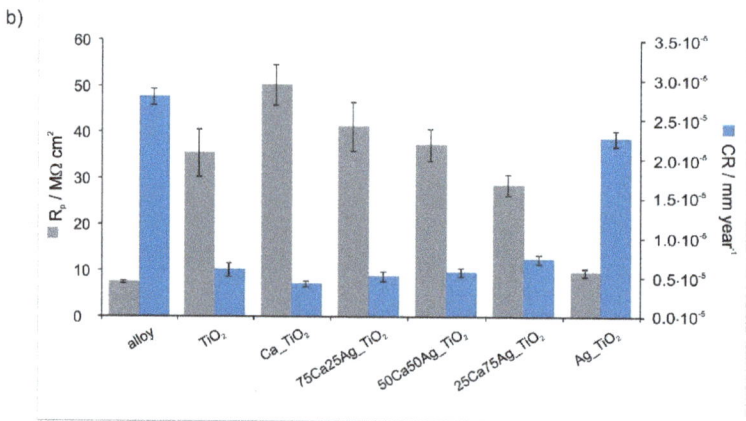

Figure 3. Values of (**a**) corrosion potential, E$_{cor}$, (**b**) polarization resistance, R$_p$, and corrosion rate, CR, determined for TiO$_2$-based coatings doped with Ca and Ag ions in different molar ratios.

It can be observed that the E$_{cor}$ value remains constant (of ca. 0.20V) for undoped TiO$_2$ coating and coatings with predominant calcium content (i.e., Ca_TiO$_2$, 75Ca25Ag_TiO$_2$, 50Ca50Ag_TiO$_2$). Whereas, when the silver content is predominant and its concentration increases in the coating, the corrosion potential progressively decreases up to 0.09V. This is attributed to the increasing amount of Ag metallic nanoparticles in the coating.

As can be seen in Figure 3b, the R_p of the Ca-doped TiO$_2$ coating is higher than that of the undoped TiO$_2$ coating, suggesting that calcium incorporation into TiO$_2$ coating has a significant effect in improving its corrosion resistance. However, as the silver addition in the films increases, the R_p of the coatings decreases gradually from 50 to 9.6 M$\Omega\cdot$cm^2. This probably means that a larger amount of silver ions is released from Ag-doped TiO$_2$ coatings with higher silver content, resulting in a higher corrosion rate (see CR diagram in Figure 3b). Analogous observations were reported by X. Zhang et al. [43]. While, some other researchers [24,35] reported opposite corrosion behavior of Ag-incorporated TiO$_2$ coatings—with an increased amount of silver content the corrosion resistance was improved. This tendency was; however, attributed to the fewer surface defects [35] or the presence of an Ag-TiO$_2$ nanocomposite [24]. Nevertheless, in terms of polarization resistance and corrosion rate, all our doubly-doped coatings act as corrosion protective—the samples with those coatings exhibit better corrosion resistance than the uncoated alloy substrate. According to the R_p and CR results, the Ca_TiO$_2$ coating provides the best anticorrosion protection for the steel substrate.

Pitting corrosion resistance of M30NW alloy samples coated with TiO$_2$-based coatings was examined through the potentiodynamic anodic polarization. The potentiodynamic curves of undoped TiO$_2$- and Ca,Ag-doped coatings were recorded in PBS solution within the wide anodic potential range (up to ca. 1.7V) in order to study the passivation and breakdown behavior of all types of coatings, and are shown in Figure 4. Table 3 gives values of corrosion quantities determined from potentiodynamic curves: current density in passive range (at arbitrary chosen potential of 0.2V) and breakdown potential E_b.

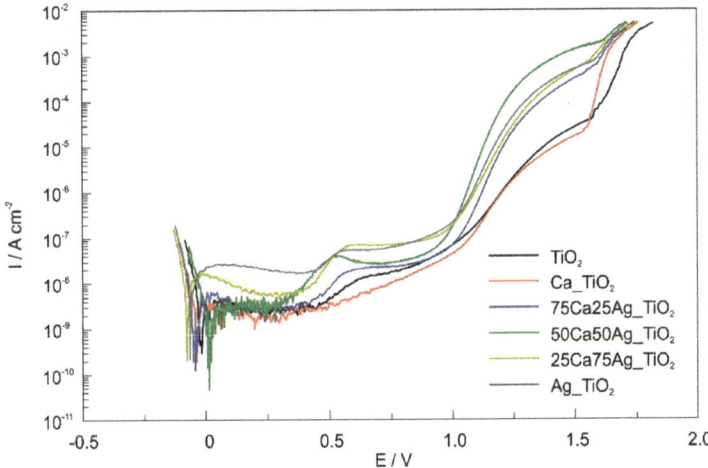

Figure 4. Potentiodynamic polarization curves of TiO$_2$-based coatings in phosphate buffered saline (PBS) solution (scan rate 1 mV$\cdot$s^{-1}).

Table 3. Values of corrosion quantities determined from potentiodynamic characteristics.

Coating	$i_{0.2}$/nA$\cdot$cm^{-2}	E_b/V
TiO$_2$	3.3 ± 1.1	1.600 ± 0.013
Ca_TiO$_2$	3.0 ± 2.2	1.524 ± 0.029
75Ca25Ag_TiO$_2$	3.8 ± 1.8	1.579 ± 0.024
50Ca50Ag_TiO$_2$	3.2 ± 3.3	1.613 ± 0.003
25Ca75Ag_TiO$_2$	6.2 ± 1.7	1.569 ± 0.054
Ag_TiO$_2$	22.9 ± 2.7	1.593 ± 0.021

Based on the potentiodynamic characteristics shown in Figure 4 and data presented in Table 3, it can be stated that an increasing amount of silver in the TiO_2 coatings results in higher electrochemical activity of the coatings. For the sample with Ca/Ag molar ratio of 1:3 (25Ca75Ag_TiO_2), the current density in passive range is two times higher when compared to the Ag-free coatings (TiO_2, Ca_TiO_2) and coatings with predominant or equal calcium content (75Ca25Ag_TiO_2, 50Ca50Ag_TiO_2). However, for the coating doped only with silver ions (Ag_TiO_2), the value of current density in the passive range is the highest, and is about 23 nA/cm^2, which is about seven times higher than for the undoped coating. This fact can be related to the previously found higher porosity of the coatings containing silver ions. The deep pores present in the coating can facilitate the penetration of the corrosion solution through the coating toward the substrate and thus increase the reactivity of the sample.

The breakdown potential E_b value was determined as the potential at which there is a sharp increase in current on the potentiodynamic curve. As shown in Figure 4 and Table 3, prepared materials are characterized by relatively high values of E_b potential of ca. 1.6 V regardless of doped ions. In order to confirm the veracity of such high E_b values, an additional experiment was also performed for each sample, and polarization was stopped at 1.5 V, just after the earlier increase in current recorded on the characteristic curve. Nevertheless, post-polarization microscopic analysis showed no pits on the surface, which proves that, in the case of investigated samples, pitting corrosion occurs at potentials higher than 1.5 V. Such high values of E_b may result from the surface finishing degree (polishing to mirror surface), passivation in mixture of HF and HNO_3 acids, as well as the nature of the titanium dioxide. Based on the corrosion tests results, it can be stated that the M30NW alloy samples with TiO_2-based coatings doped with calcium and silver ions belong to the group of high pitting corrosion-resistant materials.

On the surface of all tested TiO_2-based coatings, anodic polarization caused the formation of corrosion damages as pits, differing in morphology, depth, and width. In many cases, these pits were spherical-ish in shape and covered with corrosion sludge. Moreover, for coatings containing the addition of silver ions (75Ca25Ag_TiO_2, 50Ca50Ag_TiO_2, 25Ca75Ag_TiO_2, Ag_TiO_2), the destruction of the coating in close proximity to the pits can be observed. This is most likely the result of the greater reactivity of these samples. The SEM images shown in Figure 5 indicate that the pitting mechanism starts with the breakdown of the coating, followed by under-film corrosion (dissolution) of the substrate material. In subsequent stages, the damaged fragments of the coating wrap become detached and reveal the corroded substrate (pit). The interiors of the pits reveal the dissolved intergranular edges of the alloy grains.

3.3. Immersion Test

Studies on in vitro bone-bonding ability (referred to as bioactivity) of materials were started by Kokubo and co-workers dozens of years ago [44]. They proposed that the essential requirement for an artificial material to bond to living bone is the formation of bone-like apatite on its surface when implanted in the living body. In laboratory conditions, this ability can be assessed by immersion test in a simulated body fluid (SBF) with ion concentrations nearly equal to those of human blood plasma [39]. The evidence of the bioactive properties of the biomaterial is the formation of an apatite layer as a result of exposure to the SBF solution. Thus, an immersion test in SBF allows for the prediction of the material's in vivo bone bioactivity.

In order to study the effect of calcium and silver ions doping on bioactivity of TiO_2-based coatings, the M30NW alloy samples with five types of coatings, undoped TiO_2, Ca_TiO_2, 75Ca25Ag_TiO_2, 50Ca50Ag_TiO_2, 25Ca75Ag_TiO_2 and Ag_TiO_2, were immersed in SBF solution for 28 days, and then the samples' surfaces were examined using SEM-EDS. Figure 6 shows SEM micrographs (magnitude of 50,000×) of apatite formed on undoped TiO_2 (a), Ca_TiO_2 (b), 75Ca25Ag_TiO_2 (c), and Ag_TiO_2 (d) after soaking for 28 days in SBF solution. All these SEM images are in the same scale, thus direct comparison of the results is possible. In addition, the results of qualitative elemental analysis in the form of Ca/P molar ratios estimated for these samples are given as insets in Figure 6.

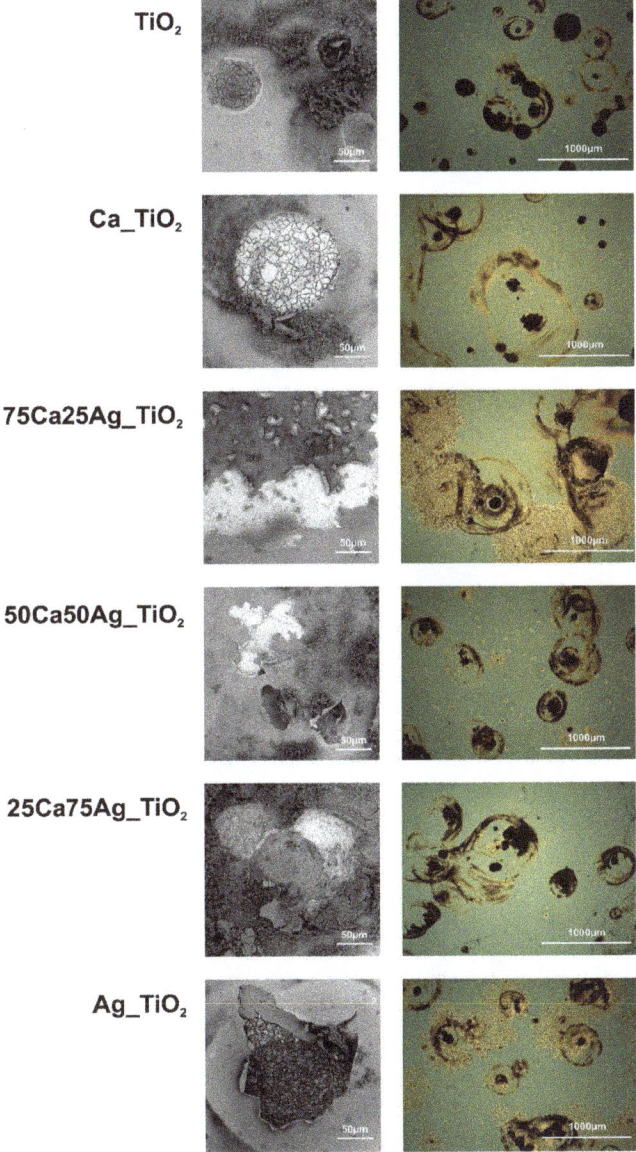

Figure 5. Post-polarization SEM images (1000× mag., bar 50 μm) and optical microscopic images (50× mag., bar 1000 μm).

SEM analysis revealed new particles of different morphologies existing on the samples' surfaces after 28 days of exposure to SBF. In case of undoped TiO_2 coating the randomly distributed agglomerates with Ca/P molar ratio of ca. 1.2 can be observed (Figure 6a). A deposit of a completely different morphology can be seen in Figure 6b for the coating doped with calcium ions (Ca_TiO_2). In this case, needle-like particles, fully covering the surface, with Ca/P molar ratio of ca. 1.5 were formed. As the content of calcium ions in the coating decreased the Ca/P ratio also decreased, and it was ca. 1.2 for 75Ca25Ag_TiO_2, ca. 1.2 for 50Ca50Ag_TiO_2, ca. 1.0 for 25Ca75Ag_TiO_2, and finally ca. 1.1 for Ag_TiO_2

sample. Apart from the low values of Ca/P molar ratio for coatings doped with both calcium and silver ions, the resulting deposits did not completely cover the surfaces; on the part of the surface they formed large agglomerates, while the remaining part of the surface was uncovered. The size and morphology of the particles deposited onto Ag-doped coating (Ag_TiO$_2$) were larger in average diameter and more interconnected comparing with undoped and Ca-Ag co-doped surfaces. No apatite aggregates were observed on Ag_TiO$_2$ surface.

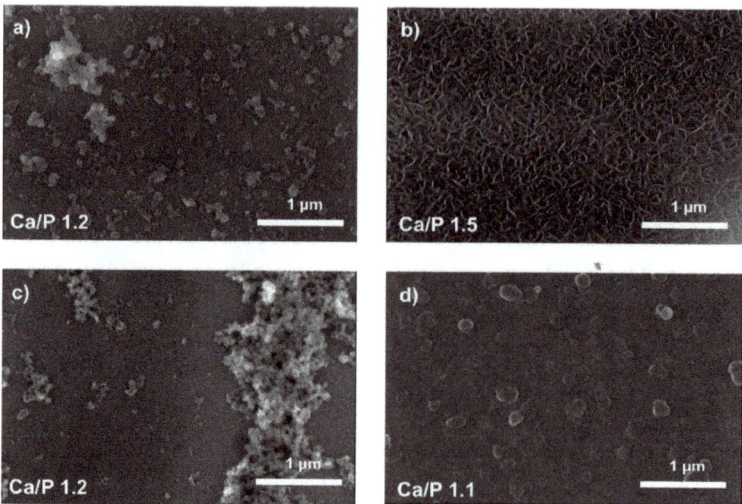

Figure 6. SEM micrographs of apatite formed on: (**a**) undoped TiO$_2$, (**b**) Ca_TiO$_2$, (**c**) 75Ca25Ag_TiO$_2$, and (**d**) Ag_TiO$_2$ after soaking for 28 days in simulated body fluid (SBF) solution (detector TLD, mag 50,000×, bar 1 μm).

According to Zhang et al. [45], the apatite particles nucleate spontaneously onto bioactive surfaces by consuming calcium and phosphate ions from the SBF solution (Equations (2) and (3)):

$$10Ca^{2+} + 8OH^- + 6HPO_4^{2-} \rightarrow Ca_{10}(PO_4)_6(OH)_2 + 6H_2O \tag{2}$$

$$10Ca^{2+} + 2OH^- + 6PO_4^{3-} \rightarrow Ca_{10}(PO_4)_6(OH)_2 \tag{3}$$

In case of coatings doped with calcium, the apatite nucleation is enhanced due to the presence of positively-charged calcium ions in the coating, which react with phosphate anions to form an amorphous calcium phosphate. Since this phase is metastable, it is eventually transformed into stable crystalline bone-like apatite [45]. Results obtained in this study indicate that the more Ca^{2+} that is incorporated in TiO$_2$ coating, the easier and quicker is the apatite nucleation on the surface. The Ca/P ratio of the apatite formed on most of the coatings fabricated in this study is in the range 1.1–1.2, which indicates that the apatites are calcium-deficient HA [33]. Only for that coating, the Ca/P ratio was approaching the value of 1.67 characteristic of the stoichiometric hydroxyapatite. It is a very important factor for orthopedic implants, since literature data indicate that the newly formed bone tissue closely adheres to the implanted element when the Ca/P molar ratio is in the range of 1.67–2.0.

3.4. Wettability

Wettability plays an important role in biological performances of materials. Hydrophilicity of the titania-based coatings on M30NW biomedical alloy samples was evaluated by measuring their water contact angles Θ (Figure 7).

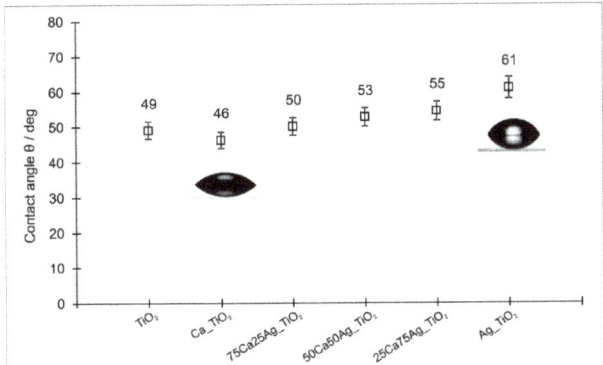

Figure 7. Contact angles Θ of the titania-based coatings on M30NW alloy samples depending on doped ions.

The TiO$_2$ and Ca_TiO$_2$ coatings have similar Θ values of 49° and 46°, respectively. The addition of silver ions causes an increase in the contact angle. As the content of silver ions in the coating increases, the value of the contact angle increases up to 61° for the highest Ag content. On the basis of wetting measurements, it was found that the surface free energy decreases with increasing concentration of silver ions in the coatings (Figure 8).

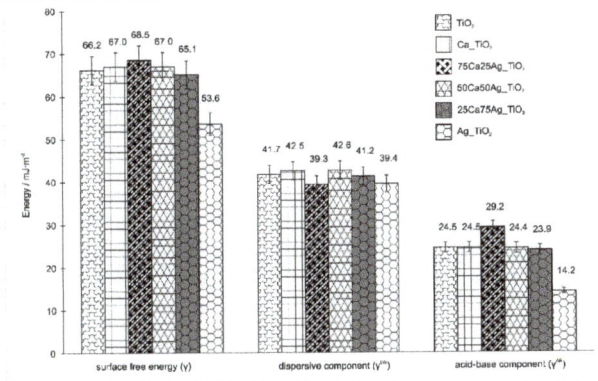

Figure 8. Surface free energy γ values (with distinction between dispersive and polar components) of the titania-based coatings doped with Ca and Ag ions in different molar ratios.

As it is known, the surface free energy (γ) is a sum of the dispersive (γ^{LW}) and acid-base (γ^{AB}) components (Equation (4)), which both determine this value.

$$\gamma = \gamma^{LW} + \gamma^{AB}. \tag{4}$$

Figure 8 gives the values of the γ, with distinction between polar (γ^{AB}) and dispersive (γ^{LW}) components. In our study, the surface free energy value is mainly influenced by the γ^{AB} component, since the value of the γ^{LW} component is similar for all surfaces and amounts ca. 40 mJ/m^2. It can be clearly seen that both surface free energy γ and its polar component γ^{AB} decrease with increasing Ag concentration in the coatings. It is related to the formation of silver oxide particles on the surface of the coatings. Silver atoms present on the coatings' surface are exposed to the atmosphere and are free to bond with other atoms, especially with oxygen and water [46]. Therefore, as the concentration of silver in the coating increases, the number of Ag-O bonds increases, which, in consequence, changes the properties of surface.

It can be concluded that all investigated TiO_2-based coatings are hydrophilic regardless of the type and molar ratio of the dopants. The higher surface wettability results in better adhesion and proliferation of the eukaryotic cells [47]. That can be beneficial for biological applications, especially for use in the circulatory system. According to the literature [48], the hydrophilic titania coating reduces adsorption of proteins and minimizes adherence of blood platelets on the surface.

3.5. Biological Evaluation: Cell Viability and Proliferation Ability Assays

Figure 9a presents the results obtained from the live/dead test and the determined amounts of cells viability after direct contact with the examined surfaces. The highest amount of live cells (i.e., above 98%) was observed for the sample 50Ca50Ag_TiO_2. The biggest percentage of dead cells was noted for the sample 25Ca75Ag_TiO_2 (~22% of all collected cells). Statistical significance on the level of $p < 0.05$ was noted between sample 25Ca75Ag_TiO_2 and TiO_2, Ag_TiO_2, 75Ca25Ag_TiO_2 for live cells as well as for the dead ones. Nevertheless, none of the examined materials is cytotoxic. All the samples fulfill the requirements defined in the ISO 10993-5—the viability is above 70%. That states that prepared sol-gel coatings, regardless of doped element—Ca, Ag or their mixture—are biocompatible materials. Taking into consideration the average amount of cells, in the case of a cells' proliferation (Figure 9b), it was observed that the trend of obtained results is similar to the trend for results obtained for contact angle measurement. It can be stated also that cells' proliferation slightly increases with decreasing surface free energy—the highest is for Ag_TiO_2 sample, which has the lowest γ values. Calcium addition does not influence the cells, most proliferate onto the samples doped only with Ag (Ag_TiO_2) and their amount is higher than for uncoated basic sample. Nevertheless there are no statistically significant differences between all evaluated coatings. It can state that, in the case of doped sol-gel coatings, their biological response depends mainly on surface topography and wettability of the samples, and to a very small extent is the effect of the content of individual elements included in the coating.

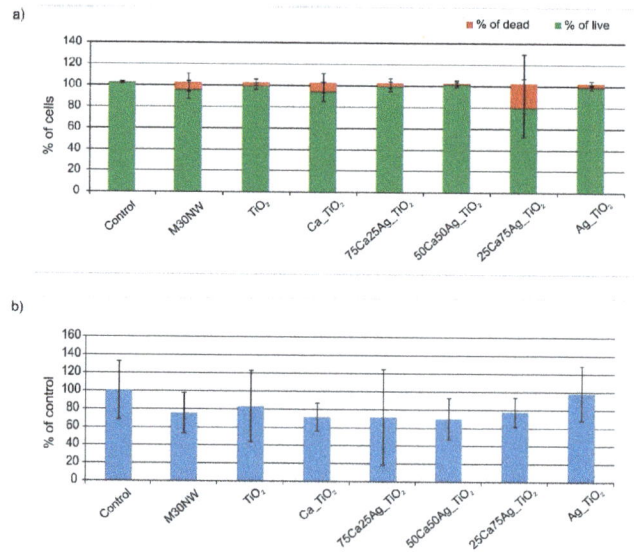

Figure 9. The results of (**a**) the live/dead test and (**b**) the cells' proliferation evaluation for all the examined coatings after 48 h of direct contact (conducted according to the protocol of ISO 10993-5: Tests for Cytotoxicity—In Vitro Methods).

The biocompatibility is usually defined as "the ability of a material to perform with an appropriate host response in a specific application" [49]. Interactions between biological system and biomaterial surface run in the following order: In the first few nanoseconds, the water molecules and proteins reach the surface, being followed by the cells [50]. The interaction of proteins and cells with the surface is driven by the specific surface features: Surface chemistry, topography, roughness, wettability, and crystallinity. Cells can sense the chemistry and topography of the surface to which they adhere. Cell behavior is different on different nanosurfaces, because nanomorphology of the material may significantly influence protein and cell adhesion. In general, cells show good spreading, proliferation, and differentiation on hydrophilic surfaces. Nevertheless, the major factor determining the nature of the cells' interaction with biomaterials is the composition and conformation of the proteins adsorbed on the surface. The adhesion and behavior of cells are affected by adsorption of serum and extracellular matrix proteins [51]. Therefore, the observed difference in the proliferation and viability of osteoblast cells may be caused by the difference in the absorption of proteins responsible for the cell colonization process. The adsorption of proteins responsible for the cell colonization and their activity may be affected by one or more interactions between proteins and surfaces, including van der Waal's interactions, electrostatic interactions, hydrogen bonding, and hydrophobic interactions [52–56]. According to the literature, generally higher surface wettability results in better adhesion and proliferation of the eukaryotic cells [57]. Although, when the surface wettability is very high, water adsorbs preferentially on the surface [58] and thus can reduce adsorption of the proteins. It has been shown by the study of Xu et al. that surfaces with $\theta > \sim 60°-65°$ show stronger adhesion forces for proteins than the surfaces with $\theta < 60°$ [57]. Generally, hydrophobic surfaces are considered to be more protein adsorbent than hydrophilic surfaces, due to strong hydrophobic interactions occurring at these surfaces [59] in direct contrast to the repulsive solvation forces arising from strongly bound water at the hydrophilic surface [53]. As proteins determine the cell proliferation results, for the sample with the highest amount of Ag having the highest contact angle (above 61°), the average number of proliferated cells is the highest—almost at the same level as a control. For the other doped samples, with the changing molar ratios of calcium ions Ca^{2+} and silver ions Ag^+ (Ca/Ag 3:1, 1:1, 1:3) in TiO_2 sol, contact angle is on the level of ~50°–55°. The surface wettability is on a very similar level for all doped TiO_2 coatings, so the results obtained from the live/dead assay confirmed this dependence—osteoblast proliferation for all doubly-doped coatings is at the comparable level. However, it should be noted that the differences between proliferation results for all coatings are not significant and are in the range of experimental error. Therefore, it could be concluded that, in general, osteoblast cells growth is promoted on all coated surfaces, regardless of the increase of particular component elements as Ag and Ca, nor differences in nanotopography. Similarly, no morphological differences for the osteoblast-like cells were reported in the literature for the Ca-Ag coexisting nano-structured titania layer on Ti metal surface [34], as well as for Ag-Sr co-doped hydroxyapatite/TiO_2 nanotube bilayer coatings [33]. These reports prove that incorporating a secondary bioactive compound (e.g., Ca or Sr ions) not only improves bioactivity of the coating, but it is also effective in lessening Ag cytotoxicity and optimally preserving its antibacterial properties.

According to T.T Liao el at. [60], less ordered phases in a coating results in a lower adsorption of proteins and cells on the surface, while increasing crystallinity of the coating improves cell colonization. For the coatings examined in this work, the XRD results showed that the crystallinity of the anatase in TiO_2 coating increases with increasing amount of Ag and the highest was obtained for the Ag_TiO_2 sample. At the same time, for this sample, the highest value of R_q = 7.88 nm was observed. Surface topography plays an important role in providing three-dimensionality of cells [61]. For instance, the topography of the collagen fibers, with repeated 66 nm binding, has shown to affect cell shape [62]. Focal adhesion interacting with the surface is established by cell filopodia (which are 0.25–0.5 μm wide and 2–10 μm long) [63]. Filopodia can interact with the surface due to surface features, which are either arranged randomly or in some geometrical order and have dimensions from the micro to the nanometer range [61]. Beyond micrometers, it has been shown that nanometric (1–500 nm) features can elicit specific cell responses [61,62]. In case of our experiment, as the surfaces of the coatings show different nanostructures, we noted that osteoblast cells react differently on the surface revealing dissimilar morphology (Figure 10). The results we got are consistent with the observations made by S. Lee et al. [64]. In their results they noticed that as micropore size increases, cell number is reduced and cell differentiation and matrix production is increased. Their study demonstrated that the surface topography plays an important role for phenotypic expression of the MG63 osteoblast-like cells. In our case, we observed that cell shape and proliferation level are different as the coatings' topography in nanoscale is different. For most porous surfaces (Ca_TiO$_2$, 75Ca25Ag_TiO$_2$) we observed fewer cells and increased matrix production, although their number is still relatively high. For surfaces without pores, osteoblasts are more elongated (more natural morphology), although other factors decrease their number compared to control.

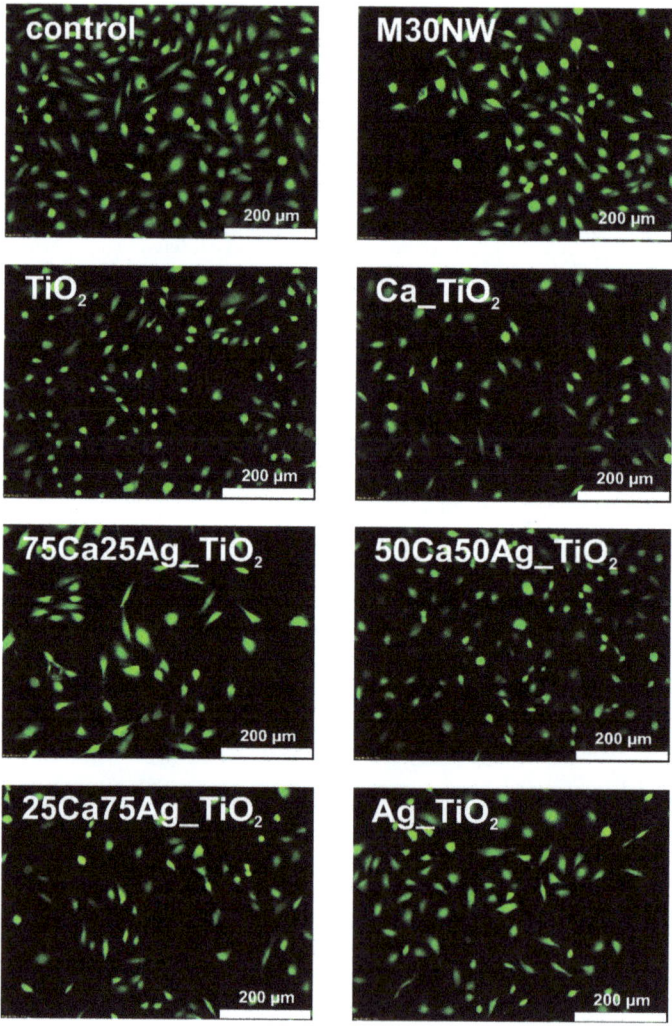

Figure 10. The osteoblast cells images (fluorescent stained, live cells—green colour, bar 200 µm) after 48 h of growing on samples (direct contact) modified by coatings with different composition and nanostructure.

4. Conclusions

Application of the sol-gel method made it possible to obtain homogeneous titanium dioxide coatings co-doped with calcium and silver ions in a different molar ratio. The thermal treatment at 450 °C allowed crystalline coatings of anatase structure to be obtained at a thickness of approx. 70–80 nm, regardless of the amount of each dopant. It was found that the amount of crystalline anatase phase increases with increasing silver content. All sol-gel TiO_2-based coatings investigated in this study have hydrophilic properties regardless of the type of dopant. Thus, each of the produced coatings is beneficial to the adsorption of osteoblast cells and thus for bone-bonding properties of implants. The doping with calcium and silver ions affects the topography of the TiO_2-based coatings. Undoped coatings and coatings doped only with calcium ions are characterized by small surface development. Along with the increase of silver content in the coating, its surface development increases as well, due to pores existing in the coating. The corrosion tests confirmed anticorrosive properties of TiO_2-based coatings. The best protective (anticorrosive) properties were registered for the coating doped with calcium ions. Analysis of corrosion results showed that the increase in silver content resulted in increasing of the electrochemical activity of the investigated samples in PBS solution. The immersion test in the SBF solution confirmed the bioactivity of the tested coatings—the apatite layer was found on the samples' surface. It was found that, with the increase in calcium ion content in the coating, the Ca/P ratio increases and approaches the value of 1.67, characteristic of stoichiometric hydroxyapatite.

Silver is recognized as an antibacterial element, but at the same time can be also cytotoxic for cells. Calcium is very well known as an element that can improve the osseointegration processes. Our results showed that coatings containing Ca and Ag particles, independently of their molar ratio in TiO_2 coating, are biocompatible and do not significantly reduce the proliferation ability of the osteoblast cells, compared to the pure material, as the M30NW steel is. The lack of toxicity and viability of cells above 80% for all coatings may indicate that these doubly-doped coatings meet at least one of the requirements to define them as biocompatible materials. As the other results indicated, they may also show the ability for bone-like apatite formation and significantly improved corrosion resistance in comparison to the steel biomaterial.

Author Contributions: B.B.: Conceptualization, methodology, investigations (surface characterization and corrosion investigations), results analysis, visualization, writing, editing, and supervising the manuscript; P.O.: Resources, corrosion investigations, and data analysis; D.B. (Damian Batory): Surface characterization, data and results analysis, discussion, and editing the manuscript; M.C.: Wetting investigations, data and results analysis, discussion; M.K.: biological investigations, data analysis, and discussion; D.B. (Dorota Bociaga): Biological investigations, data and results analysis, discussion, writing, and editing the manuscript. All authors have read and agreed to the published version of the manuscript.

Funding: This research received no external funding.

Conflicts of Interest: The authors declare no conflicts of interest.

References

1. Fu, T.; Wen, C.S.; Lu, J.; Zhou, Y.M.; Ma, S.G.; Dong, B.H.; Liu, B.G. Sol-gel derived TiO_2 coating on plasma nitrided 316L stainless steel. *Vacuum* **2012**, *86*, 1402–1407. [CrossRef]
2. Velten, D.; Biehl, V.; Aubertin, F.; Valeske, B.; Possart, W.; Breme, J. Preparation of TiO_2 layers on cp-Ti and Ti6Al4V by thermal and anodic oxidation and by sol gel coating techniques and their characterization. *J. Biomed. Mater. Res.* **2002**, *59*, 18–28. [CrossRef] [PubMed]
3. Wang, G.; Li, J.; Lv, K.; Zhang, W.; Ding, X.; Yang, G.; Liu, X.; Jiang, X. Surface thermal oxidation on titanium implants to enhance osteogenic activity and in vivo osseointegration. *Sci. Rep.* **2016**, *6*, 31769. [CrossRef] [PubMed]
4. Wojcieszak, D.; Mazur, M.; Indyka, J.; Jurkowska, A.; Kalisz, M.; Domanowski, P.; Kaczmarek, D.; Domaradzki, J. Mechanical and structural properties of titanium dioxide deposited by innovative magnetron sputtering process. *Mater. Sci. Pol.* **2015**, *33*, 660–668. [CrossRef]

5. Dreesen, L.; Cecchet, F.; Lucas, S. DC magnetron sputtering deposition of titanium oxide nanoparticles: Influence of temperature, pressure and deposition time on the deposited layer morphology, the wetting and optical surface properties. *Plasma Process. Polym.* **2009**, *6*, S849–S854. [CrossRef]
6. Altmayer, J.; Barth, S.; Mathur, S. Influence of precursor chemistry on CVD grown TiO_2 coatings: Differential cell growth and biocompatibility. *RSC Adv.* **2013**, *3*, 11234–11239. [CrossRef]
7. Sobczyk-Guzenda, A.; Pietrzyk, B.; Szymanowski, H.; Gazicki-Lipman, M.; Jakubowski, W. Photocatalytic activity of thin TiO_2 films deposited using sol–gel and plasma enhanced chemical vapor deposition methods. *Ceram. Int.* **2013**, *39*, 2787–2794. [CrossRef]
8. Sobczyk-Guzenda, A.; Owczarek, S.; Fijalkowski, M.; Batory, D.; Gazicki-Lipman, M. Morphology, structure and photowettability of TiO_2 coatings doped with copper and fluorine. *Ceram. Int.* **2018**, *44*, 5076–5085. [CrossRef]
9. Zhang, Q.; Li, C. Pure anatase phase titanium dioxide films prepared by mist chemical vapor deposition. *Nanomaterials* **2018**, *8*, 827. [CrossRef]
10. Alam, M.J.; Cameron, D.C. Preparation and characterization of TiO_2 thin films by sol-gel method. *J. Sol-Gel Sci. Technol.* **2002**, *25*, 137–145. [CrossRef]
11. Piwoński, I. Preparation method and some tribological properties of porous titanium dioxide layers. *Thin Solid Film.* **2007**, *515*, 3499–3506. [CrossRef]
12. Burnat, B.; Dercz, G.; Błaszczyk, T. Structural analysis and corrosion studies on an ISO 5832-9 biomedical alloy with TiO_2 sol–gel layers. *J. Mater. Sci.: Mater. Med.* **2014**, *25*, 623–634. [CrossRef] [PubMed]
13. Owens, G.J.; Singh, R.K.; Foroutan, F.; Alqaysi, M.; Han, C.M.; Mahapatra, C.; Kim, H.W.; Knowles, J.C. Sol–gel based materials for biomedical applications. *Prog. Mater. Sci.* **2016**, *77*, 1–79. [CrossRef]
14. Carrera-Figueiras, C.; Pérez-Padilla, Y.; Estrella-Gutiérrez, M.A.; Uc-Cayetano, E.G.; Juárez-Moreno, J.A.; Avila-Ortega, A. Surface science engineering through sol-gel process. *Appl. Surf. Sci.* **2019**. [CrossRef]
15. Viitala, R.; Jokinen, M.; Peltola, T.; Gunnelius, K.; Rosenholm, J.B. Surface properties of in vitro bioactive and non-bioactive sol–gel derived materials. *Biomaterials* **2002**, *23*, 3073–3086. [CrossRef]
16. Park, J.W.; Park, K.B.; Suh, J.Y. Effects of calcium ion incorporation on bone healing of Ti6Al4V alloy implants in rabbit tibiae. *Biomaterials* **2007**, *28*, 3306–3313. [CrossRef]
17. Park, J.W.; Kim, Y.J.; Jang, J.H. Surface characteristics and in vitro biocompatibility of a manganese-containing titanium oxide surface. *Appl. Surf. Sci.* **2015**, *258*, 977–985. [CrossRef]
18. Park, J.W. Increased bone apposition on a titanium oxide surface incorporating phosphate and strontium. *Clin. Oral Implant. Res.* **2011**, *22*, 230–234. [CrossRef]
19. Anne Pauline, S.; Kamachi Mudali, U.; Rajendran, N. Fabrication of nanoporous Sr incorporated TiO_2 coating on 316L SS: Evaluation of bioactivity and corrosion protection. *Mater. Chem. Phys.* **2013**, *142*, 27–36. [CrossRef]
20. Burnat, B.; Robak, J.; Batory, D.; Leniart, A.; Piwoński, I.; Skrzypek, S.; Brycht, M. Surface characterization, corrosion properties and bioactivity of Ca-doped TiO_2 coatings for biomedical applications. *Surf. Coat. Technol.* **2015**, *280*, 291–300. [CrossRef]
21. Burnat, B.; Robak, J.; Leniart, A.; Piwoński, I.; Batory, D. The effect of concentration and source of calcium ions on anticorrosion properties of Ca-doped TiO_2 bioactive sol-gel coatings. *Ceram. Int.* **2017**, *43*, 13735–13742. [CrossRef]
22. Visai, L.; De Nardo, L.; Punta, C.; Melone, L.; Cigada, A.; Imbriani, M.; Arciola, C.R. Titanium oxide antibacterial surfaces in biomedical devices. *Int. J. Artif. Organs* **2011**, *34*, 929–946. [CrossRef] [PubMed]
23. Albert, E.; Albouy, P.A.; Ayral, A.; Basa, P.; Csik, G.; Nagy, N.; Roualdes, S.; Rouessac, V.; Safran, G.; Suhajda, A.; et al. Antibacterial properties of Ag–TiO_2 composite sol–gel coatings. *RSC Adv.* **2015**, *5*, 59070–59081. [CrossRef]
24. Cotolan, N.; Rak, M.; Bele, M.; Cör, A.; Muresan, L.M.; Milošev, I. Sol-gel synthesis, characterization and properties of TiO_2 and Ag-TiO_2 coatings on titanium substrate. *Surf. Coat. Technol.* **2016**, *307*, 790–799. [CrossRef]
25. Heidenau, F.; Mittelmeier, W.; Detsch, R.; Haenle, M.; Stenzel, F.; Ziegler, G.; Gollwitzer, H. A novel antibacterial titania coating: Metal ion toxicity and in vitro surface colonization. *J. Mater. Sci. Mater. Med.* **2005**, *16*, 883–888. [CrossRef] [PubMed]

26. Haenle, M.; Fritsche, A.; Zietz, C.; Bader, R.; Heidenau, F.; Mittelmeier, W.; Gollwitzer, H. An extended spectrum bactericidal titanium dioxide (TiO$_2$) coating for metallic implants: In vitro effectiveness against MRSA and mechanical properties. *J. Mater. Sci. Mater. Med.* **2011**, *22*, 381–387. [CrossRef] [PubMed]
27. Wang, Y.; Xue, X.; Yang, H. Modification of the antibacterial activity of Zn/TiO$_2$ nano-materials through different anions doped. *Vacuum* **2014**, *101*, 193–199. [CrossRef]
28. Pérez-Jorge Peremarch, C.; Perez Tanoira, R.; Arenas, M.A.; Matykina, E.; Conde, A.; De Damborenea, J.J.; Barrena, E.G.; Esteban, J. Bacterial adherence to anodized titanium alloy. *J. Phys. Conf. Ser.* **2010**, *252*, 012011. [CrossRef]
29. Pérez-Jorge, C.; Conde, A.; Arenas, M.A.; Pérez-Tanoira, R.; Matykina, E.; de Damborenea, J.J.; Gómez-Barrena, E.; Esteban, J. In vitro assessment of Staphylococcus epidermidis and Staphylococcus aureus adhesion on TiO$_2$ nanotubes on Ti–6Al–4V alloy. *J. Biomed. Mater. Res. A* **2012**, *100*, 1696–1705. [CrossRef]
30. Arenas, M.A.; Pérez-Jorge, C.; Conde, A.; Matykina, E.; Hernández-López, J.M.; Pérez-Tanoira, R.; de Damborenea, J.J.; Gómez-Barrena, E.; Esteba, J. Doped TiO$_2$ anodic layers of enhanced antibacterial properties. *Colloids Surf.* **2013**, *105*, 106–112. [CrossRef]
31. Choi, J.; Park, H.; Hoffmann, M.R. Combinatorial doping of TiO$_2$ with platinum (Pt), chromium (Cr), vanadium (V), and nickel (Ni) to achieve enhanced photocatalytic activity with visible light irradiation. *J. Mater. Res.* **2010**, *25*, 149–158. [CrossRef]
32. Zhou, J.; Zhao, L. Multifunction Sr, Co and F co-doped microporous coating on titanium of antibacterial, angiogenic and osteogenic activities. *Sci. Rep.* **2016**, *6*, 29069. [CrossRef] [PubMed]
33. Huang, Y.; Zhang, X.; Zhang, H.; Qiao, H.; Zhang, X.; Jia, T.; Han, S.; Gao, Y.; Xiao, H.; Yang, H. Fabrication of silver-and strontium-doped hydroxyapatite/TiO$_2$ nanotube bilayer coatings for enhancing bactericidal effect and osteoinductivity. *Ceram. Int.* **2017**, *43*, 992–1007. [CrossRef]
34. Rajendran, A.; Vinoth, G.; Nivedhitha, J.; Iyer, K.M.; Pattanayak, D.K. Ca-Ag coexisting nano-structured titania layer on Ti metal surface with enhanced bioactivity, antibacterial and cell compatibility. *Mater. Sci. Eng.* **2019**, *99*, 440–449. [CrossRef] [PubMed]
35. Lv, C.; Yang, M.; Lu, J.; Zhao, X.; Wang, G. Effect of Ca and Ag doping on structure, in-vitro mineralization, and anti-bacterial properties of TiO$_2$ coatings. *J. Chin. Ceram. Soc.* **2019**, *47*, 692–700.
36. Yetim, A.F.; Yildiz, F.; Alsaran, A.; Celik, A. Surface modification of 316L stainless steel with plasma nitriding. *Kov. Mater.* **2008**, *46*, 105–115.
37. Junping, Y.; Wei, L.; Keya, S. Study of Ce-modified antibacterial 316L stainless steel. *China Foundry* **2012**, *9*, 307–312.
38. Burnat, B.; Błaszczyk, T.; Leniart, A. Effects of serum proteins on corrosion behavior of ISO 5832-9 alloy modified by titania coatings. *J. Solid State Electrochem.* **2014**, *18*, 3111–3119. [CrossRef]
39. Kokubo, T.; Takadama, H. How useful is SBF in predicting in vivo bone bioactivity? *Biomaterials* **2006**, *27*, 2907–2915. [CrossRef]
40. Lide, D.R. *CRC Handbook of Chemistry and Physics*, 75th ed.; CRC Press: Boca Raton, FL, USA, 1994.
41. García-Serrano, J.; Gómez-Hernández, E.; Ocampo-Fernández, M.; Pal, U. Effect of Ag doping on the crystallization and phase transition of TiO$_2$ nanoparticles. *Curr. Appl. Phys.* **2009**, *9*, 1097–1105. [CrossRef]
42. *ASTM G102-89(2015)e1, Standard Practice for Calculation of Corrosion Rates and Related Information from Electrochemical Measurements*; ASTM International: West Conshohocken, PA, USA, 2015.
43. Zhang, X.; Li, M.; He, X.; Huang, X.; Hang, R.; Tang, B. Effects of silver concentrations on microstructure and properties of nanostructured titania films. *Mater. Des.* **2015**, *65*, 600–605. [CrossRef]
44. Kokubo, T. Bioactive glass ceramics: Properties and applications. *Biomaterials* **1991**, *12*, 155–163. [CrossRef]
45. Zhang, W.; Du, K.; Yan, C.; Wang, F. Preparation and characterization of a novel Si-incorporated ceramic film on pure titanium by plasma electrolytic oxidation. *Appl. Surf. Sci.* **2008**, *254*, 5216–5223. [CrossRef]
46. de Rooij, A. The Oxidation of Silver by Atomic Oxygen. *ESA J.* **1989**, *13*, 363–392.
47. Bociaga, D.; Sobczyk-Guzenda, A.; Komorowski, P.; Balcerzak, J.; Jastrzebski, K.; Przybyszewska, K.; Kaczmarek, A. Surface characteristics and biological evaluation of Si-DLC coatings fabricated using magnetron sputtering method on Ti6Al7Nb substrate. *Nanomaterials* **2019**, *9*, 812. [CrossRef] [PubMed]
48. Chun, Y.; Levi, D.S.; Mohanchandra, K.P.; Carman, G.P. Superhydrophilic surface treatment for thin film NiTi vascular applications. *Mater. Sci. Eng.* **2009**, *29*, 2436–2441. [CrossRef]
49. Williams, D.F. *Advances in Biomaterials*; Elsevier: Amsterdam, The Netherlands, 1988.

50. Junkar, I. Chapter Two—Interaction of cells and platelets with biomaterial surfaces treated with gaseous plasma. In *Advances in Biomembranes and Lipid Self-Assembly*; Iglic, A., Kulkarni, C.H., Rappolt, M., Eds.; Elsevier: Amsterdam, The Netherlands, 2016; Volume 23, pp. 25–59.
51. Keselowsky, B.G.; Collard, D.M.; García, A.J. Surface chemistry modulates fibronectin conformation and directs integrin binding and specificity to control cell adhesion. *J. Biomed. Mater. Res. Part A* **2003**, *66*, 247–259. [CrossRef]
52. Heynes, C.A.; Norde, W. Globular proteins at solid/liquid interfaces. *Colloids Surf. B: Biointerfaces* **1994**, *2*, 517–566. [CrossRef]
53. Israelachvili, J.; Wennerstrom, H. Role of hydration and water structure in biological and colloidal interactions. *Nature* **1996**, *379*, 219–225. [CrossRef]
54. Sit, P.S.; Marchant, R.E. Surface-dependent differences in fibrin assembly visualized by atomic force microscopy. *Surf. Sci.* **2001**, *491*, 421–432. [CrossRef]
55. Kidoaki, S.; Matsuda, T. Mechanistic aspects of protein/material interactions probed by atomic force microscopy. *Colloids Surf. B Biointerfaces* **2002**, *23*, 153–163. [CrossRef]
56. Xu, L.C.; Logan, B.E. Interaction forces between colloids and protein-coated surfaces measured using an atomic force microscope. *Environ. Sci. Technol.* **2005**, *39*, 3592–3600. [CrossRef] [PubMed]
57. Xu, L.C.; Siedlecki, C.A. Effects of surface wettability and contact time on protein adhesion to biomaterial surfaces. *Biomaterials* **2007**, *28*, 3273–3283. [CrossRef] [PubMed]
58. Ogwu, A.A.; Okpalugo, T.I.T.; Ali, N.; Maguire, P.D.; McLaughlin, J.A.D. Endothelial cell growth on silicon modified hydrogenated amorphous carbon thin films. *J. Biomed. Mater. Res. Part B Appl. Biomater.* **2008**, *85*, 105–113. [CrossRef]
59. Kongdee, A.; Bechtold, T.; Teufel, L. Modification of cellulose fiber with silk sericin. *J. Appl. Polym. Sci.* **2005**, *96*, 1421–1428. [CrossRef]
60. Liao, T.T.; Zhang, T.F.; Li, S.S.; Deng, Q.Y.; Wu, B.J.; Zhang, Y.Z.; Zhou, Y.J.; Guo, Y.B.; Leng, Y.X.; Huang, N. Biological responses of diamond-like carbon (DLC) films with different structures in biomedical application. *Mater. Sci. Eng.* **2016**, *69*, 751–759. [CrossRef] [PubMed]
61. Dalby, M.J.; Riehle, M.O.; Johnstone, H.; Affrossman, S.; Curtis, A.S.G. In vitro reaction of endothelial cells to polymer demixed nanotopography. *Biomaterials* **2002**, *23*, 2945–2954. [CrossRef]
62. Curtis, A.; Wilkinson, C. New depths in cell behaviour: Reactions of cells to nanotopography. *Biochem. Soc. Symp.* **1999**, *65*, 15–26.
63. Burridge, K.; Chrzanowska-Wodnicka, M. Focal adhesions, contractility, and signaling. *Annu. Rev. Cell Dev. Biol.* **1996**, *12*, 463–518. [CrossRef]
64. Lee, S.J.; Choi, J.S.; Park, K.S.; Khang, G.; Lee, Y.M.; Lee, H.B. Response of MG63 osteoblast-like cells onto polycarbonate membrane surfaces with different micropore sizes. *Biomaterials* **2004**, *25*, 4699–4707. [CrossRef]

© 2020 by the authors. Licensee MDPI, Basel, Switzerland. This article is an open access article distributed under the terms and conditions of the Creative Commons Attribution (CC BY) license (http://creativecommons.org/licenses/by/4.0/).

Article

Thin SiNC/SiOC Coatings with a Gradient of Refractive Index Deposited from Organosilicon Precursor

Hieronim Szymanowski [1], Katarzyna Olesko [1], Jacek Kowalski [1], Mateusz Fijalkowski [2], Maciej Gazicki-Lipman [1] and Anna Sobczyk-Guzenda [1],*

1. Institute of Materials Science and Engineering, Lodz University of Technology, Stefanowskiego 1/15 Str., 90-924 Lodz, Poland; hieronim.szymanowski@p.lodz.pl (H.S.); katarzyna.olesko@gmail.com (K.O.); jm.kowalski@wp.pl (J.K.); Maciej.Gazicki-Lipman@p.lodz.pl (M.G.-L.)
2. Institute for Nanomaterials, Advanced Technologies and Innovation, Technical University of Liberec, 2 Studentska St., 461 17 Liberec, Czech Republic; mateusz.fijalkowski@tul.cz
* Correspondence: anna.sobczyk-guzenda@p.lodz.pl; Tel.: +48-42-631-30-73

Received: 13 July 2020; Accepted: 11 August 2020; Published: 17 August 2020

Abstract: In this work, optical coatings with a gradient of the refractive index are described. Its aim was to deposit, using the RF PECVD method, films of variable composition (ranging from silicon carbon-oxide to silicon carbon-nitride) for a smooth change of their optical properties enabling a production of the filter with a refractive index gradient. For that purpose, two organosilicon compounds, namely tetramethyldisilazane and hexamethyldisilazane, were selected as precursor compounds. The results reveal better optical properties of the materials obtained from the latter source. Depending on whether deposited in pure oxygen atmosphere or under conditions of pure nitrogen, the refractive index of the coatings amounted to 1.65 and to 2.22, respectively. By using a variable composition N_2/O_2 gas mixture, coatings of intermediate magnitudes of "n" were acquired. The optical properties were investigated using both UV-Vis absorption spectroscopy and variable angle spectroscopic ellipsometry. The chemical structure of the coatings was studied with the help of Fourier transform infrared and X-ray photoelectron spectroscopies. Finally, atomic force microscopy was applied to examine their surface topography. As the last step, a "cold mirror" type interference filter with a gradient of refractive index was designed and manufactured.

Keywords: thin films; silicon carbon-nitride; silicone carbon-oxide; PECVD method; inhomogeneous optical filters; gradient interference filters; organosilicon precursors

1. Introduction

Due to their wide-ranging applications in optics, interference filters constitute a group of broadly developed and manufactured optical devices [1]. Among others, they are applied as coatings on lenses, camera and projector optics [2], telescope components [3], solar panels [4], and car and aircraft windows [5].

Initially, interference filters were composed of stacks of alternated homogeneous films characterized by the low and high magnitude of refractive index. These systems exhibited numerous shortcomings, one example being an occurrence of secondary harmonic bands in their spectra, substantially lowering optical quality of a filter [6]. Another often encountered problem was poor adhesion between the subsequent layers with the resulting interfacial penetration of water frequently leading to complete filter destruction [7]. This is a reason for which a new generation of interference filters was designed, assuming a continuous periodical (sinusoidal) change of their index of refraction [8]. A gradient inhomogeneous structure of a coating allows one to find a photonic forbidden band similar to that of

Bragg's mirror, but one that is narrower (by a coefficient of π/4) than the width of the quarter-wave band. This is one reason these structures are characterized by better optical properties, with perfectly suppressed secondary and higher harmonic bands [7]. However, one needs to remember that the more complex the coating structure, the higher the difficulty of the technological problems connected with its deposition.

Despite numerous advantages of inhomogeneous interference filters, there is a rather limited number of scientific publications dealing with these devices present in the literature. One of the principal reasons for that is the fact that the manufacture of these filters requires very precise control of their optical parameters, which is extremely difficult (if not impossible) in the case of such conventional deposition techniques as the sol–gel method [9]. The realization of this type of project requires the availability of optical materials of variable magnitude of refractive index, controlled by the continuous (and high-precision) change of their elemental composition. The deposition techniques most often used for the manufacture of optical coatings include co-sputtering [10], co-evaporation [11], ion-assisted deposition [12], and glancing angle deposition [13]. In all these cases, both a precise control of the coating composition and safeguarding of its high optical quality require excessive financial investment.

Among the different materials used for the purpose of manufacturing inhomogeneous optical filters, perhaps silicon and its derivatives are the most popular. In recent years, there have been several articles published dealing with the technology of porous silicon for these applications. The first paper on that subject was published in 1997 by Berger [14]. Later, in 2002, Cunin et al. showed an application potential for the filter in screening biomolecular studies [15]. This type of filter, produced by Chhasatia et al. with the help of a thermal hydrosilylation method, was used in the construction of an insulin detection biosensor [16]. Regrettably, the above-mentioned filters are not transparent to visible light, which substantially limits their applications. It was Zhang et al. who produced antireflective narrow line-width filters with a refractive index gradient using silicon oxide (SiO_2) as a low (1.50)-index material and niobium oxide (Nb_2O_5) as a high (2.14)-index material, applying the glancing angle deposition technique for that purpose [17]. The magnetron sputtering technique is also used to manufacture rugate type optical filters. With the help of the reactive magnetron sputtering method, Bartzsch et al. produced silicon oxynitride (SiON) antireflection filters, wherein index of refraction varied within a relatively narrow range of 1.46 to 1.99 [18].

The application of the RF PECVD technique for that purpose appears to be an interesting and fruitful solution [19,20]. The technique allows one to selectively control the chemical vapor deposition processes by means of appropriate adjustment of energy and intensity of substrate bombarding ions. By using this method, Larouche et al. obtained stable filters with very good optical characteristics, stripped of harmonic reflexes. These filters were made of titanium dioxide/silicon oxide (TiO_2/SiO_2) coatings with their refractive index varying between 1.5 and 2.35, depending on the TiO_2/SiO_2 ratio [20]. Rats et al., on the other hand, produced inhomogeneous silicon oxide/silicon nitride (SiO_2/Si_3N_4) types of filters using RF/MW PE CVD technique with silane as a source of silicon and N_2, NH_3 and N_2O gas mixture as a source of nitrogen and oxygen [21]. Lin et al. obtained inhomogeneous optical SiON filters by using a PE CVD method and a helium diluted mixture of nitrogen and nitrogen suboxide. With silane as a source of silicon, the resulting magnitude of refractive index varied between 1.5 and 1.85 [22]. In the majority of works dealing with the deposition of SiON types of coatings with the help of the RF PE CVD method, silane is used as a source of silicon. It is a flammable and explosive gas and working with it requires special safety precautions. Therefore, a precious alternative for this precursor is presented by organosilicon connections. There are several literature positions dealing with a use of this group of compounds in the deposition processes of silicon-containing films [23–25]. It should be stressed, however, that these works usually concern either single homogeneous coatings [23] or stacks of a few homogeneous layers [25]. As far as an application of these precursors in the production of "rugate" filters is concerned, there is practically no information available.

The present work reports a process of deposition of silicon carbon-oxynitride (SiONC) coatings of variable composition for the purpose of the construction of a "rugate" optical filter. To do that, the RF

PE CVD technique was applied with either hexamethyldisilazane (HMDSN) or tetramethyldisilazane (TMDSN) used as a source of silicon. A mixture of gaseous N_2 and O_2 of continuously changing composition was used as a working gas. The coatings were characterized in detail with respect to their optical properties as well as their composition, chemical structure and surface morphology. Since better optical properties were characteristic of the films deposited from the HMDSN, this precursor was finally selected as a starting material for the production of a inhomogeneous SiOC/SiONC/SiNC films with the gradient of refraction index. The solution constitutes a subject of national patent procedure with several PL423097 (A1) PL233603 (B1). An application of that system as a "cold mirror" type of rugate interference filter is also described in this work.

2. Materials and Methods

2.1. Materials

Semicon p-type silicon wafers, 525 µm thick, of an <111> orientation, resistivity of 9–12 Ω·cm and dimensions of $15 \times 15 \times 0.5$ mm^3 were used as substrates for FTIR *(Fourier Transform Infrared Spectroscopy)*, XPS *(X-ray Photoelectron Spectroscopy)* and ellipsometric investigations.

Boron-silicon glass slides and fused silica plates of the dimensions $25 \times 25 \times 0.2$ mm^3 300–700 were used as substrates for UV-Vis *(Ultraviolet-Visible)*, AFM *(Atomic Force Microscopy)* and SEM *(Scanning Electron Microscopy)* studies as well as for preparation of the "cold mirror" type of interference filter.

Prior to deposition, all substrates were subjected to ultrasonic rinsing in 99.8% methanol for 10 min in order to remove surface fat and microparticle contamination.

As a source of silicon:

- Sigma Aldrich hexamethyldisilazane (HMDSN) with a purity 99% and
- abcr GmbH tetramethyldisilazane (TMDSN) with a purity 97% were used.

As working gases:

- Linde Gas Poland, Ltd. oxygen with a purity 99.999% and
- Linde Gas Poland, Ltd. nitrogen with a purity 99.999% were applied.

2.2. Deposition of the Coatings

In the present work, an RF PE CVD reactor (a schematic representation of which is given in reference [26]) was used for the purpose of deposition of SiNC, SiNOC and SiOC films. The stainless-steel jar-type deposition chamber of this reactor is furnished with the RF electrode, with the counter-electrode being the entire jar. The chamber volume amounts to approximately 50 dm^3. Working gases and precursor vapors were introduced to the reactor through separate lines terminated with shower-type distributors. Organosilicon compound container as well as (ca. 40 cm long) vapor supply lines were heated, with the former remaining at 303 K and the latter being subjected to a temperature gradient of 308 hw–413 K, in order to avoid condensation. Precursor flow rate was constant, and amounted to 6 sccm (standard cubic centimeters per minute). Initial pressure was equal 0.5 Pa and the self-bias voltage amounted to −880 V. Both the role and the amount of self-bias voltage were optimized as described in one of our earlier works [26]. The most important deposition process operational parameters are presented in Table 1, below.

Table 1. The settings of deposition parameters.

	Parameter
Initial pressure	0.5 Pa
Working pressure	20 Pa
Self-bias voltage	−880 V
Temperature of organosilicon compound	303 K
Precursor flow rate	6 sccm
Total flow rate of working gases	20 sccm

The coatings were deposited in the working atmosphere of oxygen and nitrogen with the total pressure being set constant. For that purpose, an especially designed valve system was constructed, allowing for pressure equalization in all the reactor elements. This solution is a subject of national patent procedure of several PL424592 (A1). With the help of that valve system, the reactor pressure was maintained at 20 Pa, independent of the composition of the working gas, which was changed between 100% O_2 and 100% N_2 at a constant total flow rate of 20 sccm. Each gas flow rate was independently regulated by a separate flow rate controller. Changes in working gas composition were performed step-wise, with the steps being not longer than 5 s, which ensured a smooth flow regulation. In the direction from oxide to nitride, deposition time amounted to 240 s, while that in the opposite direction was equal 60 s. Flow changes of working gas components were controlled with the MONITOR software. The acronyms of the samples used in this work are presented in Table 2, below.

Table 2. The acronyms of the samples used in this work.

Gas Composition	Tetramethyldisilazane	Hexamethyldisilazane
20 sccm N_2	T-20 N_2	H-20 N_2
12 sccmN_2/8 sccm O_2	T-12 N_2/8 O_2	H-12 N_2/8 O_2
10 sccmN_2/10 sccm O_2	T-10 N_2/10 O_2	H-10 N_2/10 O_2
8 sccmN_2/12 sccm O_2	T-8 N_2/12 O_2	H-8 N_2/12 O_2
20 sccm O_2	T-20 O_2	H-20 O_2

2.3. Optical Properties

Optical properties of the coatings, such as index of refraction (n) and extinction coefficient (k), as well as their thickness, were determined using a J.A. Woollam variable angle spectroscopic ellipsometer (VASE) and the related software. All the measurements were performed in the spectral range of 260–1000 nm for three different angles of incidence (65°, 70°, 75°), with the measurement step equal 5 nm. The measurements were made in the reflectance mode and optical properties of one component coatings, either SiOC or SiNC, were modelled with a Cauchy model layer with Urbach absorption represented by the following mathematical expression:

$$n(\lambda) = A + \frac{B}{\lambda^2} + \frac{C}{\lambda^4} \tag{1}$$

where A, B and C are parameters describing the dispersion of the refractive index $n(\lambda)$. The extinction coefficient $k(\lambda)$ was modelled by an exponential absorption tail. Fitting the procedure of A, B, and C parameters gave a mean square error of an order of magnitude of MSE = 20 (mean square error) for all the films examined. For each sample, average values of thickness and refractive index were computed from (at least) three ellipsometric measurements performed at different sites of the sample. Mean thickness values varied within the range of 260–1000 nm.

Gradient optical thin films, i.e., films with refractive index changing along their thickness, were analyzed using a graded layers algorithm supplied with the WVASE32 software by the ellipsometer manufacturer. In the gradient model, the real layer is divided into homogeneous sub-layers whose refractive indices n_i change slightly for a particular layer (characteristic jumps of the value of n).

The refractive index n profiles were determined taking into account the thickness of the film and the number of sublayers determined for the best fit of the model to the experimental data.

Transmittance of the coatings within the range of 190–1000 nm was studied with the help of ThermoScientific™ Evolution 220 UV-Vis systems. Absorbance spectra, on the other hand, were used to determine the magnitude of optical gap using the Tauc model and the following equation:

$$\alpha h\upsilon = B(h\upsilon - E_g)^m \tag{2}$$

where:
   α—denotes the absorption coefficient,
   h—denotes the Planck constant,
   υ—denotes photon frequency,
   E_g—denotes optical energy band gap,
   B—denotes a constant
   The value of m coefficient amounted to $1/2$.

2.4. Surface Topography

To determine the surface morphology of the coatings, a Bruker multimode AFM microscope, equipped with a Nanoscope V controller, was used. This enabled a collection of images 10×10 µm^2 in size. The MicroMash OTESPA type of probe had the following parameters: nominal probe radius of 7 nm, elasticity constant of 26 N/m and resonance sampling frequency of 300 kHz. All data acquisition was performed with the help of Brüker Nanoscope 7.3 software with the images being processed using the Bruker Nanoscope Analysis 1.5 application. Each measurement series was conducted on three different coatings deposited under identical conditions, with the arithmetic mean and standard deviation values being calculated from the results acquired.

2.5. XPS Spectroscopic Analysis of Elemental Composition and Chemical Structure

XPS analysis was conducted with the help of Kratos AXIS Ultra XPS spectrometer equipped with a monochromatic Al Kα source of X-rays of excitation energy of 1486.6 eV. Spectra were collected from an area of 300×700 µm^2, with the anode power amounting to 150 W. Transfer energy of a semicircular analyzer was equal 20 eV. Due to a non-conductive character of the sample surface, an additional charge neutralizer was applied. All narrow chemical shifts for each element were calibrated with respect to the most intensive spectrum component, i.e., Si (2p) maximum positioned at 101.8 eV and assigned to a Si-N bond. Sample etching was carried out down to different depths at nine different sites of the coating. The following were the etching parameters: beam energy of 4 keV, etching time of 90 s, etching rate of 26.706×0.85 nm^2/min, surface area etched of 2×2 mm^2 and surface area analyzed of 200×200 µm^2.

2.6. FTIR Analysis of Chemical Bonding

FTIR Analysis was carried out with the help of ThermoScientific model Nicolet iS50 FTIR spectrometer, working in absorbance mode and using a MCT/B beam splitter. Spectra were collected within the range of 4000 to 400 cm^{-1}, with a resolution of 4 cm^{-1}. Several scans in a single measurement cycle amounted to 128.

2.7. Design of a "Cold Mirror" Type of a Gradient Interference Filter

The gradient interference system was simulated with the help of the TFCalc™ 3.5 software from Software Spectra. The material database was updated with the optical properties of the layers obtained from HMDS mixed with pure oxygen or nitrogen, which had refractive indices of 1.65 and 2.22, respectively. In the next step, the database was enriched with materials obtained for intermediate mixtures of reactive gases. A complete set of materials served as an input to create the basic elements of a stack formula with a gradual change of optical properties. Such a package can be used for more complex systems by repeating and setting common optical thickness values. Filter design can be further developed and optimized by setting continuous or discrete target values for optical properties of the filter.

In the present work, a basic stack of 10 multiplications was designed without adaptations on the single layer level. Environmental details used for simulation, such as illumination, detector and medium properties, and light incident angle, were aligned with ThermoScientific™ Evolution 220 UV-Vis equipment, which was used to measure the optical properties of a real filter. The filter designed belongs to the category of edge filters. A longpass (LP) filter is an optical interference filter that attenuates shorter wavelengths and transmits (passes) longer wavelengths over the active range of the target spectrum (ultraviolet, visible, or infrared). Such filters are known as the "cold mirror" type. Longpass filters (also referred to as edge filters) have a very sharp slope and are described by the cut-on wavelength at 50 percent of the peak transmission. In the designed filter, its value was set at 290 nm.

2.8. The Cross-Section of the "Cold Mirror" Gradient Optical Filter

For the assessment of the film surface morphology, a Carl Zeiss ULTRAPlus scanning electron microscope (SEM), supported with the SmartSEM software, was used. The microscope was equipped with an FEG-type cathode, allowing observations to be made at low values of accelerating voltage of 0.5–30 kV. Surface charge scattering was performed using a Carl Zeiss Car Charge Compensator mechanism. For the magnification of 60,000 times, an accelerating voltage of 2.5 kV at the distance of 6 mm was applied.

3. Results

3.1. Films of a Constant Magnitude of Refractive Index

3.1.1. Optical Properties

In the case of thin solid films deposited for optical applications, the most significant tests comprise light transmission measurements in the UV-Vis range and ellipsometric examinations. Together, they supply the principal information concerning such optical parameters of the material as levels of light absorption or transmission as well as values of optical gap (E_g), index of refraction (n) and extinction coefficient (k). In addition, they also allow one to assess film thickness (d) and surface roughness (S_r). Most of these parameters can be relatively easily correlated with the structure of the coatings. This is a reason both tests were selected as simple and fast criteria for the selection of an appropriate precursor of gradient films on one hand, and for the optimization of the deposition process on the other.

Light transmission measurements were carried out within the wavelength range of 200–1000 nm. The results for the coatings prepared from tetramethyldisilazane and from hexamethyldisilazane are presented in Figure 1, below. These coatings were deposited under conditions of different composition of working gases in order to assess the effect of nitrogen-to-oxygen ratio on their optical parameters. The total flow rate of the gas mixture was constant and amounted to 20 sccm with the actual composition of that mixture being represented by the ratio of respective flow rates also expressed in sccm units.

An analysis of the results shows that all the coatings are characteristic of good optical properties revealed by an appropriate system of interference maxima. Independent of the type of precursor, films deposited under conditions of an excess of nitrogen in the reaction mixture exhibit sharper interference maxima compared to those prepared at the excess of oxygen. This is a consequence of the fact that silicon nitrides and carbonitrides are characterized by higher values of refractive index than those of the respective oxides.

A comparison of the spectra presented in Figure 1 allows one to conclude that the coatings deposited from tetramethyldisilazane are substantially thicker that those obtained from hexamethyldisilazane, since they exhibit a larger number of interference maxima. One should remember that the time of deposition was the same in all cases, and amounted to one minute. The thickness difference is a result of a substantially higher content of carbon incorporated into the coating prepared from TMDSN [26]. An increased content of carbon in the films deposited from this precursor is also revealed by their values of absorption threshold shifted to the infrared direction. The absorption threshold of the films is also dependent on the composition of the gas mixture—the films deposited at the excess of nitrogen are characteristic of a shift towards larger wavelengths (lower energy). In addition, on the contrary, the coatings prepared under conditions of oxygen excess exhibit an absorption threshold shift towards lower wavelengths and higher energy. It is interesting to note, however, that this tendency, generally shown by the films deposited from both precursors, is substantially stronger in the case of the coatings made from HMDSN.

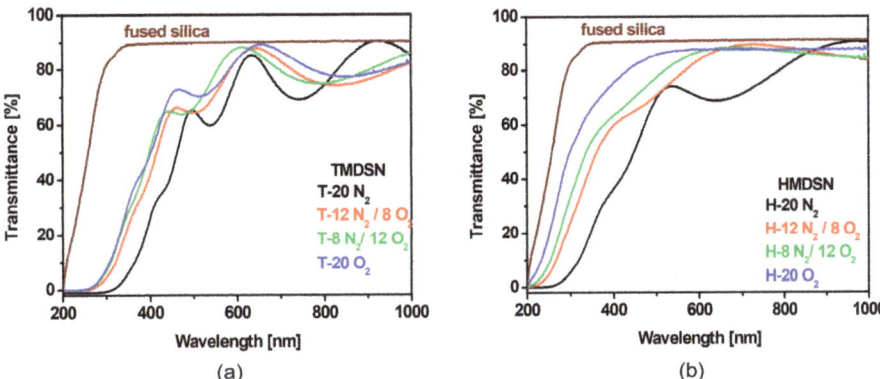

Figure 1. UV-Vis transmission spectra of the coatings deposited from TMDSN (**a**) and HMDSN (**b**) at different N_2/O_2 proportions of the working gas mixture.

The effect described above finds its confirmation in the magnitudes of the optical gap E_g calculated for the respective coatings. For that purpose, the Tauc formalism was applied to the results of absorption measurements within the wavelength range of 200–1000 nm. The resulting E_g values for the coatings prepared from both precursors at different compositions of the working gas are presented in Table 3.

Table 3. The magnitudes of optical gap E_g, index of refraction n, roughness S_r and thickness d of the coatings deposited from different organosilicon precursor compounds under conditions of three different compositions of N_2/O_2 working gas.

Gas Composition	TMDSN				HMDSN			
	E_g [eV]	n	S_r [nm]	d [nm]	E_g [eV]	n	S_r [nm]	d [nm]
20 sccm N_2	2.93	2.31	4.18	309.8	2.71	2.22	2.83	171.9
10 sccm O_2/10 sccm N_2	3.06	2.13	3.50	277.2	2.91	1.92	1.75	143.0
20 sccm O_2	3.11	1.99	3.21	247.9	2.95	1.65	0.26	136.0

In general, literature reports reveal higher magnitudes of optical gap of SiO_2 coatings than those of silicon nitride [27–29]. As seen in Table 2, the E_g value for the films deposited from TMDSN under pure oxygen conditions amounts to 3.11 eV. This result is close enough to the literature data showing the highest values of optical gap being typical for silicon oxides and remaining in the range of 3.5–9.3 eV [30,31]. The coatings made of silicon nitride are characterized by lower magnitudes of optical gap. In our case, the E_g value for the films deposited from TMDSN in pure nitrogen equals 2.93 eV, which remains in agreement with the literature data, where values within the range of 2.4–4.75 eV are reported [32]. One has to remember that the coatings presented in this work contain substantial amounts of carbon in their structure. Therefore, the respective values of silicon carbide should also be considered, and they are still lower than those of silicon nitride. Taking all the above arguments into account, one should state that a supplement of carbon lowers the magnitude of optical gap in both silicon oxide and silicon nitride coatings and the values recorded in this work remain well within the ranges reported in the literature [33–35].

An application of VASE spectroscopic ellipsometry allows one to determine the values of n and k coefficients as well as thickness and surface roughness of the material investigated. The respective data for the coatings deposited from both precursor compounds under different compositions of working gas are presented in Table 2, below. All the VASE measurements were conducted within the spectral range of 200–1000 nm, using three different angles of incidence of 65°, 70° and 75°. These values were selected in such a manner that they comprise the Brewster angle for the coatings produced. As a basis for the calculations, the Cauchy's model, often used for this type of materials [36–38], was assumed. An example of ψ angle measurement results together with the fitting of the above model for a coating prepared in pure oxygen is presented in Figure 2. When considering the relationship presented, one can observe a slight departure of the model from the experimental data within the 250–280 nm range, resulting from higher absorption in this range. Further studies have shown a responsibility for that absorption of carbon containing chemical moieties. Despite the above difference, the MSE fitting error is small and it amounts to 12.

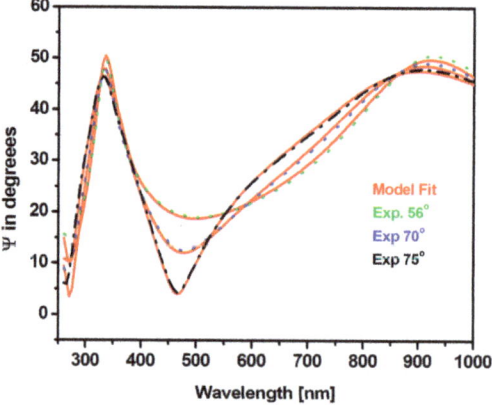

Figure 2. Experimental and calculated values of tan Ψ for the coating obtained from HMDSN under pure oxygen conditions.

The results acquired from ellipsometric measurements conducted for the coatings deposited from both TMDSN and HMDSN at different oxygen concentrations are presented in Figure 3. They show unambiguously that, by changing the composition of the gaseous reaction mixture, one is able to deposit coatings characterized by low index of refraction when depositing in pure oxygen atmosphere and those of high index of refraction when using a pure nitrogen atmosphere. When, however, a mixture of oxygen and nitrogen is used as the working gas, then depending on their proportions, a coating characterized by a refractive index within the range of 2.31–1.99 for TMDSN precursor and within the range of 2.22–1.65 for HMDSN precursor is obtained. It was thus proven in these studies that, by using a gaseous reaction mixture of a changing composition, coatings of variable optical parameters can be obtained.

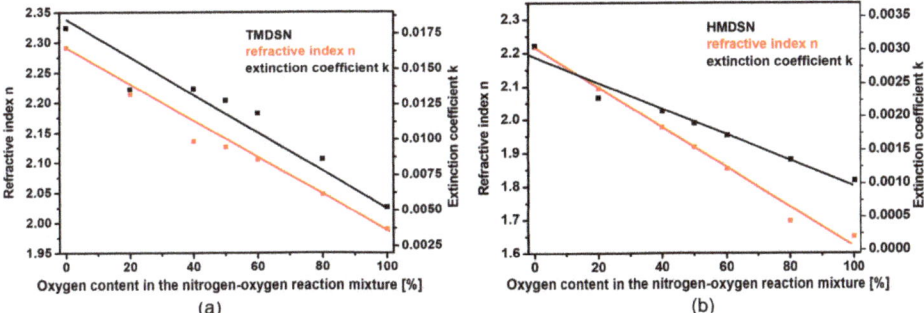

Figure 3. Dependence of optical parameters of the coatings deposited from TMDSN (**a**) and from HMDSN (**b**) on the composition of working gas mixture (oxygen–nitrogen).

The relationships presented above reveal high magnitudes of extinction coefficient of the coatings. They are still higher in the case of the films obtained under high nitrogen content conditions and lower for those deposited at high oxygen concentrations. In thin solid films, the extinction coefficient k usually depends on two factors. One is light absorption. As revealed by the UV-Vis transmission measurements, the materials under investigation primarily absorb in the 200–400 nm range, with substantially higher absorption exhibited by the coatings produced at higher nitrogen concentrations. This effect is a result of a presence of carbon in these films. Another important phenomenon is surface roughness. The rougher the surface is, the larger the fraction of radiation that undergoes scattering, thus affecting the magnitude of extinction coefficient. Ellipsometric measurements allow one to assess the roughness S_r of a given sample. The results obtained for the coatings under investigation are presented in Table 2 below. As seen in the table, surface roughness of the coatings deposited from TMDSN is far higher than that of the films obtained from HMDSN. This finding has been confirmed by surface topography examination with the help of AFM microscopy.

Ellipsometric measurements also provide data concerning sample thickness. The results presented in Table 2 show that, despite the fact that for all the samples deposition time amounts to 60 s, the coatings deposited from TMDSN are ca. two times thicker than those obtained from HMDSN. In both precursor molecules, there are two atoms of silicon. At the same time, there are four atoms of carbon in TMDSN and six atoms of carbon in HMDSN. The thickness and deposition rate differences result from the deposition mechanism, in which an important role is played by silyl radicals [39]. Due to the differences in bond energy, these radicals are much easier to form by a cleavage of Si-H bonds (E = 298 kJ/mole) than by a decomposition of Si-C bonds (E = 435 kJ/mole) [40]. A presence of the former bonds in a TMSDN molecule substantially enhances deposition, at the same time leading to an unfortunate incorporation of larger amounts of carbon.

From the results presented above, one can conclude that the optical quality of the coatings strongly depends on the composition of the reaction mixture. Those deposited in pure nitrogen atmosphere are characterized by high values of refractive index, amounting to 2.3 and to 2.2 for TMDSN and HMDSN, respectively. For samples deposited in pure oxygen, the same values amount to 1.99 and 1.65. In addition, the coatings deposited from TMDSN are of a lower optical quality than those obtained from HMDSN. An index of refraction of stoichiometric silica equals 1.45 [41]. To design and manufacture optical interference filters, one needs two materials: one of a high (n_H) and one of a low (n_L) value of refractive index. These two values should considerably differ from one another. In the case of the coatings deposited from TMDSN, that difference amounts to 0.32, and it is equal 0.57 for those obtained from HMDSN. In addition, the films deposited from TMDSN are characterized by higher absorption in the range of 200–400 nm.

3.1.2. Surface Topography Studies

To project surface topography of the coatings, AFM microscopic observations were carried out. Surface images of the samples deposited from both precursor compounds at different compositions of working gas mixture are presented in Figure 4. The images presented reveal a relatively homogeneous structure stripped of large hills and other inequities.

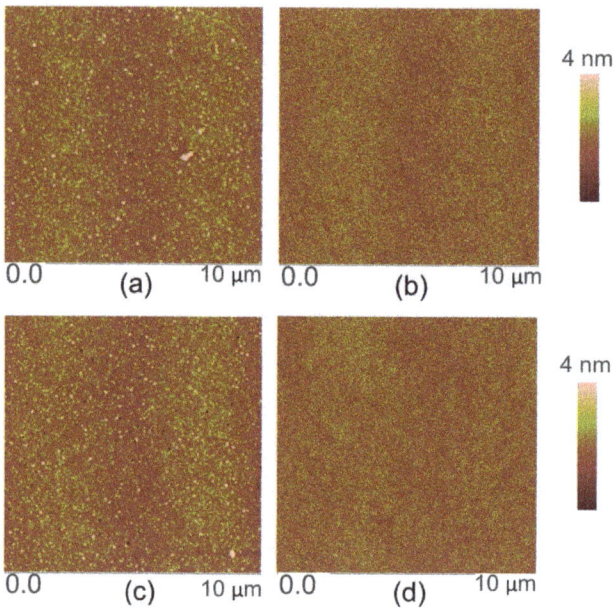

Figure 4. Surface topography of the coatings: T-20 N_2 (**a**), T-20 O_2 (**b**), H-20 N_2 (**c**), and H-20 O_2 (**d**).

Based on the AFM observations, measurements of surface roughness have been performed, and their results, such as arithmetic mean of the profile departure from the average line (R_a) and mean squared deviation of the surface roughness (R_q), are presented in Table 4 below. As seen in the table, a coating becomes smoother and smoother with an increasing concentration of oxygen. This tendency was observed in all samples, independent of whether TMDSN or HMDSN was used as a precursor compound. In addition, slightly lower values of R_a i R_q parameters were observed in the case of the coatings deposited from HMDSN, which corresponds well with the results of optical roughness measurements performed with the help of ellipsometry.

Table 4. R_a i R_q surface roughness parameters of the coatings deposited from different organosilicon precursors at different N_2/O_2 composition of the working gas mixture.

Gas Composition	TMDSN		HMDSN	
	R_a [nm]	R_q [nm]	R_a [nm]	R_q [nm]
20 sccm N_2	0.340	0.501	0.299	0.439
10 sccm O_2/10 sccm N_2	0.279	0.389	0.256	0.289
20 sccm O_2	0.187	0.241	0.156	0.203

Taking into account the results of optical measurements as well as those of surface topography, one can conclude that the coatings deposited from HMDSN precursor are characterized by better quality and higher homogeneity. It was due to these reasons that the further studies were confined to the coatings obtained from HMDSN as a precursor more suitable for a successful design and manufacture of an interference optical filter with a gradient of refractive index.

3.1.3. Elemental Composition and Chemical Structure Studies Conducted with the Help of XPS Spectroscopy

XPS studies allowed for an assessment of qualitative as well as quantitative composition of the coatings deposited under conditions of different composition of the working gas mixture. An example of a wide scan XPS spectrum of a coating deposited in pure nitrogen atmosphere is shown in Figure 5, while Table 5 presents elemental composition of the films compared with that of the HMDSN precursor compound.

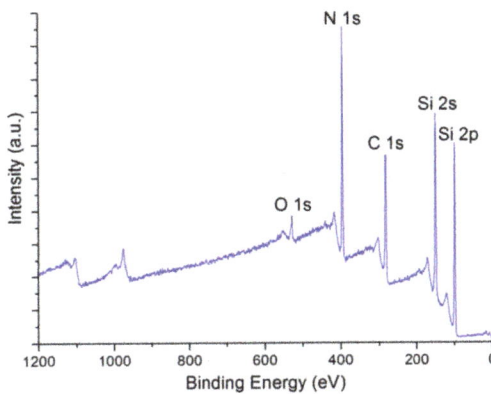

Figure 5. An example of a wide scan XPS spectrum of the H-20 N_2 coating.

Table 5. Elemental composition of the films deposited from HMDSN at different composition of the working gas mixture as well as that of the HMDSN precursor compound.

Gas Composition	Si [at.%]	N [at.%]	C [at.%]	O [at.%]
20 sccm N_2	22.7 ± 0.2	46.5 ± 0.1	25.8 ± 0.1	5.0 ± 0.2
10 sccm O_2/10 sccm N_2	23.9 ± 0.3	20.5 ± 0.2	19.9 ± 0.1	35.7 ± 0.2
20 sccm O_2	26.4 ± 0.2	5.8 ± 0.1	13.3 ± 0.2	54.4 ± 0.3
HMDSN	22.2	11.1	66.67	0.00

An analysis of the above data reveals a close relationship between the elemental composition of the coatings and the N_2/O_2 proportion in the working atmosphere. The silicon content remains at a relatively stable level of 22.7–26.4 at.% for all the samples, with a weak increasing tendency with increasing concentration of oxygen in the working gas. A similar result was reported by Chang et al. [42], with a difference being a lack of reactive atmosphere. Both in their studies and in our work, the concentration of silicon in the films remains close to its content in the precursor compound.

The amount of nitrogen in HMDSN equals 11.1 at.%. In the coating deposited in pure nitrogen atmosphere, that content increases to 46 at.%. This result strongly indicates that, under plasma conditions, nitrogen becomes a reactive gas able to form chemical bonds with other elements. In the sample prepared in the 1:1 N_2/O_2 atmosphere, the content of nitrogen drops down to 20.5 at.%, while the coating deposited under pure oxygen conditions contains 5.8 at.% of nitrogen only. In the latter case, the entire amount of this element very likely originates from the precursor and its presence in the coating may be a result of secondary processes.

Oxygen is characterized by a high affinity towards silicon. This is a reason this element content in a coating deposited under pure nitrogen atmosphere still amounts to ca. 5 at.%. To a large extent, this is supposed to be surface bound element resulting from a reaction of atmospheric oxygen with the coating surface after its exposure to the atmosphere. Following their etching, the oxygen content in the films, revealed by XPS measurements, drops down to 3.4 at.%, and this figure is entirely acceptable. In the coatings deposited under increasing oxygen concentration in the working gas, the content of oxygen increases dramatically to reach a value exceeding 54.4 at.%.

An interesting behavior is shown by carbon. Its content in the films is a subject of substantial variations, depending on the composition of the working gas mixture. Carbon content in the surface layer of a coating deposited under pure nitrogen conditions amounts to 25.8 at.%. When deposited in a mixed nitrogen/oxygen atmosphere, the coating contains 19.9 at.% carbon. Finally, for a coating produced in pure oxygen, C content further decreases to 13.3 at.%. This may be explained by a partial removal, under conditions of precursor fragmentation, of carbon in a form of its oxides CO_2 and CO. The larger oxygen concentration in the gas mixture, the more effective is that process. When film deposition is performed in pure nitrogen, no oxides are formed and carbon is removed in a form of simple hydrocarbons, principally methane. However, this process is far less efficient, which explains larger carbon concentrations in the coatings deposited under pure nitrogen conditions. In addition, it has to be noted that the content of carbon in the coatings, independent of whether deposited in the presence of oxygen or in the presence of nitrogen, is much lower that its concentration in the precursor compound. This effect points to a substantial fragmentation of the precursor molecules taking place in plasma, wherein the final content of carbon to a large extent results from secondary reactions.

Similar is the situation of nitrogen in the coatings investigated. In the case of materials deposited under pure nitrogen conditions, its content amounts to 46.5 at.%, and is over four times higher than that of the precursor compound. It convincingly indicates a reactive character of this element, thus forming Si-N bonds under plasma conditions applied. When, however, oxygen is used as the deposition process working atmosphere, the content of nitrogen in the resulting coatings deceases down to 5.8 at.%, a figure that is lower than that of the precursor compound.

Further characteristics of the coatings chemical structure is based on deconvolution of the Si 2p, N 1s and C 1s XPS spectral bands, therefore enabling a detailed analysis of chemical bonding present in that structure as a function of oxygen content in the working atmosphere. Such an analysis was performed for three different samples: one deposited under pure nitrogen condition (H-20 N_2, Figure 6), one obtained in an equimolar nitrogen/oxygen mixture (H-10 O_2/10 N_2, Figure 6) and one prepared in pure oxygen (H-20 O_2, Figure 6).

Deconvolution of the Si 2p band points to a dominant role of a Si-N bond in the samples deposited under pure nitrogen conditions. Si-O bonds are also present, confirming oxygen affinity towards silicon. Finally, the spectrum also indicates the presence of Si-C bonding. In the coatings deposited under mixed N_2/O_2 atmosphere, the amount of Si-O bonds increases and a band characteristic for SiO_2 appears, all that at the expense of decreasing contents of Si-N and Si-C bonds. When deposited under pure oxygen atmosphere, the coatings are characterized by dominant XPS bands corresponding to both Si-O and SiO_2 chemical bonding.

Deconvolution process of the XPS N 1s band confirms a dominant role of a Si-N bond in the samples deposited under pure nitrogen conditions. Along an increasing amount of oxygen in the working atmosphere, the content of both C-N and N-O bonds in the coatings also rises. In fact, an increasing amount of N-O bonds, absent in the precursor molecule, points to plasma energetic conditions allowing for a substantial fragmentation of that molecule.

On the basis of XPS results one can state that, with an increasing concentration of oxygen in the working atmosphere, the amount of silicon-oxygen bonding (102.8 eV) in the coatings increases, and that of carbon-nitrogen bonding (399.1 eV) declines. In terms of optical properties of these materials, it indicates a drop of their index of refraction "n".

Deconvolution of the C 1s XPS band reveals a presence of the following carbon bonds: Si-C (283 eV), C-C (C-H) (284.8 eV), C-N (285.5 eV), as well as C-O (285.5 eV) and C=O (287.8 eV). Here, too, one can observe a decrease of the amount of Si-C and Si-N chemical bonds present in the coatings along an increasing concentration of oxygen in the working gas mixture. A decreasing content of carbon in these films makes an additional proof of carbon removal (in the form of carbon oxides) from an oxygen rich working atmosphere.

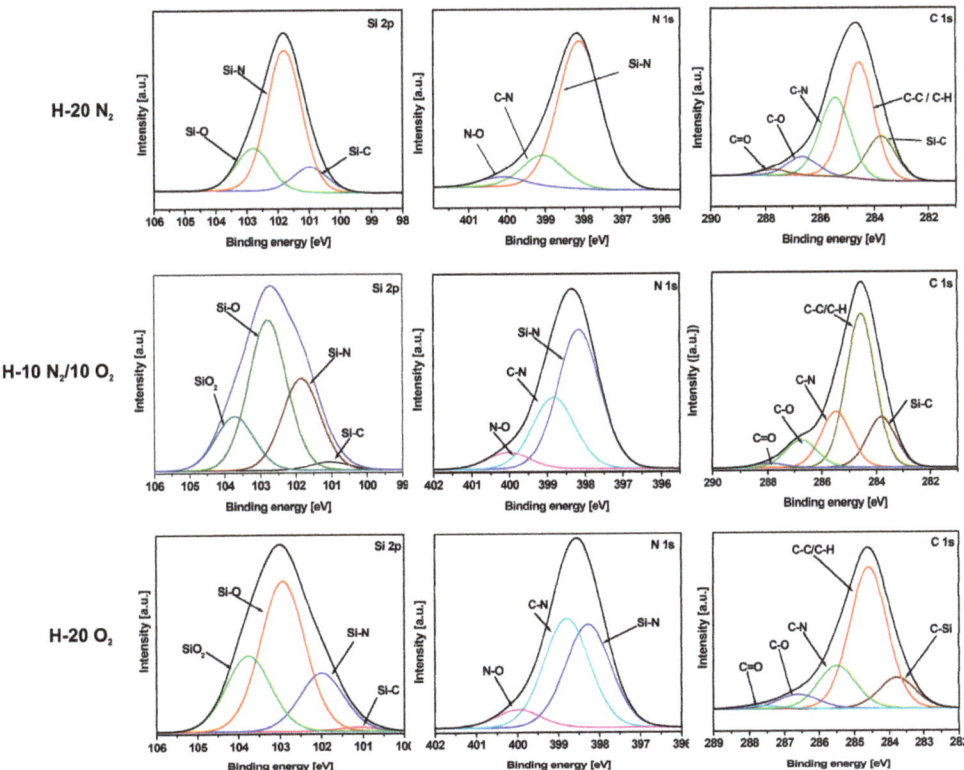

Figure 6. The Si 2p, N 1s and C 1s XPS narrow scans of films deposited from HMDSN precursor at different working gas compositions.

3.1.4. Chemical Composition Studies by FTIR Spectroscopy

Structural studies were performed on selected samples of SiONC coatings deposited from HMDSN precursor at negative self-bias potential of −880 V and three different compositions of the working atmosphere: pure oxygen, 1:1 oxygen/nitrogen mixture and pure nitrogen. The respective FTIR spectra, recorded within the spectral range of 4000–400 cm^{-1}, are presented in Figure 7. To enhance their clarity, the spectra have been divided into two ranges: 4000–1400 cm^{-1} (a) and 1300–400 cm^{-1} (b).

In the spectra presented in Figure 7a, characteristic bands corresponding to the following molecular vibrations can be identified:

- 3440–3150 cm^{-1}—N–H bond stretching vibrations,
- 3000–2800 cm^{-1}—symmetric and asymmetric stretching vibrations of C-H bonds belonging to CH$_2$ and CH$_3$ groups,
- 2200 cm^{-1}—C≡N bond stretching vibrations
- 2170–2020 cm^{-1}—Si–H bond stretching vibrations,
- 1880 cm^{-1}—C=O bond stretching vibrations,
- 1640 cm^{-1}—C=N bond stretching vibrations,
- 1550 cm^{-1}—N–H group deformation vibrations,

- 1475–1440 cm^{-1}—deformation vibrations of CH$_2$ and CH$_3$ groups [43,44].

It appears from the FTIR spectra that addition of nitrogen to the working atmosphere results in an enrichment of the coating structure with organic CH$_3$ and CH$_2$ groups originating from the precursor. In addition, a proportional increase of the amount of Si–H bond is also noted. A lowering of the C–H bond content in the films deposited in the oxygen atmosphere is interpreted as a result of a formation of volatile CO$_2$ and H$_2$O oxides in the course of the PE CVD process. These by-products are easily removed from the environment and they do not contribute to the process of deposition. As a result, the coating thickness drops down. In the case of N–H bonds, absorbing within the spectral range of 3440–3150 cm^{-1}, their existence is also observed in the case of the samples deposited under pure nitrogen atmosphere. It is connected to the presence of nitrogen in the precursor molecule. Certainly, the amount N-H bonds in the coating structure increases with an increasing concentration of nitrogen in the reaction mixture. A shift of a 2200 cm^{-1} band, corresponding to the vibrations of C–N bonds towards higher wavenumbers, is also observed for the coatings deposited under oxygen atmosphere.

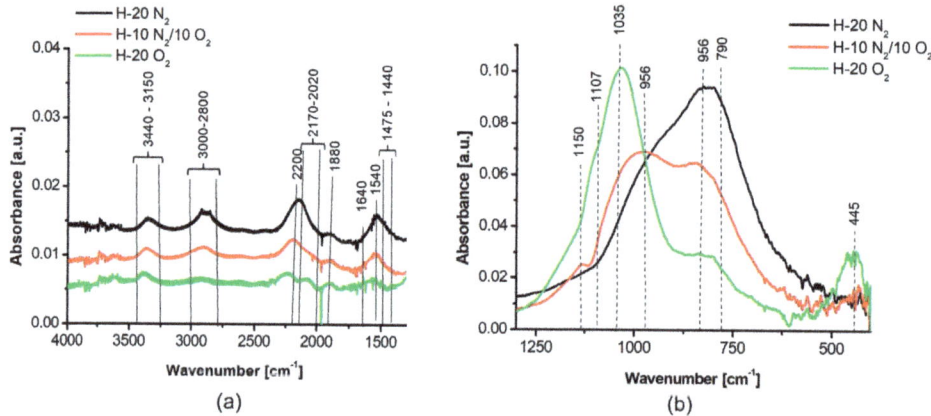

Figure 7. FTIR spectra of the coatings deposited at three different compositions of the working atmosphere and recorded within the range of 4000–1400 cm^{-1} (**a**), and 1300–400 cm^{-1} (**b**).

Within the spectral range of 1300–400 cm^{-1}, shown in Figure 7b, several bands corresponding to chemical bonds of elements having significant effect on the coating optical properties have been separated. They were subjected to a detailed deconvolution procedure with the resulting surface areas of particular band components presented in Table 6. In the case of the coatings deposited under pure oxygen atmosphere, absorption bands corresponding to the following chemical moieties were selected: Si–O–(CH$_x$) groups, as well as stoichiometric SiO$_2$ and non-stoichiometric SiO$_x$ silicon oxide bonds. The strongest band, recorded at 1035 cm^{-1} and corresponding to stretching vibrations of stoichiometric silicon dioxide Si-O bonds, is accompanied by a substantial maximum at 445 cm^{-1}, resulting from rocking vibrations of that group. A presence of such a large amount of oxygen in the coating is responsible for a low value of its refractive index equal 1.65. An increase of nitrogen content in the working environment results in an increasing number of Si–NH–Si bonds absorbing at 953 cm^{-1} as well as, typical for ceramic materials, Si–N bonds resonant with the radiation of the wavenumber of 830 cm^{-1}. Additionally, Si–C bonds, typical for silicon carbide, are formed under these conditions. A presence of both these bond structures is responsible for an increase of the coatings refractive index up to 2.2. This means that, by a strict control of nitrogen content in the working N$_2$/O$_2$ atmosphere, one should be able to substantially affect chemical structure of the films and, therefore, to regulate, within a certain range, the magnitude of refractive index n of the resulting coating.

Table 6. Surface areas of particular components of a broad IR absorption band recorded at 1300–700 cm^{-1} for selected coatings deposited from HMDSN (key: s—stretching vibrations; r—rocking vibrations).

Type of Bond	Wavenumber [cm^{-1}]	Area [a.u.]		
		H-20 N$_2$	H-10 N$_2$/10 O$_2$	H-20 O$_2$
Si–O–(CH$_x$)$_x$ (s) [45]	1150	0.6	1.5	3.0
Si–O (s) [43,45]	1107	-	-	0.8
Si–O (s) in SiO$_2$ [43,45]	1035	-	5.1	12.1
Si–NH–Si (s) [45]	956	10.9	3.3	0.7
Si–N (s) [45]	830	13.2	9.2	0,1
Si–C (s) [43,45]	790	3.1	2.0	0.2
Si–O (r) [43,45]	445	-	0.1	0.6

3.2. Coatings Characterized by a Gradient of Refractive Index

3.2.1. Optical Properties

The results presented thus far concerned coatings deposited under different, but constant, proportions of nitrogen to oxygen in the working atmosphere. This also means a constant composition of these coatings. In the present chapter, results obtained for materials deposited under conditions of a variable composition of that atmosphere are reported. The composition of the working gas was changed every 5 s, with the subsequent O$_2$/N$_2$ proportions presented in Table 7. These proportions were selected in such a manner that in every case the total flow rate of the working gas was equal to 20 sccm. Such an arrangement resulted in a step-wise alteration of the coating chemical composition and, consequently, introduced a gradient of refractive index. Several sequences of a coating were produced, from a single nitrogen–oxygen sequence up to the most complex nitrogen–oxygen–nitrogen–oxygen–nitrogen sequence. The single nitrogen–oxygen sequence resulted in a film thickness of approximately 110 nm, while proportionally larger thickness values were acquired for the multiple sequences.

Table 7. Oxygen/nitrogen proportions applied in subsequent steps of a deposition of a gradient refractive index coating.

Flow rate O$_2$ [sccm]	0	4	8	12	16	20
Flow rate N$_2$ [sccm]	20	16	12	8	4	0

Profiles of refractive index and extinction coefficient for particular coatings were determined with the help of spectroscopic ellipsometry. In the model applied for the calculations, respective values of these optical parameters obtained for different (but constant) compositions of deposition working atmosphere, were used. The resulting profiles for a coating deposited under the working atmosphere changing from 100% nitrogen to 100% oxygen are presented in Figure 8 below.

On the basis of the models proposed, one can judge that the shape of the real profile does not substantially depart from that theoretically assumed. In the model, the magnitude of the refractive index varies in the range of 1.65–2.22, while ellipsometric data show respective values from the 1.65–2.38 range, with the course of changes remaining relatively close to that assumed.

In a similar way to the above data, the results of a two-fold sequence of a gradient coating deposition exhibit a satisfactory fitting of the proposed model. In this case, the refractive index was assumed to vary according to the model change of 2.25–1.65–2.25, while ellipsometric measurements showed that index to follow the 2.38–1.67–2.28 pattern. The difference may indicate the contribution of residual oxygen, lowering the refractive index, in the deposition process.

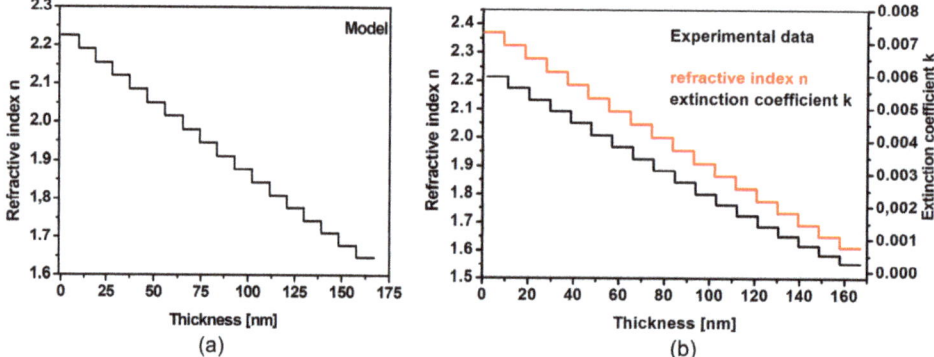

Figure 8. Profiles of refractive index and extinction coefficient of a SiONC coating, deposited under conditions of variable nitrogen-to-oxygen proportions in the working atmosphere, as a function of the film thickness: theoretical model (**a**) and the real profile recorded by ellipsometric measurements (**b**).

A similar approach was applied in the case of the coating characterized by a four–fold nitrogen–oxygen–nitrogen–oxygen–nitrogen sequence. The profiles of theoretical and real values of refractive index of that coating are presented in Figure 9. In all instances, a good agreement between the model assumed and the measurement results have been observed, thus pointing to a high stability of the deposition process.

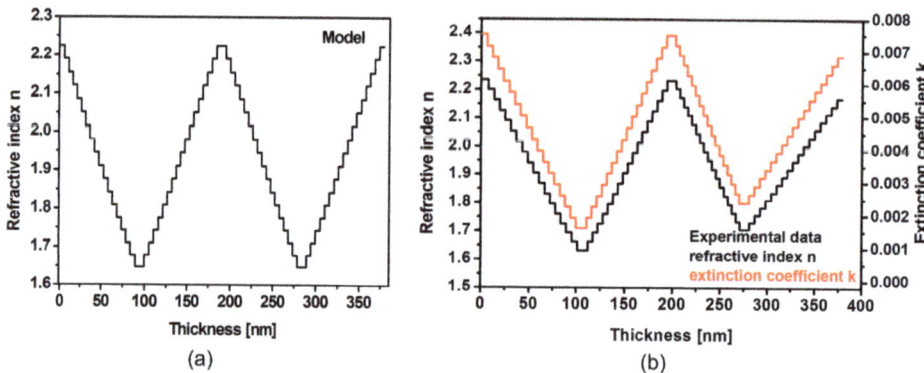

Figure 9. Model (**a**) and real profiles of refractive index and extinction coefficient changes (**b**) for a coating deposited in a N_2–O_2–N_2–O_2–N_2 sequence.

Materials with a gradient of refractive index were also investigated by means of UV-Vis spectroscopy. Figure 10 presents transmission spectra of the coatings deposited with a one-fold, two-fold and fourfold change of the composition of working atmosphere.

It appears from Figure 10 that the effect of optical interference is recorded even in the case of the coatings deposited with a one-fold N–O change of working atmosphere. As far as materials produced with the two-fold N–O–N and four-fold N–O–N–O–N processes are concerned, an evident cutback of absorption at low wavelengths is observed in this case. What is important to note, is the fact that light interference takes place despite a lack of separate layers characterized by different magnitudes of refractive index. In fact, the effect develops from a gradient of refractive index. This finding constitutes proof of application potential of the coatings investigated in the technology of monolayer optical filters with a gradient of refractive index.

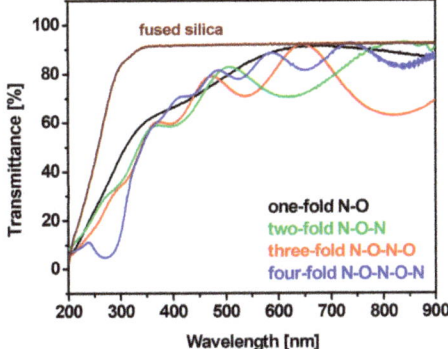

Figure 10. UV-Vis transmission spectra of the coatings deposited with a one-fold N–O, two-fold N–O–N and fourfold N–O–N–O–N change of working atmosphere.

3.2.2. Elemental Composition Profiles of the Coatings

XPS studies of the coatings allow one to follow changes of elemental composition of these materials as a function of their thickness. Such elemental composition profiles of a coating deposited with a two-fold N–O–N change of working atmosphere are presented in Figure 11, below.

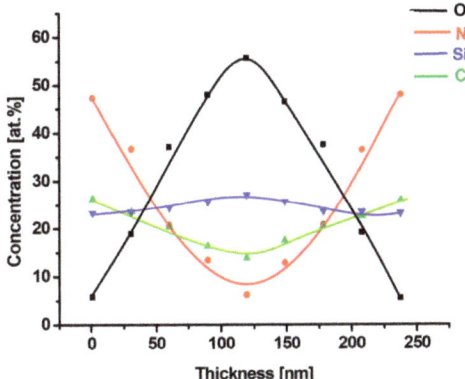

Figure 11. Elemental composition in-depth profiles of a coating deposited with a two-fold N–O–N change of working atmosphere.

It appears from the figure that a change of working gas composition, although having a significant impact on the chemical structure of the resulting material, does not affect the concentration of silicon in the coating, which continues to stay at the level of 24–28 at.% along the entire thickness of the sample. In contrast to that, the contents of the remaining elements are subjected to considerable changes. The amount of both nitrogen and oxygen in the coating is fundamentally dependent on the respective gas concentration in the working atmosphere. While being ca. 46 at.% for the best samples deposited under pure nitrogen conditions, the content of nitrogen drops down to 5.8 at.% in the case of a material obtained in pure oxygen. A return to the 100% N_2 atmosphere results in the reestablishment of high nitrogen content in the films. One can conclude that the concentration of this element in the coatings varies in a periodic manner, following periodic alterations of the N_2/O_2 proportion in the working atmosphere. Similar is the behavior of oxygen. In the case of the best samples deposited under pure nitrogen conditions, its amount in the coating totals 2.8 at.%, only to increase up to 54 at.% for samples obtained in pure oxygen. Just as is the case with nitrogen, the concentration of oxygen in the coatings varies in a periodic manner, following periodic alterations of the O_2/N_2 ratio in the working atmosphere.

The concentration of carbon in the films follows the nitrogen-to-oxygen proportions in the working gas. However, these changes are not so evident, and they result from the periodicity of chemical reactions taking place in the glow discharge. As discussed earlier, the coatings deposited under oxygen-rich conditions are stripped of carbon because of a swift removal of volatile carbon oxides from the reaction mixture.

3.3. "Cold Mirror" Interference Optical Filter with a Gradient of Refractive Index

The effect of optical interference taking place in the coating deposited with the four-fold exchange of the working atmosphere led us to believe that it was possible to design and to construct the entire interference filter with a gradient change of refractive index. The design of a "cold mirror" type filter was accomplished with the use of a TFCalc.™ 3.5 (*transmission filter calculation*) specialized software. An initial stage of that design comprised the construction of a database containing all the magnitudes of refractive index and extinction coefficient of the respective materials obtained at different compositions of the working atmosphere. It was then assumed that the filter of interest will be of a band type with the absorption threshold set at 300 nm. The calculations showed that it takes a ten-fold exchange of the working gas atmosphere along the nitrogen–oxygen–nitrogen sequence to obtain of desired filter structure. The designed gradient change of refractive index for such a filter is presented below, in Figure 12.

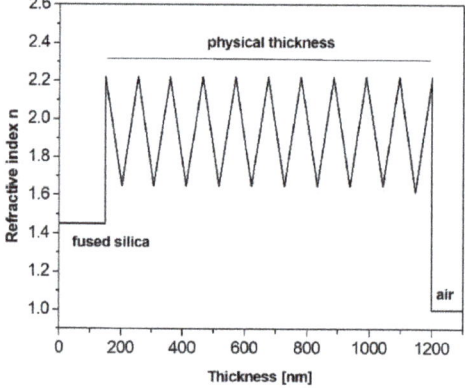

Figure 12. A model gradient change of refractive index of the "cold mirror" interference filter.

Reflection spectra of the filter model as well as of the real device are presented in Figure 13. It appears from the figure that the optical characteristics of the real filter are not much different from those of the model. According to the design, the 50% reflection level should be observed at the wavelength of 290 nm, while in reality it is recorded at 313 nm. This difference is likely to be a result of deposition rate fluctuations taking place in the course of the changes of the working atmosphere. It may also be a consequence of a moderate temperature increase in the reactor during deposition. In any case, a manufacture of the presented filter constituted a test of both the deposition equipment and the technology developed. This proves the high stability and high repeatability of the process, thus opening a path to depositing much more complex optical coatings.

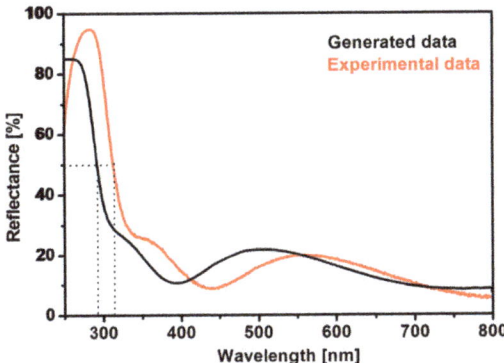

Figure 13. Reflection spectra of the "cold mirror" type of interference filter: theoretical model, and real device.

The coatings with a continuous change of refractive index resulting from gradients of their chemical composition, constitute a subject of studies carried out in numerous laboratories. Some publications deal with the methods using two different sources of material. There are reports of gradient coating deposition from solid SiO_2, Ta_2O_5, or Nb_2O_5 targets using sputtering techniques. Optical properties of the coatings acquired are good [9,46,47]. However, the deposition technique itself is somewhat problematic, because of different sputtering rates of different targets. Retaining appropriate proportions between particular materials require a continuous control of the magnetron sputtering power. It is also possible to sputter materials with a use of either electron or ion beams, but these techniques are very costly both in the capital investment phase and in the course of operation. Use of two different precursor compounds is also practiced in the PE CVD technologies. In the work cited, HSDSN and $TaCl_5$ were used as sources of low and high refractive index materials, respectively. The gradient coatings were deposited with the help of plasma enhanced technology using oxygen as one of the reactive mixture components [48]. Aggressive chlorine-containing chemicals are obtained as by-products in this technique, which enforces the application of highly (chemically) resistant materials in the construction of deposition equipment.

In the literature, there are reports concerning coatings with a gradient of physical properties based on a single precursor compound. These publications usually deal with silane SiH_4 playing the role of that precursor [22,49]. As working gases in these works, mixtures of oxygen and nitrogen of different proportions are applied, with the resulting structure of the PE CVD coatings varying from silicon oxide to silicon nitride through intermediate materials of a SiO_xN_y composition. Silane, however, is a very volatile, flammable and explosive chemical, and its use requires extra safety precautions. The method of deposition of gradient coatings described in the present work removes the above-mentioned inconveniences. Instead of a hazardous gas, easy to handle liquid organosilicon compounds are used as precursors. As far as literature reports are concerned, thus far there has been none whatsoever that refers to gradient coatings deposited from these precursors.

The optical "cold mirror" filter, whose model and real transmission spectra are presented in Figure 13, has also been investigated with the help of scanning electron microscopy. The SEM image of the manufactured filter cross-section is presented in Figure 14.

In the micrograph presented above, one can identify subsequent "sublayers" present in the optical coating. The thickness of the coating equals ca. 1045 nm. From the design assumptions, that thickness should amount to 1050 nm. That agreement is quite satisfactory, and it proves the quality of the technology developed. That technology, and the method of deposition of single-layer optical coatings with the gradient of refractive index using organosilicon precursor and oxygen-nitrogen mixture as working atmosphere in particular, constitutes a subject of patent application number PL423097 (A1), PL233603 (B1).

Figure 14. SEM image of the cross-section of the optical interference filter manufactured.

4. Conclusions

In the work presented, a novel method of deposition of optical coatings with a gradient of refractive index, using the RF PECVD technology, is introduced. Compared to the existing literature reports, this work is unique in the sense that it presents a successful endeavor to obtain a complex "optical system with a gradient of refractive index by means of the plasma-chemical technique, using a single organometallic precursor and a variable oxygen-to-nitrogen ratio in the working atmosphere. The unquestioned advantage of this work is an elimination, as precursor compounds, of explosive silane and corrosive silicon halogenides and their replacement with safe and easy to handle liquid organosilicon compounds. Two organosilicon precursors, namely TMDSN and HMDSN, were used as precursor compounds. As the reaction atmosphere, a mixture of oxygen and nitrogen was applied. An application of pure components resulted in a formation of SiOC films of low refractive index in the case of oxygen and SiN:C films of high refractive index in the case of nitrogen. The coatings deposited from TMDSN were characterized by refractive index magnitudes of 1.99 and 2.31 for the SiOC and for the SiNC materials, respectively. Similar data for the films obtained from HMDSN amounted to 1.65 and to 2.22. On the basis of the above results, HMDSN was selected as the more promising precursor of the gradient coatings, characterized by both chemical composition and refractive index varying with their thickness. Those gradients were acquired by an application of the variable composition of reaction atmosphere. Sinusoidal alterations of that composition resulted in a deposition of the coatings with sinusoidal changes in their optical properties. The method developed was used to manufacture a "cold mirror" type of interference filter. The filter is characterized by a good agreement between its real parameters and those assumed in the model, which reveals a good stability of the deposition process established.

5. Patents

PL423097 (A1) PL233603 (B1), K. Olesko, A. Sobczyk-Guzenda, H. Szymanowski, S. Owczarek, Method for producing one-layered optical filters with the light refractive index gradient, 23 April 2019.

PL424592 (A1), K. Olesko, A. Sobczyk-Guzenda, A. Nosal, Method for supplying with vapours of the precursor of reactors intended for applying coatings by vacuum methods and the system for supplying with vapours of the precursor of reactors intended for applying coatings by vacuum methods, 26 August 2019.

Author Contributions: Conceptualization, H.S. and A.S.-G., methodology, H.S., K.O., and A.S.-G.; validation, K.O. and A.S.-G.; formal analysis, K.O. and A.S.-G.; investigation, K.O., H.S., M.F., A.S.-G., and J.K.; resources, H.S., K.O., and A.S.-G.; data curation, K.O. and J.K.; writing—original draft preparation, H.S. and A.S.-G.; writing—review and editing, H.S., A.S.-G., and M.G.-L.; visualization, H.S. and A.S.-G.; supervision, H.S. and A.S.-G.; project management, A.S.-G. All authors have read and agreed to the published version of the manuscript.

Funding: The research was founded by the National Science Centre (NCN), Poland, Grant No 2014/13/B/ST8/04293.

Conflicts of Interest: The authors declare no conflict of interest.

References

1. Ejigu, E.K. Simulating radiation thermometer temperature measurement error from theperformance change of an interferencefilter due to polarization effect. *Measurement* **2020**, *114*, 471–477. [CrossRef]
2. Li, H.; Wang, K.; Qian, L. Tunable color filter with non-subwavelength grating at oblique incidence. *Optik* **2020**, *207*, 164432. [CrossRef]
3. Hussain, T.; Jie, T.; Fung, W.J.; Nabeel, M.; Xiao, H.D. UV radiation protection for space telescope FPA using cerium. *Radiat. Phys. Chem.* **2018**, *153*, 159–163. [CrossRef]
4. Syafiq, A.; Pandey, A.K.; Adzman, N.N.; Rahim, N.A. Advances in approaches and methods for self-cleaning of solar photovoltaic panels. *Sol. Energy* **2018**, *162*, 597–619. [CrossRef]
5. Gralewicz, G.; Owczarek, G. Analysis of the selected optical parameters of filters protecting against hazardous infrared radiation. *Int. J. Occup. Saf. Ergon.* **2016**, *22*, 305–309. [CrossRef]
6. Macleod, H.A. *Thin-Film Optical Filters*, 3rd ed.; Institute of Physics Publishing: Bristol, VA, USA, 2001; pp. 1–5.
7. Jun-Chao, Z.; Ming, F.; Yu-Chuan, S.; Yun-Xia, J.; Hong-Bo, H. The synchronization of a fractional order hyperchaotic system based on passive control. *Chin. Phys.* **2011**, *20*, 1–4.
8. Vernhes, R.; Zabeida, O.; Klemberg-Sapieha, J.E.; Martinu, L. Single-material inhomogeneous optical filters based on microstructural gradients in plasma-deposited silicon nitride. *Appl. Opt.* **2004**, *43*, 97–103. [CrossRef]
9. Leitel, R.; Stenzel, O.; Wilbrandt, S.; Gäbler, D.; Janicki, V.; Kaiser, N. Optical and non-optical characterization of Nb_2O_5–SiO_2 compositional graded-index layers and rugate structures. *Thin Solid Film.* **2006**, *497*, 135–141. [CrossRef]
10. Ouellette, M.F.; Lang, R.W.; Yan, K.L.; Bertram, R.W.; Owles, R.S.; Vincent, D. Experimental studies of inhomogeneous coatings for optical applications. *J. Vac. Sci. Technol. A* **1991**, *9*, 1188–1992. [CrossRef]
11. Redinger, A.; Mousel, M.; Djemour, R.; Gütay, L.; Valle, N.; Siebenritt, S. $Cu_2ZnSnSe_4$ thin film solar cells produced viaco-evaporation and annealing including a$SnSe_2$ capping layer. *Prog. Photovolt. Res. Appl.* **2014**, *22*, 51–57. [CrossRef]
12. Donovan, E.P.; van Vechten, D.; Kahn, A.D.F.; Carosella, C.A.; Hubler, G.K. Near infrared rugate filter fabrication by ion beam assisted deposition of $Si_{(1-x)}N_X$ films. *Appl. Opt.* **1989**, *28*, 2940–2944. [CrossRef] [PubMed]
13. Kaminska, K.; Brown, T.; Beydaghyan, G.; Robbie, K. Vacuum evaporated porous silicon photonic interference filters. *Proc. SPIE Int. Soc. Opt. Eng.* **2003**, *4833*, 633–639. [CrossRef] [PubMed]
14. Berger, M.G.; Arens-Fischer, R.; Thönissen, M.; Krüger, M.; Billat, S.; Lüth, H.; Hilbrich, S.; Theiss, W.P. Grosse, Dielectric filters made of PS: Advanced performance by oxidation and new layer structures. *Thin Solid Film.* **1997**, *297*, 237–240. [CrossRef]
15. Cunin, F.; Schmedake, T.A.; Link, J.R.; Li, Y.; Koh, J.; Bhatia, S.; Sailor, M. Biomolecular screening with encoded porous-silicon photonic crystals. *Nat. Mater.* **2002**, *1*, 39–41. [CrossRef] [PubMed]
16. Chhasatiaa, R.; Sweetman, M.J.; Prieto-Simon, B.; Voelckerd, N.H. Performance optimisation of porous silicon rugate filter biosensor for the detection of insulin. *Sens. Actuators B Chem.* **2018**, *273*, 1313–1322. [CrossRef]
17. Zhang, J.; Fang, M.; Jin, Y.; Hongbo, H. Narrow line-width filters based on rugate structure and antireflection coating. *Thin Solid Film* **2012**, *520*, 5447–5450. [CrossRef]
18. Bartzsch, H.; Lange, S.; Frach, P.; Goedicke, K. Graded refractive index layer systems for antireflective coatings and rugate filters deposited by reactive pulse magnetron sputtering. *Surf. Coat. Technol.* **2004**, *180*, 616–620. [CrossRef]

19. Lim, S.; Ryu, J.H.; Wager, J.F.; Casas, L.M. Inhomogeneous dielectrics grown by plasma-enhanced chemical vapor deposition. *Thin Solid Film* **1993**, *236*, 64–66. [CrossRef]
20. Larouche, S.; Szymanowski, H.; Klemberg-Sapieha, J.E.; Martinu, L.; Gujrathi, S.C. Microstructure of plasma-deposited SiO_2/TiO_2 optical films. *J. Vac. Sci. Technol. A* **2004**, *22*, 1200–1207. [CrossRef]
21. Rats, D.; Poitras, D.; Soro, J.M.; Martinu, L.; von Stebut, J. Mechanical properties of plasma-deposited silicon-based inhomogeneous optical coatings. *Surf. Coat. Technol.* **1999**, *111*, 220–228. [CrossRef]
22. Lim, S.; Shih, S.; Wager, L.F. Design and fabrication of a double bandstop rugate filter grown by plasma-enhanced chemical vapor deposition. *Thin Solid Film* **1996**, *277*, 144–146. [CrossRef]
23. Schäfer, J.; Hnilica, J.; Šperka, J.; Quade, A.; Kudrle, V.; Foest, R.; Vodák, J.; Zajíčková, I. Tetrakis(trimethylsilyloxy)silane for nanostructured SiO_2-like films deposited by PECVD at atmospheric pressure. *Surf. Coat. Technol.* **2016**, *295*, 112–118. [CrossRef]
24. Mahajan, A.M.; PatilJ, L.S.; Bange, P.; Gautam, D.K. TEOS-PECVD system for high growth rate deposition of SiO_2 films. *Vacuum* **2005**, *79*, 194–202. [CrossRef]
25. García-Valenzuela, A.; Butterling, M.; Liedke, M.O.; Hirschmann, E.; Trinh, T.T.; Attallah, A.G.; Wagner, A.; Alvarez, R.; Gil-Rostra, J.; Rico, V.; et al. Positron annihilation analysis of nanopores and growth mechanism of oblique angle evaporated TiO_2 and SiO_2 thin films and multilayers. *Microporous Mesoporous Mater.* **2020**, *295*, 109968. [CrossRef]
26. Sobczyk-Guzenda, A.; Oleśko, K.; Gazicki-Lipman, M.; Szymanski, W.; Balcerzak, J.; Wendler, B.; Szymanowski, H. Chemical structure and optical properties of $Si_xN_yC_z$ coatings synthesized from two organosilicone precursors with the RF PECVD technique—A comparative study. *Mater. Res. Express.* **2019**, *6*, 016410.1-21.
27. Brinkmann, N.; Sommer, D.; Micard, G.; Hahn, G.; Terheiden, B. Electrical, optical and structural investigation of plasma-enhanced chemical-vapor-deposited amorphous silicon oxynitride films for solar cell applications. *Sol. Energy Mater. Sol. C* **2013**, *108*, 180–188. [CrossRef]
28. DiStefano, T.H.; Eastman, D.E. The band edge of amorphous SiO_2 by photoinjection and photoconductivity measurements. *Solid State Commun.* **1971**, *9*, 2259–2261. [CrossRef]
29. Weinberg, Z.A.; Rubloff, G.W.; Bassous, E.; Watson, T. Transmission, photoconductivty, and the experimental band gap of thermally grown SiO2 films. *Phys. Rev. B* **1979**, *19*, 3107–3118. [CrossRef]
30. Efimov, A.; Weber, M.J. *Handbook of Optical Materials*; CRC Press: Boca Raton, FL, USA; London, UK; New York, NY, USA; Washington, DC, USA, 2003; p. 68.
31. Sakamoto, N.; Umezu, I.; Maeda, K. A comparative study on structural and electronic properties of PECVD a-SiO_x with a-SiN_x. *J. Non Cryst. Solids* **1995**, *187*, 287–290.
32. Deshpande, S.V.; Gulari, E.; Brown, S.W.; Rand, S.C. Optical properties of silicon nitride films deposited by hot filament chemical vapor deposition. *J. Appl. Phys.* **1995**, *77*, 6534–6541. [CrossRef]
33. Cheng, Y.; Huang, X.; Du, Z.; Xiao, J. Effect of sputtering power on the structure and optical band gap of SiC thin films. *Opt. Mater.* **2017**, *73*, 723–728. [CrossRef]
34. Momeni, A.; Pourgolestani, M.; Taheri, M.; Mansour, N. Enhanced red photoluminescence of quartz by silicon nanocrystals thin film deposition. *Appl. Surf. Sci.* **2018**, *434*, 674–680. [CrossRef]
35. Khatami, Z.; Wilson, P.R.J.; Wojcik, J.; Mascher, P. The influence of carbon on the structure and photoluminescence of amorphous silicon carbonitride thin films. *Thin Solid Film* **2017**, *622*, 1–10. [CrossRef]
36. Tompkins, H.G.; Hilfiker, J.N. *Spectroscopic Ellipsometry: Practical Application to Thin Film Characterization*; Momentum Press: New York, NY, USA, 2015.
37. Jaglarz, J.; Dulian, P.; Karasinski, P. Thermo-optical properties of porous silica thin films produced by sol-gel metod. *Mater. Chem. Phys.* **2020**, *243*, 122603. [CrossRef]

38. Gómez-Varela, A.I.; Castro, Y.; Durán, A.; de Beule, P.A.A.; Flores-Arias, M.T.; Bao-Varela, C. Synthesis and characterization of erbium-doped SiO$_2$-TiO$_2$ thin films prepared by sol-gel and dip-coating techniques onto commercial glass substrates as a route for obtaining active GRadient-INdex materials. *Thin Solid Film* **2015**, *583*, 115–121.
39. Blaszczyk-Lezak, I.; Wrobel, A.M.; Aoki, T.; Nakanishi, Y.; Kucinska, I.; Tracz, A. Remote nitrogen microwave plasma chemical vapor deposition from a tetramethyldisilazane precursor. 1. Growth mechanism, structure, and surface morphology of silicon carbonitride films. *Thin Solid Film* **2006**, *497*, 24–34. [CrossRef]
40. Cottrell, T.L. *The Strengths of Chemical Bonds*, 2nd ed.; Butterworth: London, UK, 1958.
41. Gracia, F.; Yubero, F.; Holgado, J.P.; Espinos, J.P.; Gonzalez-Elipe, A.R.; Girardeau, T. SiO$_2$/TiO$_2$ thin films with variable refractive index prepared by ion beaminduced and plasma enhanced chemical vapor deposition. *Thin Solid Film* **2006**, *500*, 19–26. [CrossRef]
42. Chang, W.; Chang, C.; Leu, J. Optical properties of plasma-enhanced chemical vapor deposited SiCxNy films by using silazane precursors. *Thin Solid Film* **2017**, *636*, 671–679. [CrossRef]
43. Lambert, J.B.; Shurvell, H.F.; Graham Cooks, R. *Introduction to Organic Spectroscopy*, 1st ed.; Macmillan: New York, NY, USA, 1987; pp. 174–177, ISBN 0023673001.
44. Silverstein, R.M.; Bassler, G.C.; Morrill, T.C. *Spectrometric Identification of Organic Compounds*, 5th ed.; Wiley: New York, NY, USA, 1991; pp. 72–118, ASIN B00CIFTXXK.
45. 45. Anderson, D.R. *Analysis Silicones*; Lee Smith, A., Ed.; Wiley-Interscience: New York, NY, USA, 1974; Chapter 10; ISBN 10: 0471800104.
46. Juškevičius, K.; Audronis, M.; Subačius, A.; Kičas, S.; Tolenis, T.; Buzelis, R.; Drazdys, R.; Gaspariūnas, M.; Kovalevskij, V.; Matthews., A.; et al. Fabrication of Nb$_2$O$_5$/SiO$_2$ mixed oxides by reactive magnetron co-sputtering. *Thin Solid Film* **2015**, *589*, 95–104.
47. Liu, H.; Chen, S.; Ma, P.; Pu, Y.; Qiao, Z.; Zhang, Z.; Wei, Y.; Liu, Z. Ion beam sputtering mixture films with tailored refractive indices. *Opt. Laser Technol.* **2014**, *55*, 21–25. [CrossRef]
48. Bauer, S.; Klippe, L.; Rothhaar, U.; Kuhr, M. Optical multilayers for ultra-narrow bandpass filters fabricated by PICVD. *Thin Solid Film* **2003**, *442*, 189–193. [CrossRef]
49. Linkens, D.A.; Abbod, M.F.; Metcalfe, J.; Nichols, B. Modeling and fabrication of optical interference rugate filters. *ISA Trans.* **2001**, *40*, 3–16. [CrossRef]

© 2020 by the authors. Licensee MDPI, Basel, Switzerland. This article is an open access article distributed under the terms and conditions of the Creative Commons Attribution (CC BY) license (http://creativecommons.org/licenses/by/4.0/).

Article

Deposition Mechanism and Thickness Control of CVD SiC Coatings on Nextel™440 Fibers

Yi Wang [1,*], **Jian Sun** [1], **Bing Sheng** [1] **and Haifeng Cheng** [2]

1 Unit 96901 of People's Liberation Army, Beijing 100094, China; jackgfkd@163.com (J.S.); chenjsaea@163.com (B.S.)
2 Science and Technology on Advanced Ceramic Fibers and Composites Laboratory, National University of Defense Technology, Changsha 410073, China; zyjcfc@163.com
* Correspondence: wycfcnudt@163.com; Tel.: +86-10-6634-5471

Received: 17 February 2020; Accepted: 15 April 2020; Published: 20 April 2020

Abstract: SiC coatings were successfully synthesized on Nextel™440 fibers by chemical vapor deposition (CVD) using methyltrichlorosilane as the original SiC source at 1373 K. After deposited, the fibers were fully surrounded by uniform coatings with some bulges. The X-ray diffraction (XRD), X-ray photoelectron spectroscopy (XPS) and high-resolution transmission electron microscopy (HR-TEM) results indicated that the coatings were composed of β-SiC and free carbon. Moreover, thickness control of the coatings could be carried out by adjusting the deposition time. The coating thickness rose exponentially, and the exterior of the coatings became looser as the deposition time increased. The thickness of about 1.5 µm was obtained after depositing for 4 h. The coating thickness was also theoretically calculated, and the result agreed well with the measured thickness. Finally, the related deposition mechanism is discussed and a deposition model is built.

Keywords: SiC coatings; oxide fibers; chemical vapor deposition; deposition mechanism; thickness control

1. Introduction

SiC ceramic has attracted extensive attention due to its excellent performance, such as appropriate high-temperature strength, relatively high oxidation, corrosion and thermal-shock resistance [1–4], etc. SiC ceramic can be used in coating materials, especially in fiber-reinforced ceramic matrix composites (FRCMCs) to improve the bond strength between the fiber and matrix. After introducing SiC interphases, weak fiber/matrix interfaces can be obtained in FRCMCs and several toughening mechanisms like crack deflection, fiber debonding, bridging and subsequent pullout can occur, all of which contribute to damage tolerance [5–7]. Our previous works proved that SiC interphases were suitable for the strength improvement of high-temperature structural and functional materials due to its oxidation tolerance and relatively temperature-stable dielectric characteristics [8,9].

SiC coatings have been successfully deposited by CVD on silicon carbide fibers, carbon fibers, oxide fibers and more [10–12]. Most research has focused on the coatings' effects on microstructure and mechanical properties of those fibers, the strength improvement of the fiber-reinforced ceramic matrix composites, or the degradation mechanism of the coatings [13,14]. However, research on the deposition mechanism and thickness control of CVD SiC coatings on oxide fibers is rare.

In the present study, SiC coatings were synthesized on Nextel™440 fibers by CVD using the gas system of CH_3SiCl_3–H_2–Ar. The coating thickness was adjusted by varying the deposition time, and the deposition mechanism was investigated by the aid of microstructure and composition analysis.

2. Experimental Procedure

2.1. Materials and Deposition of Coatings

Nextel^{TM}440 fiber fabrics (BF-30, 3M, St. Paul, MN, USA) were employed as the sample for deposited coating. The fiber, with a diameter of about 11 μm, was composed of Al_2O_3 (70 wt.%), SiO_2 (28 wt.%) and B_2O_3 (2 wt.%) [15]. Methyltrichlorosilane (MTS, CH_3SiCl_3) was used as the SiC precursor, with hydrogen as carrier gas and argon as diluent and protective gas, respectively. As shown in Figure 1, the deposition process of SiC coating was carried out in a hot-wall vertical reactor. Prior to the deposition, the fiber fabrics were desized, ultrasonically cleaned in acetone and dried in an oven. The deposition was performed at 1373 K under a total pressure of 5.0 kPa, with hydrogen and argon flow rates of 200 and 75 sccm, respectively. The coatings were deposited for 1, 2 and 4 h, respectively.

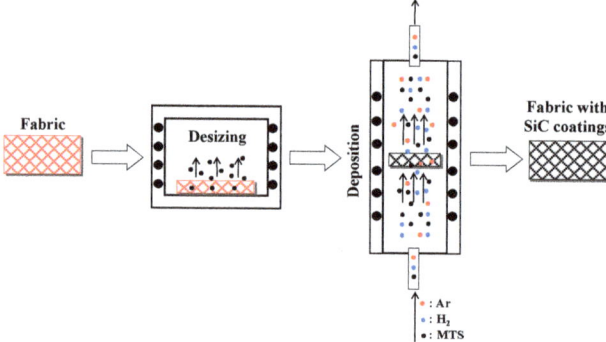

Figure 1. Deposition process of SiC coatings on Nextel^{TM}440 fibers.

2.2. Characterization

The microstructure of the fibers without and with coatings was characterized by scanning electron microscope (SEM) equipped with energy dispersed spectroscopy (EDS) (HITACHI FEG S4800, HITACHI, Tokyo, Japan). The phase composition of the fibers without and with coatings was analyzed by XRD equipped with a D8 ADVANCE diffractometer using monochromatic Cu Kα radiation (Bruker, Hamburg, Germany). The chemical composition of the fibers without and with coatings was traced by XPS using Al Kα radiation of energy 1486.6 eV (Thermo ESCALAB 250, Thermo Scientific, Waltham, MA, USA). TEM analysis was carried out on Tecnai F20 operating at 200 kV, and the samples were prepared according to [16].

3. Results and Discussion

3.1. Characteristics of the CVD SiC Coatings

The surface of the fibers without and with SiC coatings was observed by SEM. As shown in Figure 2, the as-received fibers showed a smooth and homogeneous surface, except for a few white sheet alumina micro-grains. After depositing for 2 h with SiC coatings, the fiber surface was no longer smooth and a large number of micro-bulges with the largest diameter of about 1 μm appeared. The coatings were generally dense, without any micro-cracks or pores.

The phase composition of the fibers without and with SiC coatings was analyzed by XRD. As shown in Figure 3a, diffraction peaks of mullite and γ-Al_2O_3 were detected for the as-received fibers. After being coated with SiC coatings (Figure 3b), three accessional peaks (2θ = 35.60°, 60.06° and 71.83°) were observed, which respectively belonged to the (111), (220) and (311) crystal planes of β-SiC. It is clear that the coatings were mainly composed of β-SiC phase [17].

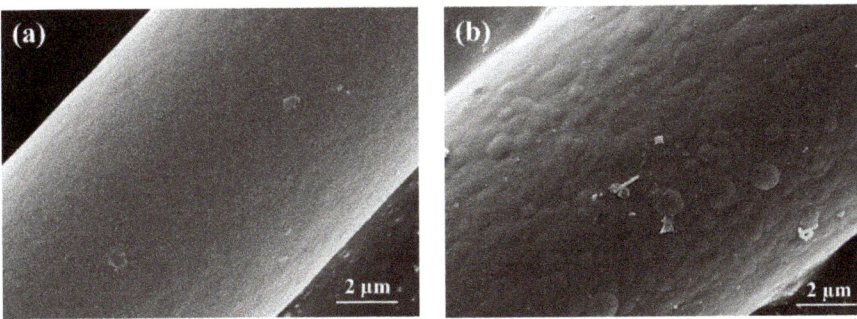

Figure 2. Surface morphology of the fibers without (**a**) and with (**b**) coatings.

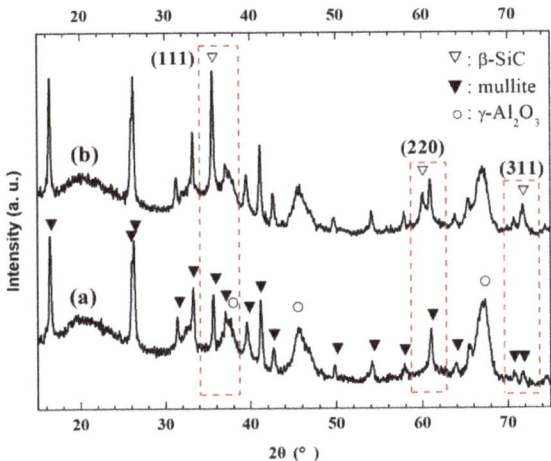

Figure 3. XRD patterns of the fibers without (**a**) and with (**b**) SiC coatings.

The chemical composition of the fibers without and with SiC coatings was analyzed by XPS. As shown in the survey scan (Figure 4a), the peaks of Al 2s and Al 2p disappeared after depositing with SiC coatings. Element analysis of the coated fibers revealed that the content of C increased while the contents of Al, Si and O decreased greatly. Moreover, the Cl 2$p_{3/2}$ (~198 eV) peak and Cl 2s (~271 eV) peak were detected [18], indicating the absorption of HCl. As shown in Figure 4b, the C 1s peak could be fitted into two sub peaks, C–Si bond (282.6 eV) and C–C bond (283.2 eV) [19], which was attributed to SiC and carbon, respectively. As shown in Figure 4c, the Si 2p peak positioned at 100.9 eV was also attributed to Si–C bonding. In conclusion, the coatings were composed of β-SiC and free carbon.

An HR-TEM image of SiC coatings deposited for 2 h on the fibers is displayed in Figure 5a. It could be found that the coatings consisted of large quantities of turbostratic carbon and some cubic β-SiC micro-grains oriented in the (111) crystal plane with 0.250 nm spacing. The result was coincident with XPS analysis and revealed that the coatings were rich in carbon. As shown in the SAED pattern (Figure 5b), the diffraction rings could be indexed as (111), (220) and (311) planes of cubic β-SiC [20].

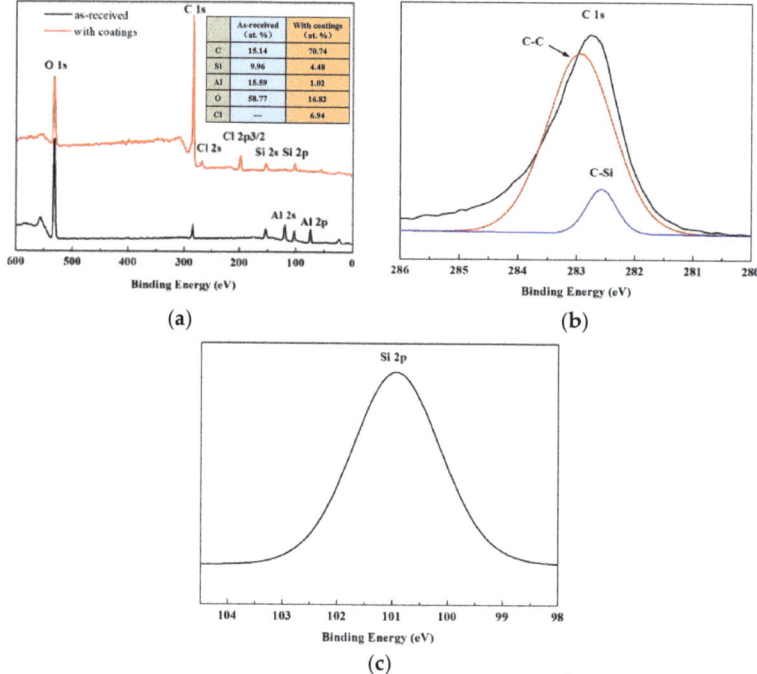

Figure 4. XPS analysis of the fibers: (**a**) survey scan of as-received ones and those with coatings, (**b**) C 1s spectra and (**c**) Si 2p spectra of those with coatings.

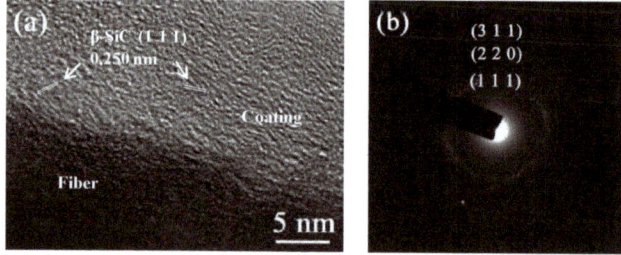

Figure 5. TEM analysis of the fibers with coatings deposited for 2 h: (**a**) HR-TEM image and (**b**) SAED pattern of coatings.

3.2. Thickness Control of the CVD SiC Coatings

The cross-sectional morphology of the samples deposited for different amounts of time is shown in Figure 6. It is apparent that the fibers were surrounded closely by SiC coatings. The coating thickness was measured 10 times at different points to obtain the average value. As shown in Table 1, the coating thickness increased with the deposition time, and the value of about 1.5 μm was obtained after depositing for 4 h. Figure 7 shows the effect of the deposition time on the coating thickness. It could be found that the thickness of the coatings deposited for 1, 2 and 4 h increased exponentially with the deposition time. However, the exterior of the coatings became looser as the deposition time increased above 2 h, owing to the reduction of the adhesion force on the coating surface, thus leading to the deposition difficulty of new raw materials [21,22].

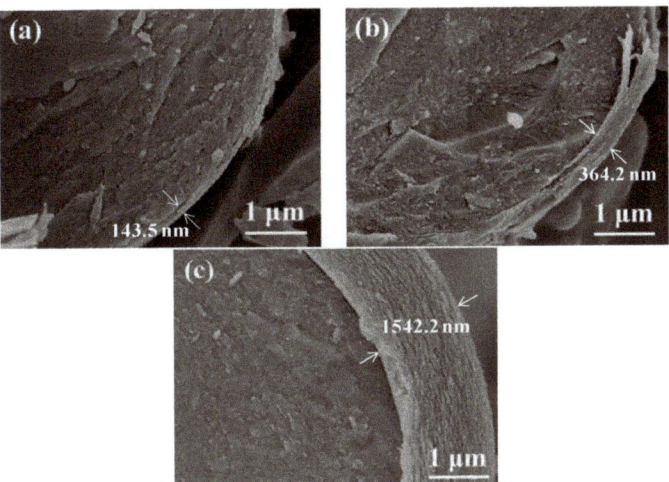

Figure 6. Cross-sectional morphology of the coatings deposited for: (**a**) 1 h, (**b**) 2 h and (**c**) 4 h.

Table 1. Measured thickness and theoretical thickness of the coatings.

Deposition Time (h)	1	2	4
Measured thickness (nm)	143.5 ± 14.2	364.2 ± 12.5	1542.2 ± 17.0
Theoretical thickness (nm)	181.5	499.3	2001.2

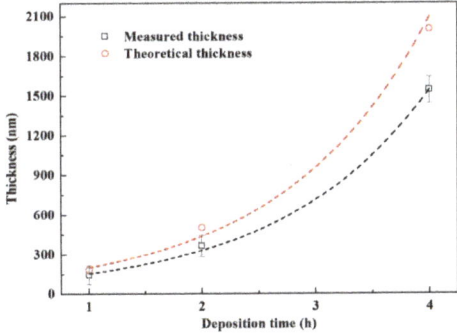

Figure 7. Measured thickness and theoretical thickness of the coatings deposited for different amounts of time.

In this study, the coating thickness (d) was also calculated according to the corresponding relationship between volume and weight for the coatings and fibers, which was summarized by Y. Zheng et al. [23]:

$$d = r_0 \left[(1 + \eta c)^{\frac{1}{2}} - 1 \right] \quad (1)$$

where r_0 is the original fiber radius; $c = (\rho_f/\rho_c)$, where ρ_c and ρ_f are the density of the coatings and fibers, respectively; and $\eta = (\omega_c/\omega_f)\%$, where ω_c and ω_f are the weight of the coatings and fibers.

The original fiber radius (r_0) was about 10 µm. The density of the pure SiC was 3.20 g/cm^3. As the coatings were rich in carbon, the coating density (ρ_c) was not constant, and should be calculated according to the volume mixing law, where the volume ratio of the carbon and SiC in the coatings can be obtained by element ratio from XPS analysis. Then, the parameter ρ_c can be expressed as:

$$\rho_c = 2.764\, e^{-0.02t} \tag{2}$$

Combining Equations (1) and (2), and adopting $\rho_f = 3.0$ g/cm³, the coating thickness could be obtained and the results are shown in Table 1:

$$d = r_0\left[\left(1 + 1.085\eta e^{0.02t}\right)^{\frac{1}{2}} - 1\right] \tag{3}$$

As shown in Figure 7, both the measured and calculated coating thickness increased exponentially with the deposition time. The results of both methods were quite consistent in general. The calculated thickness was a little larger than the measured one. There were two main reasons: one was that the volume ratio of the carbon in the coatings was not accurate due to the ladder-like decrease of the carbon with the deposition time; the other one was that the coatings were a little loose. As a result, the actual coating density was different from the theoretical one. However, the calculated thickness could still reflect the actual one, and it was meaningful for the thickness control of CVD coatings.

3.3. Deposition Mechanism Research

The chemical reaction process of CVD SiC coatings using MTS as a raw material was investigated to explain the deposition mechanism for the coatings. As the bond energy for C–Si, C–H and Si–Cl are 314, 337 and 466 J/mol, respectively, the breakdown of the C–Si bond occurred first at high temperature. The reaction process can be summarized as follows [24,25]:

$$CH_3SiCl_3 \leftrightarrow CH_3\cdot + SiCl_3\cdot \tag{4}$$

$$CH_3\cdot \leftrightarrow \langle C\rangle + 3/2 H_2 \tag{5}$$

$$SiCl_3\cdot + 3/2 H_2 \leftrightarrow \langle Si\rangle + 3HCl \tag{6}$$

$$\langle Si\rangle + \langle C\rangle \rightarrow SiC \tag{7}$$

$$\langle Si\rangle \rightarrow Si \tag{8}$$

$$\langle C\rangle \rightarrow C \tag{9}$$

where <Si> and <C> correspond to the intermediate states of Si and C, which can form SiC as shown in Equation (7). It can be concluded from Equations (5)–(7) that hydrogen has a great effect on the formation of silicon or free carbon during the CVD process. That is to say, pure SiC coatings can hardly be prepared via CVD process, and they will always be rich in silicon or free carbon. It is reported in the literature that coatings deposited on carbon plate, fibers or composites by CVD using MTS at temperatures above 1200 °C are usually composed of SiC, and Si impurities form at lower deposition temperature [25–27]. In this study, the coatings deposited on Nextel™440 fibers by CVD using MTS at 1100 °C were composed of SiC and free carbon.

In addition, the cross-sectional morphology of the coated fibers (Figure 6a–c) revealed that the coatings were loose. In order to investigate structure and composition of the interior of the coatings, the fibers with SiC coatings deposited for 2 h were ultrasonically treated to thin the coatings from 364.2 to 97.6 nm. As shown in Figure 8a,b, many hemispherical bulges with uniform diameter of about 4 μm with a disorderly distribution on the fiber surface. The magnified SEM image (Figure 8c) shows that the plane area of the coatings was dense and smooth, without any defects. Furthermore, the EDS analysis (Figure 8e–f) revealed that after ultrasonic treatment, the fibers were totally covered by thin carbon layers with few SiC bulges. Compared to the original coatings (Figure 4b), no Cl peak was detected, indicating a dense structure for the interior of the coatings, without any absorption of HCl.

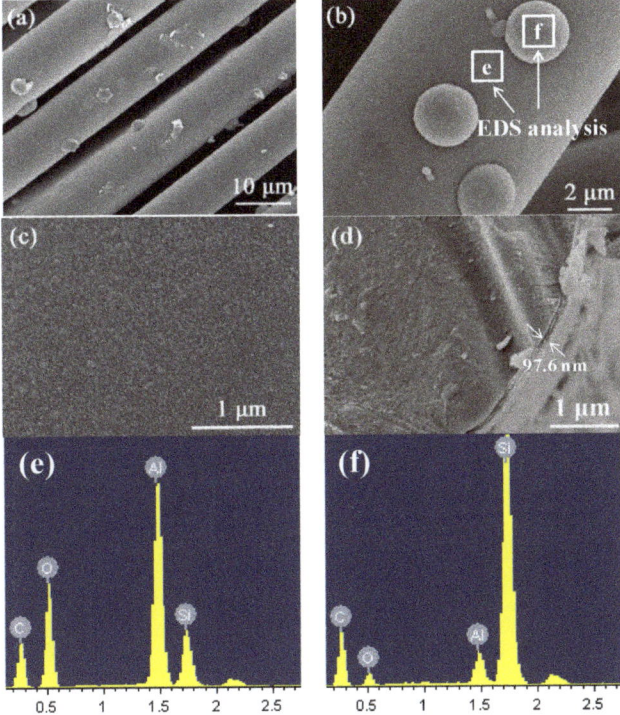

Figure 8. Surface morphology and EDS analysis of the coatings deposited for 2 h after ultrasonic treatment: (**a**) overview image, (**b**) detailed image with marked areas for the EDS measurement, (**c**) detailed image for smooth area of the fiber surface, (**d**) detailed image for the fracture surface, (**e**) EDS measurement of smooth area and (**f**) EDS measurement of bulges.

As a result, the deposition mechanism can be described. As shown in Figure 2b and schematized in Figure 9a, there were a few white sheet alumina micro-grains with disorderly distribution on the original fibers. Theses micro-grains not only increased the specific surface area but also had high catalysis activity, which led to the < Si > and < C > attaching onto them and reacting to form spherical SiC particles rapidly. At the same time, MTS was decompounding and depositing on fibers per Equations (4)–(9) to form carbon-rich SiC coatings (as shown in Figure 9b). As the deposition process continued and the coating thickness increased, the growth rate of spherical SiC particles and deposition rate of SiC coatings gradually became close to each other, and a relatively smooth surface was finally obtained (Figure 9c). As mentioned previously, SiC coatings became looser as the coating thickness increased due to the reduction of adhesion force on the coating surface.

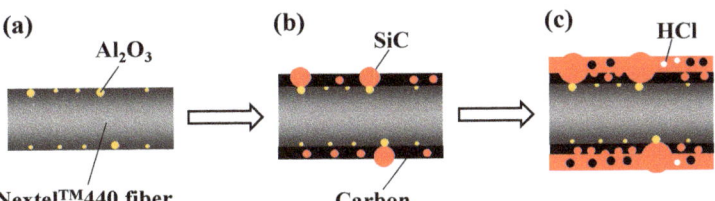

Figure 9. Schematic of the deposition mechanism for SiC coatings on Nextel^{TM}440 fibers: (**a**) the original fiber, (**b**) initial stage of deposition and (**c**) later stage of deposition.

4. Conclusions

We investigated the deposition mechanism of CVD SiC coatings on Nextel™440 fibers, and thickness control of the coatings was actualized by varying the deposition time. The main conclusions can be summarized as follows:

- Carbon-rich SiC coatings were synthesized on Nextel™440 fibers by CVD. Some traces of bulges were observed on the coating surface.
- CVD SiC coatings with different thickness were obtained by varying the deposition time. As the deposition time increased from 1 to 4 h, the coating thickness increased from 143.5 nm to 1.5 μm accordingly. An empirical formula was put forward to calculate the coating thickness, and the calculated thickness was quite coincident with the measured one.
- The deposition mechanism of the coatings was discussed. Spherical SiC particles were formed at a high rate on alumina micro-grains at the fiber surface, and the coatings became looser with increasing thickness due to the reduction of the adhesion force on the coating surface.

Coatings composed of β-SiC and free carbon could be used for interface engineering of oxide fiber-reinforced composites. After preparation of the composites under inert atmosphere, free carbon could be oxidized to obtain porous SiC interphases, which could improve the fiber/matrix interface characteristics. Further work will focus on interface engineering of Nextel™440-reinforced mullite composite by porous SiC interphases. The damage mode and strength of the composites will be systematically investigated.

Author Contributions: Conceptualization, Y.W. and H.C.; Data curation, J.S. and B.S.; Investigation, Y.W., J.S. and H.C.; Methodology, Y.W. and B.S.; Writing—original draft, Y.W.; Writing—review and editing, H.C. All authors have read and agreed to the published version of the manuscript.

Funding: This research was funded by the National Natural Science Foundation of China (NSFC) under Grant No. 51602347.

Conflicts of Interest: The authors declare no conflicts of interest.

References

1. Naslain, R. Design, preparation and properties of non-oxide CMCs for application in engines and nuclear reactors: An overview. *Compos. Sci. Technol.* **2004**, *64*, 155–170. [CrossRef]
2. Koyanagi, T.; Katoh, Y.; Nozawa, T.; Snead, L.-L.; Kondo, S.; Henager, C.-H., Jr.; Ferraris, M.; Hinoki, T.; Huang, Q. Recent progress in the development of SiC composites for nuclear fusion applications. *J. Nucl. Mater.* **2018**, *511*, 544–555. [CrossRef]
3. Arai, Y.; Inoue, R.; Goto, K.; Kogo, Y. Carbon fiber reinforced ultra-high temperature ceramic matrix composites: A review. *Ceram. Int.* **2019**, *45*, 14481–14489. [CrossRef]
4. Shen, Z.-Z.; Chen, J.-H.; Li, B.; Li, G.-Q.; Zhang, Z.-J.; Hou, X.-M. Recent progress in SiC nanowires as electromagnetic microwaves absorbing materials. *J. Alloy. Comp.* **2020**, *815*, 152388. [CrossRef]
5. Chawla, K.-K. Interface engineering in mullite fiber/mullite matrix composites. *J. Eur. Ceram. Soc.* **2008**, *28*, 447–453. [CrossRef]
6. Zhang, M.-Y.; Li, K.-Z.; Shi, X.-H.; Tan, W.-L. Effects of SiC interphase on the mechanical and ablation properties of C/C-ZrC-ZrB$_2$-SiC composites prepared by precursor infiltration and pyrolysis. *Mater. Des.* **2017**, *122*, 322–329. [CrossRef]
7. Zhou, W.; Long, L.; Li, Y. Mechanical and electromagnetic wave absorption properties of C$_f$-Si$_3$N$_4$ ceramics with PyC/SiC interphases. *J. Mater. Sci. Technol.* **2019**, *35*, 2809–2813. [CrossRef]
8. Tian, H.; Liu, H.-T.; Cheng, H.-F. A high-temperature radar absorbing structure: Design, fabrication, and characterization. *Compos. Sci. Technol.* **2014**, *90*, 202–208. [CrossRef]
9. Wang, Y.; Cheng, H.-F.; Wang, J. Mechanical and Dielectric Properties of Mullite Fiber-Reinforced Mullite Matrix Composites with Single Layer CVD SiC Interphases. *Int. J. Appl. Ceram. Technol.* **2015**, *12*, 500–509. [CrossRef]

10. Liu, H.-T.; Cheng, H.-F.; Wang, J.; Tang, G.-P. Effects of the single layer CVD SiC interphases on the mechanical properties of the SiCf/SiC composites fabricated by PIP process. *Ceram. Int.* **2010**, *36*, 2033–2037. [CrossRef]
11. Xiang, Y.; Li, W.; Wang, S.; Chen, Z.-H. Effects of the single layer CVD SiC interphases on the mechanical properties of the C/SiC composites fabricated by PIP process. *Mater. Sci. Eng. A* **2012**, *558*, 451–455.
12. Wang, Y.; Cheng, H.-F.; Wang, J. Effects of the single layer CVD SiC interphases on mechanical properties of mullite fiber-reinforced mullite matrix composites fabricated via a sol–gel process. *Ceram. Int.* **2014**, *40*, 4707–4715. [CrossRef]
13. Yu, H.-J.; Zhou, X.-G.; Zhang, W.; Peng, H.-X.; Zhang, C.-R. Mechanical behavior of SiC$_f$/SiC composites with alternating PyC/SiC multilayer interphases. *Mater. Des.* **2013**, *44*, 320–324. [CrossRef]
14. Chen, S.-A.; Zhang, Y.-D.; Zhang, C.-R.; Zhao, D.; Hu, H.-F.; Zhang, Z.-B. Effects of SiC interphase by chemical vapor deposition on the properties of C/ZrC composite prepared via precursor infiltration and pyrolysis route. *Mater. Des.* **2013**, *46*, 497–502. [CrossRef]
15. 3M^{TM} Nextel^{TM} Ceramic Fibers and Textiles Technical Reference Guide. 2018. Available online: http://www.3M.com/ceramics (accessed on 9 February 2019).
16. Hay, R.-S.; Welch, J.-R.; Cinibulk, M.-K. TEM specimen preparation and characterization of ceramic coatings on fiber tows. *Thin Solid Films* **1997**, *308–309*, 389–392. [CrossRef]
17. Youm, M.-R.; Yun, S.; Choi, S.-C.; Park, S.-W. Synthesis of β-SiC powders by the carbothermal reduction of porous SiO$_2$–C hybrid precursors with controlled surface area. *Ceram. Int.* **2020**, *46*, 4870–4877. [CrossRef]
18. Jiamprasertboon, A.; Dixon, S.-C.; Sathasivam, S.; Powell, M.-J.; Lu, Y.; Siritanon, T.; Carmalt, C.-J. Low-Cost One-Step Fabrication of Highly Conductive ZnO:Cl Transparent Thin Films with Tunable Photocatalytic Properties via Aerosol-Assisted Chemical Vapor Deposition. *ACS Appl. Electron. Mater.* **2019**, *1*, 1408–1417. [CrossRef]
19. Zhang, C.; Qu, L.; Yuan, W.-J. Effects of Si/C ratio on the phase composition of Si-C-N powders synthesized by carbonitriding. *Materials* **2020**, *13*, 346. [CrossRef]
20. Dietrich, D.; Martin, P.-W.; Nestler, K.; Stöckel, S.; Weise, K.; Marx, G. Transmission electron microscopic investigations on SiC- and BN-coated carbon fibres. *J. Mater. Sci.* **1996**, *31*, 5979–5984. [CrossRef]
21. Górka, J.; Czupryński, A.; Żuk, M. Properties and structure of deposited nanocrystalline coatings in relation to selected construction materials resistant to abrasive wear. *Materials* **2018**, *11*, 1184. [CrossRef]
22. Czupryński, A. Flame spraying of aluminum coatings reinforced with particles of carbonaceous materials as an alternative for laser cladding technologies. *Materials* **2019**, *12*, 3467. [CrossRef] [PubMed]
23. Zheng, Y.; Wang, S.-B. Synthesis of boron nitride coatings on quartz fibers: Thickness control and mechanism research. *Appl. Surf. Sci.* **2011**, *257*, 10752–10757. [CrossRef]
24. Reznik, B.; Gerthsen, D.; Zhang, W.-G.; Hüttinger, K.-J. Microstructure of SiC deposited from methyltrichlorosilane. *J. Eur. Ceram. Soc.* **2003**, *23*, 1499–1508. [CrossRef]
25. Zhang, W.-G.; Hüttinger, K.-J. CVD of SiC from Methyltrichlorosilane. Part II: Composition of the Gas Phase and the Deposit. *Chem. Vap. Depos.* **2001**, *7*, 173–181. [CrossRef]
26. Long, Y.; Javed, A.; Chen, Z.-K.; Xiong, X.; Xiao, P. Deposition rate, texture, and mechanical properties of SiC coatings produced by chemical vapor deposition at different temperatures. *Int. J. Appl. Ceram. Technol.* **2013**, *10*, 11–19. [CrossRef]
27. Zhou, W.; Long, Y. Mechanical properties of CVD-SiC coatings with Si impurity. *Ceram. Int.* **2018**, *44*, 21730–21733. [CrossRef]

© 2020 by the authors. Licensee MDPI, Basel, Switzerland. This article is an open access article distributed under the terms and conditions of the Creative Commons Attribution (CC BY) license (http://creativecommons.org/licenses/by/4.0/).

Article

Deposition of Titanium Dioxide Coating by the Cold-Spray Process on Annealed Stainless Steel Substrate

Noor irinah Omar [1,2,*], Santirraprahkash Selvami [1], Makoto Kaisho [1], Motohiro Yamada [1], Toshiaki Yasui [1] and Masahiro Fukumoto [1]

1. Department of Mechanical Engineering, Toyohashi University of Technology,1-1, Tempaku-Cho, Toyohashi, Aichi 441-8580, Japan; santirraprakash@gmail.com (S.S.); Kaisho.makoto.ik@tut.jp (M.K.); yamada@me.tut.ac.jp (M.Y.); yasui@tut.jp (T.Y.); fukumoto@tut.jp (M.F.)
2. Faculty of Mechanical Engineering and Manufacturing Technology, Technical University of Malacca, Durian Tunggal, Melaka 76100, Malaysia
* Correspondence: noor.irinah.binti.omar.dj@tut.jp; Tel.: +81-080-9390-6329

Received: 18 September 2020; Accepted: 15 October 2020; Published: 17 October 2020

Abstract: The surface of most metals is covered with thin native oxide films. It has generally been believed that to achieve bonding, the oxide covering the surface of metallic particles or metal substrates must be broken and removed by adiabatic shear instability (ASI), whether induced at the particle–substrate interface or at the particle–particle interface. The aim of the present research is to investigate the correlation between the remaining oxide amorphous layer and substrate-deformation with the adhesion strength of cold-sprayed TiO_2 coatings towards the bonding mechanism involved. Relevant experiments were executed using stainless steel (SUS 304), subjected to various annealing temperatures and cold-sprayed with TiO_2 powder. The results indicate an increasing trend of coating adhesion strength with increasing annealed substrate temperature. The influence of substrate plastic deformation and atomic intermixing at the remaining amorphous oxide layer is discussed as the factors contributing to the increasing adhesion strength of cold-sprayed TiO_2 coatings.

Keywords: adiabatic shear instability (ASI); cold spray; titanium dioxide; bonding mechanism; adhesion strength; substrate deformation; amorphous interface layer

1. Introduction

Cold spraying is a process in which the powder particles are used to form a coating by means of ballistic impingement upon a suitable substrate. Powders range in particle sizes from 5 to 100 μm and are accelerated by injection into a high-velocity stream of gas. The particles are then accelerated by the main nozzle gas flow and impacted on the substrate. Upon impact, the solid particles deform and create a bond with the substrate [1]. When solid particles are sprayed toward a substrate, there are various phenomena that are generally observed on the substrate surface in relation to the process parameters such as substrate hardness, ductility, velocity size, incident angle, etc. However, if the velocity of particles is sufficiently fast, they can be embedded in the surface through a deposition process [2]. Therefore, substrate properties such as hardness, temperature, and degree of oxidation play a significant role in the bonding between particles and substrate [3].

It is well known that cold spraying of ceramic materials can be difficult because cold spraying requires plastic deformation of the feedstock particles for adhesion to the substrate. The challenge lies in the difficulty of plastically deforming hard and brittle ceramic materials, such as TiO_2. Previous studies have reported the possibility of cold spraying thick pure TiO_2 [4] but the bonding mechanism of cold sprayed TiO_2 is not fully understood.

Several experimental results have been published on the bonding mechanism of cold-spraying Ti or TiO_2 particles onto metal and ceramic substrates. Winnicki et al. used a low-pressure cold-spray system for cold-sprayed amorphous, anatase, and rutile TiO_2 powders with a particle size of 10–70 μm and similar shape, which were prepared by the sol-gel process. The 100 μm TiO_2 coatings were prepared on aluminum and the mechanism responsible for powder deposition was the mechanical interlocking of submicron powders with a local presence of agglomerates. They also indicate that the key parameter for the process seems to be the working gas temperature [5]. The same author also cold-sprayed amorphous and anatase TiO_2 powder on the ABS substrate with varying gas temperatures of 200 and 300 °C. The bond strength of the coating was tested using a tensile strength test machine and the highest value was 2 MPa for TiO_2 amorphous+anatase at 200 and 300 °C gas temperature [6]. Hajipour et al. cold sprayed two types of TiO_2 powder; nanocrystal particles with a particle diameter of about 100 ± 15.3 nm and agglomerating ultra-fine particle with a diameter of about 80 ± 11 μm on aluminum substrate. The thickness of the coating is about 490 nm for nanocrystal particles and 15–20 μm for the agglomerating ultra-fine particle. They indicated that for a brittle material such as TiO_2, the first layer is achieved by plastic deformation of the ductile metallic substrate when the particles are embedded into the substrate without any additional binding agent or calcination procedure. The coating/substrate interface is relatively rough when the particle hits the substrate at a high speed. As a result, the titanium dioxide particles are embedded in the Al substrate. Roughness causes mechanical entanglement that might also play an important role in the buildup stage. They also conducted a coating bond strength testing by ultrasonic cleanout of 185 W for 1 min. The adhesion was assessed according to the spalling state of the coating. They identified that there was no spalling of the coating after 1 min [7]. Schmidt et al. used 0.1–10 μm of TiO_2 particles that were cold sprayed onto the flat polished surface of the titanium substrate. They identified that the plastic deformation of the substrate leads to a large continuous contact zone between the particles and the substrate and thus to durable bonding. They also tested the bond strength of the coating by ultrasonic cleaning with a maximum intensity of 40.8 W/cm^2. No local changes in the number, positions, and volumes of the particles could be observed after the cleaning cycle, indicating that the bonding of all the TiO_2 particles to the substrate resists cleaning up to a maximum intensity of 40.8 W/cm^2 in the ultrasonic bath [8]. Kliemann et al. used 3–50 μm TiO_2 agglomerates formed from 5 to 15 nm of primary particles for continuous coating of steel, Cu, Ti, and $AlMg_3$ substrates. They identified ductile substrates that allow shear instabilities to happen as the primary bonding mechanism between the particles and the substrate [9]. Gutzmann et al. obtained the impact morphology of single TiO_2 particles and studied the deposition of different particles on substrates with different temperatures. They showed that there were concentric rings on the impacted substrates such as the shear instability zone. The deposition of a single TiO_2 particle could be achieved only when the substrate temperature was above a certain value. The softer the substrate, the higher the deposition efficiency. It has been proposed that the preheating of the substrates could make them soft and facilitate the substrate shear instability, thus helping the deposit of the coating [10]. Gardon et al. reported that the mechanism responsible for the deposition of TiO_2 on the stainless steel substrate in the cold-spraying process is the chemical bonding between the particles and the substrate. They have shown that the previous layer of titanium sub-oxide prepares the substrate with the appropriate surface roughness needed for the deposition of the TiO_2 particle. In addition, the composition of the substrate is also important for deposition as it can provide chemical affinities during the particle interaction after impact. Substrate hardness may also ease the interaction between the particles and the substrate [11]. Salim et al. prepared 400 and 150 μm TiO_2 coatings on metal and tiles, respectively. They reported that the adhesion strength changed little with the changing spraying parameters. It was discovered that the hardness and the oxidizability of the substrate affected the adhesion strength of the TiO_2 coatings. The adhesion strength of the TiO_2 coatings could be improved by altering the surface chemistry of the substrate. It was proposed that the chemical or physical bonding mechanism can the main bonding mechanism for ceramic coatings [12], and that preheating could increase the oxidizability of the substrate, thus deteriorating the adhesion strength of the coating. The result seems to contradict Gutzmann's experiment, which showed that the preheating of the substrate improved the deposition of the TiO$_2$ particle.

Clearly, there is still considerable uncertainty concerning the preheating of the substrate and the influence of any remaining amorphous oxide layer present at the interface of particle/substrate in relation to the bonding mechanism of cold sprayed TiO$_2$. Therefore, to further understand the bonding mechanism of cold-spraying TiO$_2$ onto metal substrates, in this study, we investigated the correlation between the remaining oxide amorphous phases after cold spraying, and their impacts on the particle/substrate interface toward the adhesion strength of agglomerated nano-TiO$_2$ coatings on annealed metal substrates. Stainless steel (SUS 304) was chosen as the substrates to investigate the bonding mechanism involved and subjected to heat treatment of annealing. An annealing is a heat treatment that changes the physical and chemical properties of a material in order to improve its ductility and reduce its hardness, making it more workable [13]. By increasing the ductility of stainless steel after annealing at higher temperature, we expect better bonding between the TiO$_2$ coating and the SUS 304 annealed stainless steel.

2. Materials and Methods

2.1. Process

In all coating experiments, the cold-spraying equipment with a De-Laval 24TC nozzle (CGT KINETIKS 4000; Cold Gas Technology, Ampfing, Germany) was used. Nitrogen was used as the process gas with a 500 °C operating temperature, and a 3 MPa pressure. The spray distance was 20 mm with a process traverse speed of 10 mm/s. The coatings were deposited on a grit-blasted annealed stainless steel (SUS 304). The substrates were annealed with an electric furnace to preheat the grit-blasted substrate to four different temperatures, respectively (i.e., 300, 500, 700, and 1000 °C) before spraying. In all cases, the temperature of the annealed substrate during spraying was at room temperature.

2.2. Materials

As a feedstock, we applied agglomerated TiO$_2$ powder (TAYCA Corporation, WP0097, Osaka, Japan) containing a pure anatase crystalline structure with an average particle size of about 7.55 µm, as shown in Figure 1. The material chemical composition of substrates used is presented in Table 1.

Figure 1. SEM morphology of TiO$_2$ powder.

Table 1. Material chemical composition [wt%].

Element	Fe	Cr	Ni	Mo	S	P	Mn	Si	C
SUS 304	Bal	18	11	–	0.030	0.045	2.00	1.00	0.08

2.3. Characterization

2.3.1. Tensile-Strength Testing

In accordance with JIS H 8402, specimens measuring Ø25 × 10 mm were used to assess the coatings' adhesion strength, given as the fracture load value measured by a universal testing machine (Autograph AGS-J Series 10 kN, Shimadzu, Japan). We measured the adhesion strength over an average of five specimens for each of the spraying conditions.

2.3.2. Coatings Evaluation

A scanning electron microscope (SEM: JSM-6390, JEOL, Tokyo, Japan) was used to observe the TiO_2 coating's cross-sectional microstructures on annealed substrates. The observation sample of the TiO_2 coating was prepared by embedding a 25 mm × 10 mm sample into a hardenable resin. The hardened sample embedded in the hardened resin was ground with silica papers to a #3000 grit size and finally polished with 1 and 0.3 µm alumina suspension.

2.3.3. Micro-Vickers Hardness

To investigate the relationship between the annealed substrate surface hardness and the adhesion strength of the TiO_2 coating on the annealed substrate, the substrate hardness was measured using an HMV-G micro-Vickers hardness tester (Shimadzu, Japan). The measurement showed a hardness of HV 0.1; the test load on the cross section was 98.07 mN. The final micro-hardness value was the average of five tests taken at approximately the same points for each substrate.

2.3.4. Substrate Oxide Evaluations

X-ray photoelectron spectroscopy (XPS) is a versatile surface analysis technique used for compositional and chemical state analyses. In this study, XPS analysis (ULVAC-PHI, PHI Quantera SXM-CI, Kanagawa, Japan) using a monochromatic Al Kα source (15 mA, 10 kV) was performed. Wide (0–1000 eV) and narrow scans of Fe $2p$, Cr $2p$, and O $1s$ for different annealed substrates were collected. The measured binding energies were then corrected with C $1s$ at 285.0 eV. When pre-sputtering to clean the surface was performed, the sample surface was reduced and the measurements were affected, so XPS analysis was performed without pre-sputtering. Table 2 shows the XPS analysis conditions for substrates oxide analysis.

Table 2. XPS parameter for substrate oxide layer analysis.

Measured Regions	Fe $2p$, O $1s$, Cr $2p$
Measured X-Ray output [W]	10
Probe diameter [µm]	50
Time per step [ms]	30
Pass energy	140
Cycle	30

2.3.5. Wipe Test

A CGT Kinetiks 4000 cold-spray system (Cold Gas Technology, Ampfing, Germany) with a custom-made suction nozzle was used to perform the wipe test and coating using TiO_2 powder onto the annealed 1000 °C stainless steel substrates. The wipe test was conducted to study the deformation behavior of a single particle on this substrate. Prior to deposition, the substrate was ground and polished until a mirror finish surface was obtained. The temperature of the process gas and the pressure used were 500 °C and 3 MPa, respectively. Nitrogen has been used as a process gas. The distance between the exit of the nozzle and the substrate was fixed at 20 mm. The traverse speed of the process was 2000 mm/s. Prior to spraying, the substrates were rinsed with acetone. The FEI Helios Dual

Beam 650 field emission SEM (FESEM, FEI, Oregon, USA) and focused ion beam (FIB, FEI, Oregon, USA) microscope was used to investigate the single particle TiO$_2$ deposition on mirror polished annealed substrate.

2.3.6. TEM Testing

Single titanium dioxide particles were deposited on a 25 mm × 25 mm 1000 °C annealed mirrored steel substrate (SUS 304). In this cold-spray process, nitrogen gas was used as the process gas. The distance between the exit of the nozzle and the substrate was fixed at 20 mm. The process traverse speed was 2000 mm/s. Prior to spraying, the substrates were rinsed with acetone. Thin membranes for transmission electron microscopy (TEM, JEOL, Tokyo, Japan) observation were carefully made from deposited TiO$_2$ particles onto annealed 1000 °C mirrored steel substrate by the focused ion beam (FIB) milling equipment (FEI Helios Dual Beam 650, FEI, Oregon, USA). Without further sample preparation, electron-transparent membranes were made and investigated by field emission gun (FEG) electron microscopy using a JEOL JEM-2100F FE-TEM with a scanning mode at an applied voltage of 200 kV. The elemental distribution of the membranes was acquired by means of STEM energy-dispersive X-ray spectroscopy (EDX).

3. Results

3.1. Strength of Adhesion

The adhesion strength of the cold-sprayed TiO$_2$ coating on annealed SUS 304 stainless steel are shown in Figure 2. The TiO$_2$ coating on the annealed hard substrates showed an increased trend of adhesion strength from room temperature to 1000 °C, with values from 0.51 to 2.55 MPa.

Figure 3 shows the fracture coating of TiO$_2$ on annealed SUS 304 from room temperature to 1000 °C after adhesion-strength testing. The interface fracture occurred between the coating and substrate for annealed SUS 304 substrates in all conditions, as shown in Figure 3a–e.

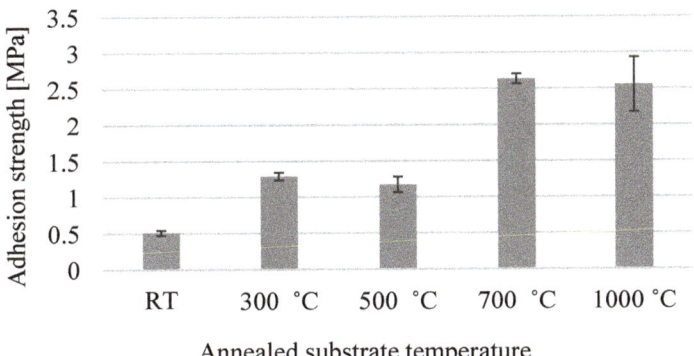

Figure 2. Adhesion strength of the TiO$_2$ coating on annealed SUS 304.

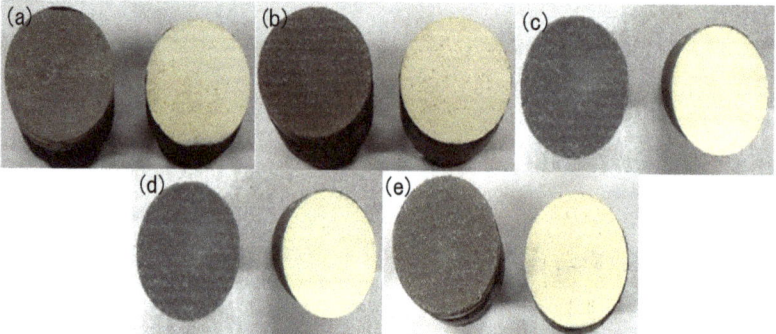

Figure 3. Fracture surface substrate and TiO$_2$ coating after tensile strength testing on SUS 304. (**a**) Room temperature; (**b**) annealed at 300 °C; (**c**) annealed at 500 °C; (**d**) annealed at 700 °C; and (**e**) annealed at 1000 °C.

3.2. SEM Cross-Section Microstructure of TiO$_2$ Coatings on Annealed SUS 304 Substrates

Figure 4a–e shows the TiO$_2$ coating cross-sectional area on SUS 304 for various substrate annealing temperatures. The figures show, in all conditions, a dense coating with a thickness of 300 μm, indicating that a critical velocity was reached for this hard material. This suggests that the TiO$_2$ coating adhered well to the annealed SUS 304 substrate from room temperature to 1000 °C annealing.

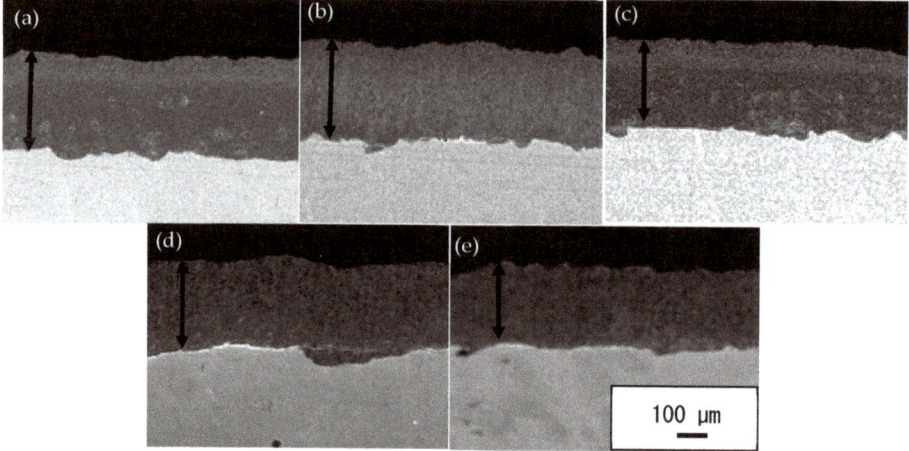

Figure 4. Cross-section microstructure of TiO$_2$ coatings on SUS 304. (**a**) Room temperature; (**b**) annealed at 300 °C; (**c**) annealed at 500 °C; (**d**) annealed at 700 °C; and (**e**) annealed at 1000 °C.

We can categorize the cold-spraying procedure into two stages: (1) Adhesion and (2) cohesion bonding. Adhesion or the formation of the interface between the substrate and particle is the first stage. The annealed substrates can clearly implement this stage, which forms the first coating layer, particularly for the hard material, SUS 304.

3.3. Substrate Vickers Microhardness

Figure 5 shows the annealing substrate hardness of SUS 304 from room temperature to 1000 °C. The stainless steel, SUS 304 showed a decreasing trend from 345.90 Hv for room temperature to 173.00 Hv for 1000 °C annealing.

Based on the iron-carbon phase diagram, when the austenite stainless steel, SUS 304 is annealed at 1000 °C, which is above the eutectoid temperature of 727 °C and slow cooled in the furnace using air medium, the phase transformation involved is austenite to pearlite (ferrite + cementite). This microstructure transformation is associated with a reduction of substrate hardness for 1000 °C annealed SUS 304 and it becomes softer [13].

The reduction of the hardness of SUS 304 may be one of the factors that contributed to the trend of the increase in the TiO_2 coating adhesion strength with increasing SUS 304 substrate annealing temperature. As the substrate becomes softer at higher annealed temperatures such as 1000 °C and when cold sprayed TiO_2 impacts the substrate surface with a high impact velocity, it is associated with substrate deformation that may contribute to the bonding. The mechanical anchoring factor is discussed later in Sections 3.5 and 3.6.

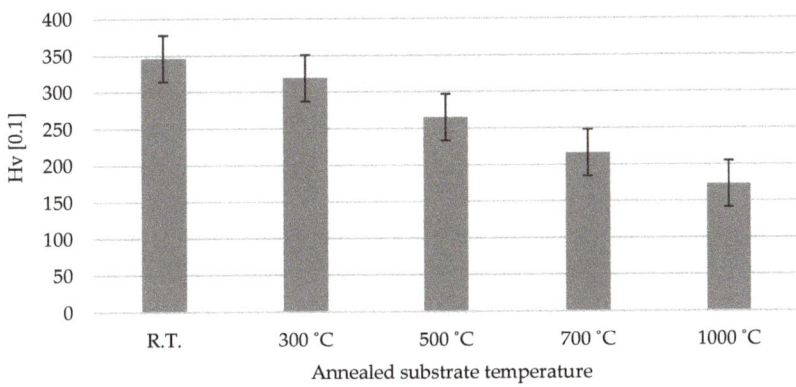

Figure 5. Annealed substrate microhardness of SUS 304 from room temperature to 1000 °C annealing.

3.4. Depth Profile of the Oxide Layer

The result of the depth analysis of room temperature substrate and annealed 700 °C by X-ray photoelectron spectroscopy for the SUS 304 substrate is shown accordingly in Figure 6a,b. The atomic composition of oxygen in the deepest part of the oxide layer increases significantly as the annealing substrate temperature increases from RT to 700 °C. This shows that the oxide layer of stainless steel grows thicker as the annealing temperature of the substrate increases. We also expect the oxide layer to be thicker on the annealed 1000 °C SUS 304 substrate.

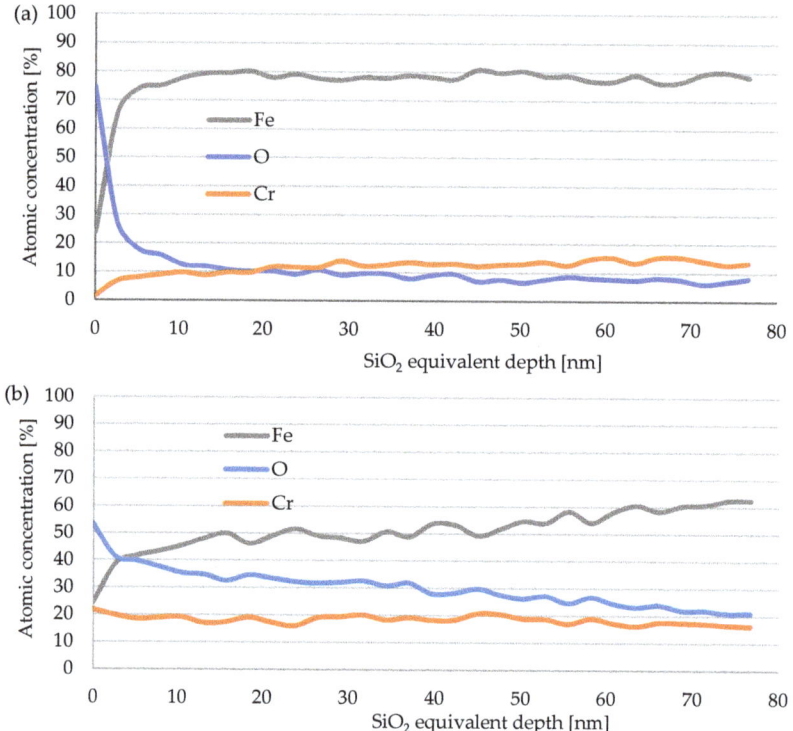

Figure 6. Depth profile analysis of SUS 304 stainless steel. (**a**) Room temperature, (**b**) annealed at 700 °C.

3.5. FIB Splat TiO$_2$ Particle on 1000 °C Annealed Substrates

Figure 7 shows the FIB result of the TiO$_2$ particles on the 1000 °C annealed SUS 304. This wipe test was conducted to further understand the bonding mechanism of the TiO$_2$ particle on the annealed SUS 304 substrate. Only the 1000 °C annealed SUS 304 was selected because it had a high adhesion strength. The results obtained revealed that the TiO$_2$ particle was found unchanged after the collision and the substrate surface of the 1000 °C annealed SUS 304 experienced a deformation due to impacting during the cold-spraying process, as shown by Figure 7. A previous study undertaken by Trompetter et al. demonstrated that for a solid particle impacting on a substrate, the substrate hardness played a significant role in the as-produced solid particles [14]. This condition can be understood by the fact that SUS 304 experienced microstructure transformation during the annealing process at 1000 °C, therefore the substrates hardness was reduced and it became softer.

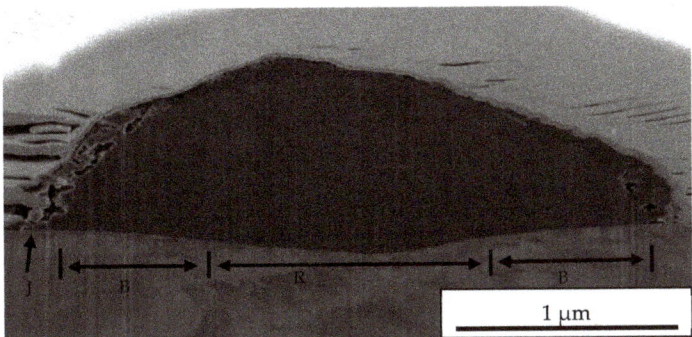

Figure 7. FIB cross-section of a single particle of TiO$_2$ on 1000 °C annealed SUS 304. J indicates the jetted-out region; B is the bonded region; R is the rebound region.

Referring to Figure 7, since the shear instability starts at a position away from the bottom center of the TiO$_2$ particle, the bottom region of the deposited particle can be divided into three regions along the particle–substrate boundary: (i) The particle jetted out region (J) generated by the severe shear plastic strain induced by adiabatic shear instability (ASI); (ii) the well-bonded region (B) where the particle and the substrate are intimately bonded; and (iii) the rebound region (R) where the shear instability did not occur and the accumulated elastic energy from the impact of a sprayed particle detached the particle from the substrate. At the boundary of B and R, ASI is accompanied by severe shear stress, and an abnormal increase in temperature can easily expel the particles, and consequently the oxide covering the surface of particle or substrate can be broken and removed [3,15–20].

The adhesion strength of the TiO$_2$ coating on annealed SUS 304 showed an increased trend as the annealed substrate temperature is increased. This indicates that substrate deformation or mechanical anchoring is one of the factors that influence the adhesion bonding of the annealed SUS 304 with TiO$_2$ at the annealing temperature of 1000 °C. This result is supported by other reports of cold-sprayed TiO$_2$ onto hard substrate such as titanium and stainless steel. Schmidt et al. used 0.1–10 μm of TiO$_2$ particles that were cold sprayed onto the flat polished surface of a titanium substrate. They identified that the plastic deformation of the substrate leads to a large continuous contact zone between the particles and the substrate and thus to a durable bonding [8]. Kliemann et al. used 3–50 μm TiO$_2$ agglomerates formed from 5 to 15 nm of primary particles for the continuous coating of steel substrate. They identified ductile substrate that allows shear instability to happen as the primary bonding mechanism between the particles and the substrate [9].

3.6. TEM Analysis on Interface Oxide Layer between TiO$_2$ Particle on 1000 °C Annealed Substrates

The TEM result is shown as the STEM image interface of the rebound region between single-particle TiO$_2$ and 1000 °C annealed SUS 304, as shown by Figure 8. It confirms the existence of the remaining interface layer after the cold-sprayed TiO$_2$ impacted, with a thickness of approximately 10 nm at the rebound region interface, R and 15 nm at the bonded region, B between single-particle TiO$_2$, and 1000 °C annealed SUS 304, as revealed in Figures 9 and 10. Kim et al. used kinetic spraying of single titanium particles on mirrored steel substrates. They showed that some portion of a thin amorphous oxide remained between the particle–substrate interface, and even a severe plastic deformation was associated with the impacts of the particles onto the substrate. The remaining oxide provided a bond between a particle or particle–substrate [21]. Our data also reported that the same, even interface thickness in the bonded region B is thicker than in the rebound region, R.

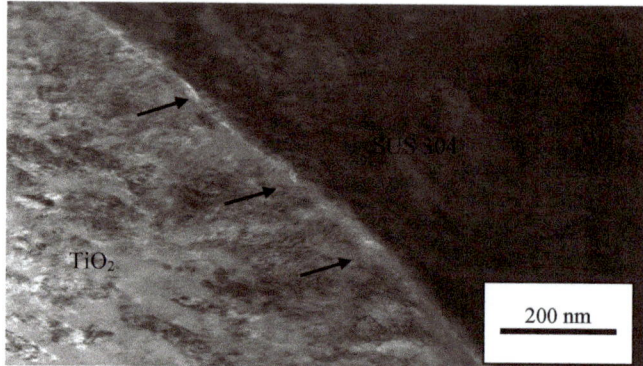

Figure 8. STEM of the TiO$_2$/1000 °C annealed SUS 304 interface at the rebound region, R.

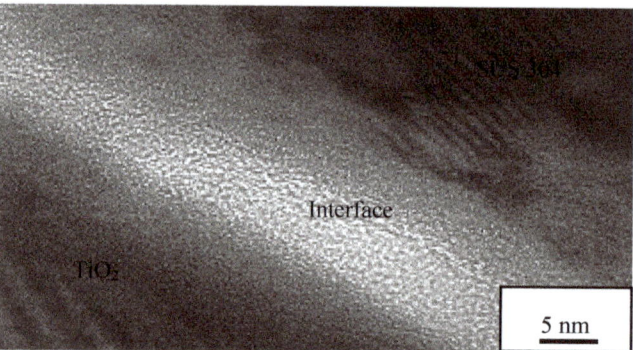

Figure 9. High-magnification images of the TiO$_2$/1000 °C annealed SUS 304 interface at the rebound region, R.

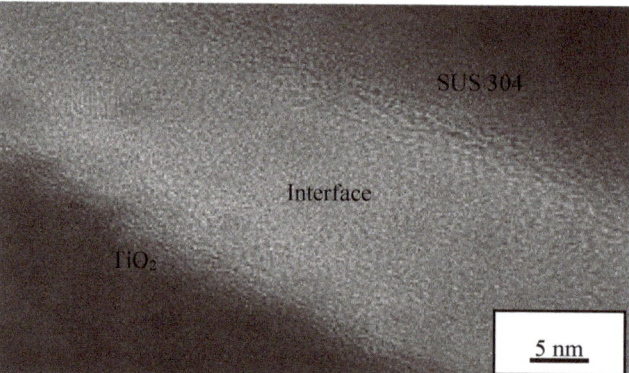

Figure 10. High-magnification images of the TiO$_2$/1000 °C annealed SUS 304 interface at the bonded region, B.

Further analysis of the EDS (JEOL, Tokyo, Japan) on the rebound region interface layer R showed the elemental composition, as shown in Table 3 and Figure 11. The results showed that the oxide layer that occurs at the interface rebound area, R of TiO$_2$ particles and 1000 °C annealed SUS 304 consist of 96.33 at% oxygen, 2.35 at% titanium, 0.95 at% iron, and 0.38 at% chromium. In addition, the EDS analysis of the interface layer in the bonded region B reveals the elemental composition, as shown in Table 4 and Figure 12. The results show that the elemental composition was oxygen at 91.88 at%, titanium at 4.22 at%, iron at 2.77 at%, chromium at 0.83 at%, and nitrogen at 0.30 at%. This EDS analysis showed that nitrogen is present in the bonded region B, which is a gas carrier that was used during the cold-spraying process. In addition, at% of titanium, iron, and chromium in the bonded region B was also slightly higher than the rebound region R.

Table 3. Chemical composition analysis by EDS for the interface at the rebound region, R of TiO_2/1000 °C annealed SUS 304.

Element	Atomic%
O	96.33
Ti	2.35
Fe	0.95
Cr	0.38
Total	100

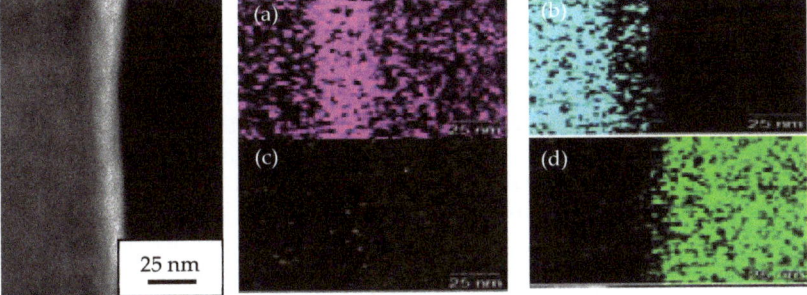

Figure 11. EDX elemental mappings of the TiO_2/1000 °C annealed SUS 304 interface at the rebound region, R: (**a**) oxygen; (**b**) titanium; (**c**) chromium; (**d**) ferum.

Table 4. Chemical composition analysis by EDS for the interface at the bonded region, B of TiO_2/1000 °C annealed SUS 304.

Element	Atomic%
O	91.88
Ti	4.22
Fe	2.77
Cr	0.83
N	0.30
Total	100

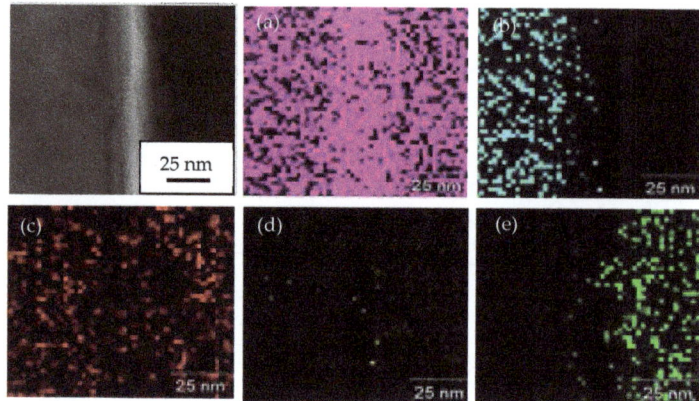

Figure 12. EDX elemental mappings at the TiO$_2$/1000 °C annealed SUS 304 interface at the bonded region, B: (**a**) oxygen; (**b**) titanium; (**c**) nitrogen; (**d**) chromium; (**e**) ferum.

Ko et al. cold-sprayed soft Al particles on the hard, but deformable substrate, Fe. They showed that the atomic intermixing of Al/Fe occurring at an amorphous 10 nm oxide-layer interface could produce a strong adhesive bond between Al and Fe due to some of the chemical adhesion forces [2]. They were also cold-sprayed Cu particles on the AlN substrate and Al particles on the ZrO$_2$. TEM images show the formation of a 10 nm-thick amorphous layer at the Cu/AlN interface and approximately 5 nm amorphous layer at the Al/ZrO$_2$ interface. Due to the restructuring of the interfacial layer upon a high-velocity particle impact, the adhesion between the malleable cold-sprayed metals (Cu and Al) on the brittle ceramic (AlN and ZrO$_2$) substrate was attributed to the high-velocity collision, instead of mechanical interlocking—resulting in limited amorphization and atomic intermixing. The degree of this type of restructuring depends on particle velocity, hardness, and mechanical deformability that is different between the particle and substrate [22]. Our experiments are consistent with the result (Ko. et al. 2016 [22]) that TEM images showed the formation of an amorphous layer approximately 10-nm from the rebound area of the interface, R and 15 nm at the bonded area, B. The EDX results confirm the elemental composition of the amorphous layer consisting of atomic mixing of Ti/Fe/Cr. Therefore, atomic intermixing also contributed to the bonding mechanism between the TiO$_2$/SUS 304 substrate due to some chemical adhesion forces. These findings provide considerable progress related to the bonding mechanism of cold-sprayed TiO$_2$ onto annealed SUS 304 at higher temperatures in terms of explaining the increased adhesion strength of the TiO$_2$ coating as the annealing temperature of the substrate also increased. Substrate deformation and atomic intermixing at the amorphous layer at interface TiO$_2$/SUS 304 are the two factors involved here.

4. Conclusions

This study investigated the correlation between the adhesion strength of cold-sprayed TiO$_2$ on the SUS 304 stainless steel—annealed at temperatures ranging from room temperature to 1000 °C. The results of the study lead to the following conclusions:

- The annealing process plays an important role in the induced ductility of the austenitic stainless steel, SUS 304 especially when annealed at a high temperature such as 1000 °C. This will lead to a decrease in the hardness of the substrate and will make it softer. When the cold-sprayed TiO$_2$ particle is impacted with a high velocity on the annealed 1000 °C SUS 304 surface, the plastic deformation of the substrate occurs and provides a large continuous contact zone between the particles and the substrate, resulting in bonding. Therefore, the adhesion strength of the TiO$_2$ coating is high on the annealed 1000 °C SUS 304 substrate.

- The oxide layer of austenitic stainless steel, SUS 304 grows thicker as the annealed temperature of the substrate increases. The TEM/EDX result shows that the existence of the remaining interface of the amorphous layer is approximately 10 nm for the rebound region, R and 15 nm for the bonded region, B between the TiO$_2$ particles and 1000 °C annealed SUS 304. Due to the restructuring of the interfacial layer upon a high-velocity particle impact, the adhesion between the brittle TiO$_2$ particle and the stainless steel SUS 304 metal was attributed to a high-velocity collision and resulted in limited amorphization and atomic intermixing. Atomic intermixing of Ti/Fe/Cr at the interface may produce a strong adhesive bond between them due to chemical adhesion forces.

Author Contributions: Conceptualization and methodology, N.I.O., S.S., and M.K.; formal analysis, investigation, data curation, and writing—original draft preparation, N.I.O.; writing—review and editing, M.Y.; supervision, T.Y. and M.F.; funding acquisition, M.Y. All authors have read and agreed to the published version of the manuscript.

Funding: This research is supported by JSPS KAKENHI, grant number JP17K06857 and was partially carried out at the Cooperative Research Facility Center at Toyohashi University of Technology.

Acknowledgments: We also acknowledge the Interface and Surface Fabrication lab, Majlis Amanah Rakyat, and the Technical University of Malacca for Noor Irinah's PhD scholarship.

Conflicts of Interest: The authors declare no conflict of interest.

References

1. Champagne, V.K. *The Cold Spray Materials Deposition Process Fundamentals and Applications*, 1st ed.; Victor, K.C., Ed.; Woodhead Publishing Limited: Cambridge, UK, 2007; Introduction; pp. 1–2.
2. Ko, K.; Choi, J.; Lee, H. Intermixing and interfacial morphology of cold-sprayed Al coatings on steel. *Mater. Lett.* **2014**, *136*, 45–47. [CrossRef]
3. Xie, Y.; Planche, M.-P.; Raoelison, R.; Liao, H.; Suo, X.; Hervé, P. Effect of Substrate preheating on adhesive strength of SS 316L cold spray coatings. *J. Therm. Spray Technol.* **2015**, *25*, 123–130. [CrossRef]
4. Yamada, M.; Isago, H.; Nakano, H.; Fukumoto, M. Cold spraying of TiO$_2$ photocatalyst coating with nitrogen process gas. *J. Therm. Spray Technol.* **2010**, *19*, 1218–1223. [CrossRef]
5. Winnicki, M.; Baszczuk, A.; Jasiorski, M.; Borak, B.; Małachowska, A. Preliminary studies of TiO$_2$ nanopowder deposition onto metallic substrate by low pressure cold spraying. *Surf. Coatings Technol.* **2019**, *371*, 194–202. [CrossRef]
6. Winnicki, M.; Jasiorski, M.; Baszczuk, A.; Borak, B. Preliminary studies of TiO2 coatings by deposition onto ABS polymer substrates by low pressure cold spraying. In Proceedings of the Les Rencontres Internationales sur la Projection Thermique—RIPT 9, Jülich, Germany, 11–13 December 2019.
7. Hajipour, H.; Abdollah-Zadeh, A.; Assadi, H.; Taheri-Nassaj, E.; Jahed, H. Effect of feedstock powder morphology on cold-sprayed titanium dioxide coatings. *J. Therm. Spray Technol.* **2018**, *27*, 1542–1550. [CrossRef]
8. Schmidt, K.; Buhl, S.; Davoudi, N.; Godard, C.; Merz, R.; Raid, I.; Kerscher, E.; Kopnarski, M.; Renno, C.M.; Ripperger, S.; et al. Ti surface modification by cold spraying with TiO$_2$ mircoparticles. *Surf. Coat. Technol.* **2017**, *309*, 749–758. [CrossRef]
9. Kliemann, J.-O.; Gutzmann, H.; Gärtner, F.; Hübner, H.; Borchers, C.; Klassen, T. Formation of cold-sprayed ceramic titanium dioxide layers on metal surfaces. *J. Therm. Spray Technol.* **2010**, *20*, 292–298. [CrossRef]
10. Gutzmann, H.; Freese, S.; Gartner, F.; Klassen, T. Cold gas spraying of ceramics using the example of titanium dioxide. In Proceedings of the nternational Thermal Spray Conference, ITSC, Hamburg, Germany, 27–29 September 2011; pp. 391–396.
11. Gardon, M.; Fernández-Rodíguez, C.; Garzón Sousa, D.; Doña-RodRíguez, J.M.; Dosta, S.; Cano, I.G.; Guilemany, J.M. Photocatalytic activity of nanostructured anatase coatings obtained by cold gas spray. *J. Therm. Spray Technol.* **2014**, *23*, 1135–1140. [CrossRef]
12. Salim, N.T.; Yamada, M.; Nakano, H.; Shima, K.; Isago, H.; Fukumoto, M. The effect of post-treatments on the powder morphology of titanium dioxide (TiO$_2$) powders synthesized for cold spray. *Surf. Coatings Technol.* **2011**, *206*, 366–371. [CrossRef]
13. Clayton, C.R. Materials science and engineering: An introduction. *Mater. Sci. Eng.* **1987**, *94*, 266–267. [CrossRef]

14. Trompetter, B.; Hyland, M.; McGrouther, D.; Munroe, P.; Markwitz, A. Effect of substrate hardness on splat morphology in high-velocity thermal spray coatings. *J. Therm. Spray Technol.* **2006**, *15*, 663–669. [CrossRef]
15. Yin, S.; Wang, X.; Suo, X.; Liao, H.; Guo, Z.; Li, W.; Coddet, C. Deposition behavior of thermally softened copper particles in cold spraying. *Acta Mater.* **2013**, *61*, 5105–5118. [CrossRef]
16. Assadi, H.; Gärtner, F.; Stoltenhoff, T.; Kreye, H. Bonding mechanism in cold gas spraying. *Acta Mater.* **2003**, *51*, 4379–4394. [CrossRef]
17. Grujicic, M.; Zhao, C.I.; De Rosset, W.S.; Helfritch, D. Adabatic shear instability based mechanism for particle/substrate bonding in the cold-gas dynamic-spray process. *Mater. Des.* **2004**, *25*, 681–688. [CrossRef]
18. Bae, G.; Xiong, Y.; Kumar, S.; Kang, K.; Lee, C. General aspects of interface bonding in kinetic sprayed coatings. *Acta Mater.* **2008**, *56*, 4858–4868. [CrossRef]
19. Yin, S.; Wang, X.-F.; Li, W.Y.; Jie, H.-E. Effect of substrate hardness on the deformation behavior of subsequently incident particles in cold spraying. *Appl. Surf. Sci.* **2011**, *257*, 7560–7565. [CrossRef]
20. Kim, K.; Li, W.; Guo, X. Detection of oxygen at the interface and its effect on strain, stress, and temperature at the interface between cold sprayed aluminum and steel substrate. *Appl. Surf. Sci.* **2015**, *357*, 1720–1726. [CrossRef]
21. Kim, K.; Kuroda, S. Amorphous oxide film formed by dynamic oxidation during kinetic spraying of titanium at high temperature and its role in subsequent coating formation. *Scr. Mater.* **2010**, *63*, 215–218. [CrossRef]
22. Ko, K.; Choi, J.; Lee, H. The interfacial restructuring to amorphous: A new adhesion mechanism of cold-sprayed coatings. *Mater. Lett.* **2016**, *175*, 13–15. [CrossRef]

Publisher's Note: MDPI stays neutral with regard to jurisdictional claims in published maps and institutional affiliations.

© 2020 by the authors. Licensee MDPI, Basel, Switzerland. This article is an open access article distributed under the terms and conditions of the Creative Commons Attribution (CC BY) license (http://creativecommons.org/licenses/by/4.0/).

MDPI
St. Alban-Anlage 66
4052 Basel
Switzerland
Tel. +41 61 683 77 34
Fax +41 61 302 89 18
www.mdpi.com

Coatings Editorial Office
E-mail: coatings@mdpi.com
www.mdpi.com/journal/coatings

www.ingramcontent.com/pod-product-compliance
Lightning Source LLC
LaVergne TN
LVHW070737100526
838202LV00013B/1257